DATE DUE FOR RETURN

NUCLEAR ORGANIZATION IN DEVELOPMENT AND DISEASE

Novartis Foundation Symposium 264

NUCLEAR ORGANIZATION IN DEVELOPMENT AND DISEASE

2005

John Wiley & Sons, Ltd

Other Wiley Editorial Offices

John Wiley & Sons Inc., 111 River Street, Hoboken, NJ 07030, USA

Jossey-Bass, 989 Market Street, San Francisco, CA 94103-1741, USA

Wiley-VCH Verlag GmbH, Boschstr. 12, D-69469 Weinheim, Germany

John Wiley & Sons Australia Ltd, 33 Park Road, Milton, Queensland 4064, Australia

John Wiley & Sons (Asia) Pte Ltd, 2 Clementi Loop #02-01, Jin Xing Distripark, Singapore
129809

John Wiley & Sons Canada Ltd, 22 Worcester Road, Etobicoke, Ontario, Canada M9W 1L1

Wiley also publishes its books in a variety of electronic formats. Some content that appears
in print may not be available in electronic books.

Novartis Foundation Symposium 264
x+290 pages, 37 figures, 4 tables

1054218946

Library of Congress Cataloging-in-Publication Data

Chadwick, Derek.
    Nuclear organization in development and disease / Derek Chadwick, Jamie Goode.
       p. cm – (Novartis Foundation symposium ; 264)
    Includes bibliographical references and index.
    ISBN 0-470-09373-0 (alk. paper)
       1. Cell nuclei–Congresses. 2. Cell nuclei–Abnormalities–Congresses. 3. Developmental
cytology–Congresses. I. Goode, Jamie. II. Title. III. Series.
    QH595.C47 2005
    571.6′6–dc22   2004062126

British Library Cataloguing in Publication Data

A catalogue record for this book is available from the British Library
ISBN 0 470 09373 0

Typeset in 10½ on 12½ pt Garamond by Dobbie Typesetting Limited, Tavistock, Devon.
Printed and bound in Great Britain by T. J. International Ltd, Padstow, Cornwall.
This book is printed on acid-free paper responsibly manufactured from sustainable forestry,
in which at least two trees are planted for each one used for paper production.

Contents

Participants

Gisèle Bonne   INSERM UR582-Institut de Myologie, Bâtiment Joseph Babinski, Groupe Hospitalier Pitié-Salpétrière, 47 Boulevard de l'Hôpital, 75 651 Paris Cédex 13, France

Brian Burke   Department of Anatomy and Cell Biology, University of Florida, 1600 SW Archer Road, Box 100235, Gainesville, FL 32610, USA

Jacqueline Capeau   INSERM U402 Faculté de Médecine Saint-Antoine, 27 rue Chaligny, 75571 Paris Cédex 12, France

Philippe Collas   Institute of Medical Biochemistry, University of Oslo, PO Box 1112, Blindern, Oslo, N-0317, Norway

Jean-Claude Courvalin   Departement de Biologie Cellulaire, Institut Jacques Monod, CNRS, Université Paris 7, 75005 Paris, France

Kay Davies   Department of Human Anatomy and Genetics, University of Oxford, South Parks Road, Oxford OX1 3QX, UK

Juliet Ellis   Randall Centre, King's College, New Hunt's House, Guy's Campus, London SE1 1UL, UK

Diane Fatkin   Molecular Cardiology Unit, Victor Chang Cardiac Research Institute, Level 6, 384 Victoria Street, Darlinghurst, NSW 2010, Australia

Susan M. Gasser   Friedrich Miescher Institute for Biomedical Research, Maulbeerstrasse 66, CH-4058 Basel, Switzerland

Larry Gerace   Departments of Cell and Molecular Biology, The Scripps Research Institute, 10550 N. Torrey Pines Road, La Jolla, CA 92037, USA

Wayne R. Giles   Departments of Bioengineering and Medicine, University of California San Diego, Whitaker Institute of Biomedical Engineering, PFGB384, 9500 Gilman Drive, La Jolla, CA 92093-0412, USA

Anne Goldman   Department of Cell and Molecular Biology, Northwestern University Medical School, Chicago, Illinois 60611-3008, USA

Robert D. Goldman (*Chair*)   Department of Cell and Molecular Biology, Northwestern University Medical School, Chicago, Illinois 60611-3008, USA

Yosef Gruenbaum   Department of Genetics, The Institute of Life Sciences, The Morris Cohen Wing, Room 2-428, The Hebrew University of Jerusalem, Givat-Ram, Jerusalem 91904, Israel

Chris Hutchison   Department of Biological Sciences, The University of Durham, South Road, Durham DH1 3LE, UK

Jean-Pierre Julien   Department of Anatomy and Physiology of Laval University, Research Centre of CHUL, 2705 Boulevard Laurier, Quebec, G1V 4G2, Canada

Richard T. Lee   Cardiovascular Division, Brigham and Women's Hospital, Partners Research Facility, Room 279, 65 Landsdowne Street, Cambridge, MA 02139, USA

Nicolas Lévy   Departement de Génétique Medicale, Laboratoire de Génétique Moléculaire, Hôpital d'enfants de la Timone, Faculté de Medecine, 13385 Marseille CEDEX 05, France

Christian Malone (*Novartis Foundation Bursar*)   University of Wisconsin, 1525 Linden Drive, Madison, WI 53706, USA

Glenn E. Morris   MRIC, PP18, North East Wales Institute, Plas Coch, Mold Road, Wrexham LL11 2AW, UK

Junko Oshima   Department of Pathology, Box 357470, Health Sciences Building, Room K-543B, University of Washington School of Medicine, 1959 NE Pacific Avenue, Seattle, WA 98195-7470, USA

Susan Shackleton   Division of Medical Genetics, Department of Genetics, University of Leicester, University Road, Leicester LE1 7RH, UK

Catherine M. Shanahan   British Heart Foundation Lecturer, Department of Medicine, ACCI, Level 6, Box 110, Addenbrooke's Hospital, Hills Road, Cambridge CB2 2QQ, UK

Dale Shumaker Department of Cell and Molecular Biology, 11-145 Ward, Goldman Laboratory, Northwestern University Medical School, 303 E Chicago Avenue, Chicago, IL 60611-3008, USA

Daniel A. Starr Center for Genetics and Development, University of California Davis, 354 Briggs Hall, Davis, CA 95616, USA

Colin L. Stewart Chief, Laboratory of Cancer and Developmental Biology, NCI at Frederick, PO Box B, Frederick, MD 21702-1201, USA

Huber R. Warner Associate Director, Biology of Aging Program, Gateway Building 2C231, National Institute on Aging, Bethesda, MD 20892, USA

Adam Wilkins BioEssays, 10/11 Tredgold Lane, Napier Street, Cambridge CB1 1HN, UK

Katherine L. Wilson Department of Cell Biology, The Johns Hopkins University School of Medicine, 725 N Wolfe Street, Baltimore, MD 21205, USA

Howard J. Worman Department of Medicine and Department of Anatomy and Cell Biology, College of Physicians and Surgeons, Columbia University, 630 West 168th Street, 10th Floor, NY 10032, USA

S. G. Young J. David Gladstone Institutes of Cardiovascular Disease, University of California San Francisco, PO Box 419100, Young Laboratory, 365 Vermont Street, San Francisco, CA 94141-9100, USA

Chair's introduction

Robert D. Goldman

Department of Cell and Molecular Biology, Feinberg School of Medicine, Northwestern University, 303 E Chicago Avenue, Chicago, IL 60611-3008, USA

I attended my first Ciba Foundation Symposium in this room in 1972. The symposium topic was cell locomotion, which was one of the most exciting areas in cell biology at that time. The most exciting and unique aspect of that symposium was the active participation by all those in attendance, not just those invited to speak. This great tradition still holds true.

I'd like to introduce the main subject of this meeting by telling you a bit about nuclear structure and how our views of it have changed over the years. Drawings of the interphase nucleus derived from direct observations during the bright-field microscopic era in the 1920s, typically showed only a spherical structure at the cell centre containing chromatin. By the early 1970s, textbook images of the nucleus, using the higher resolution afforded by transmission electron microscopy, showed that the nucleoplasm was filled with relatively amorphous structures (mainly chromatin), a nucleolus, the nuclear envelope membranes, the nuclear lamina and nuclear pore complexes. More recently, the use of high-resolution confocal imaging has shown that many of the 'amorphous' structures seen by electron microscopy are involved in splicing and gene expression.

During this meeting we will discuss the nuclear envelope/lamina complex and we will have a look at how chromatin interacts with the nuclear lamins, the major components of the lamina. We will also consider, in detail, the multitude of interesting proteins known to be associated with the inner nuclear envelope membrane, many of which are lamin binding partners. In contrast, the outer membrane of the nuclear envelope has been grossly neglected, as it is generally assumed simply to be an extension of the endoplasmic reticulum. I would like us to think about this latter point during the meeting because it is quite possible and even likely that there are specific proteins associated with the outer nuclear membrane which act as receptors for cytoplasmic components or which are involved in the molecular cross talk between the outer and inner nuclear envelope membranes. We will also discuss the finding that lamins are not restricted to the lamina subjacent to the inner nuclear envelope membrane. They are found elsewhere in the nucleus, possibly forming an extensive nucleoskeletal network. The existence of such a network may ultimately help to explain many of

1

the defects caused by the numerous lamin mutations, collectively known as the 'laminopathies', which we will learn more about this week. We will also hear a bit about studies of cytoskeletal intermediate filaments (IFs), which are closely related to the lamins, and their roles in various types of diseases.

By the end of this meeting we should be in a position to understand some of the specific functions of, and interactions amongst, the many nuclear proteins to be discussed. Hopefully our discussions will carry us beyond the definition of protein–protein interactions based only on circumstantial evidence such as co-immunoprecipitation. This is critically important as we move forward in developing a comprehensive mechanistic understanding of how the nucleus, and in particular the nuclear lamins, regulate and orchestrate a wide array of remarkably complex functions.

Nuclear lamins: building blocks of nuclear structure and function

Robert D. Goldman, Anne E. Goldman and Dale K. Shumaker

Department of Cell and Molecular Biology, Northwestern University, Feinberg School of Medicine, Chicago, IL, USA

Abstract. The cell nucleus is surrounded by a complex membranous envelope which separates the nucleoplasm from the cytoplasm. Unlike the cytoplasm, the nucleoplasm is not subdivided into membrane-bound compartments, which allows for the efficient segregation of a wide range of complex metabolic activities. In the absence of such membrane compartmentalization, the nucleus is faced with the daunting task of efficiently segregating and interconnecting an enormous array of critically important functions. These include the assembly of the large multi-component complexes or 'factories' involved in DNA replication and transcription. These structures are dynamic as they are assembled and disassembled both spatially and temporally at different times, implying the existence of an infrastructure or nucleoskeleton responsible for establishing and maintaining a complex nuclear architecture. There is increasing evidence that the nuclear lamins are essential elements of this nuclear infrastructure, and that their proper assembly and organization are required for numerous essential nuclear functions. Our goal has been to determine the roles of the nuclear lamins in vital nuclear processes including DNA replication and transcription. The hypothesis directing our investigations is that the lamins form a 3D network that courses throughout the nucleoplasm providing an infrastructure for the assembly and distribution of numerous multicomponent complexes involved in a wide range of nuclear functions.

2005 Nuclear organization in development and disease. Wiley, Chichester (Novartis Foundation Symposium 264) p 3–21

The nuclear lamins are filamentous proteins which comprise the major structural components of the nuclear lamina. Electron microscopy has revealed that the lamina is an electron-dense layer immediately subjacent to the inner nuclear envelope membrane. The lamina forms a molecular interface between the inner membrane and nucleoplasmic structures such as chromatin. Lamins are also found distributed throughout the nucleoplasm where they appear in several organizational states, including distinct foci and/or relatively uniform 'veils' that appear to course throughout the nucleoplasm (Fig. 1; Moir et al 2000a, Liu et al 2000).

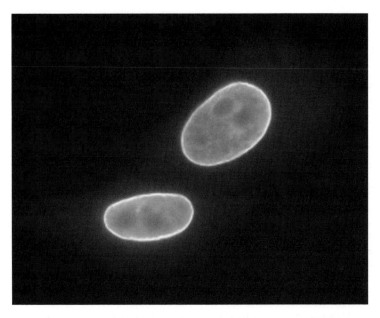

FIG. 1.   Human HeLa cells stained for indirect immunofluorescence with lamin A antibody, showing that the lamins are concentrated in the nuclear lamina at the nuclear periphery, and that they are also distributed throughout the nucleoplasm (the lamin 'veil').

Nuclear lamins are type V intermediate filament (IF) proteins, whose structure is highly conserved in invertebrate and vertebrate species (Erber et al 1999). In mammals there are several lamin genes. In humans there is one gene encoding A-type lamins (*LMNA*), and two B-type lamin genes (*LMNB1* and *LMNB2*). In humans, *LMNA* gives rise to four A/C-type proteins by alternative splicing (LA, LAΔ10, LC, LC2; Fisher et al 1986, Machiels et al 1996, Furukawa et al 1994); whereas there are three *LMNB* products LB1, LB2 and LB3 (which is also derived from *LMNB2*) (Pollard et al 1990, Biamonti et al 1992, Furukawa & Hotta 1993, Stuurman et al 1996). All vertebrate cells express at least one lamin B (LB), while A-type lamin expression is developmentally regulated and is first expressed during gastrulation. To date, LA/C expression has only been identified in differentiated cells (Rober et al 1989). These observations have led to suggestions that lamins may function in the differential expression of genes responsible for determining cell- and tissue-type specificity. Lamins also play important roles in mitosis. Lamin polymers are disassembled in the prophase to metaphase transition as the nuclear envelope breaks down (Gerace & Blobel 1980). This disassembly involves phosphorylation by p34/cdc2 (MPF) (Heald & McKeon 1990). After mitosis, lamins are dephosphorylated and they repolymerize around chromatin as the nuclear envelope is reassembled (Moir et al 2000b, Burke & Gerace 1986).

The lamins, as is the case for all IF proteins, consist of three distinct structural domains (Herrmann et al 1996). These include a highly α-helical central rod domain (McKeon et al 1986, Stuurman et al 1998) and mainly non-α-helical N- (head) and C- (tail) terminal domains. The central rod is required for the assembly of coiled-coil lamin dimers into which the two protein chains are organized so that their head and tail domains are in register. These dimers interact in a head-to-tail fashion during the polymerization of lamins into higher order structures, such as those comprising the nuclear lamina (Stuurman et al 1998, Moir et al 1991). The structure of a highly conserved portion of the tail domain has been determined by X-ray crystallography and nuclear magnetic resonance (NMR) (Krimm et al 2002, Dhe-Paganon et al 2002). The results show that there is an immunoglobulin-like fold (Ig fold) in the mid-region of this tail domain. The specific function of this domain remains unknown. However, based upon information derived from other studies, Ig folds have been shown to provide binding sites for other proteins, DNA and phospholipids (Dhe-Paganon et al 2002). Little is known about the detailed ultrastructure of lamin polymers *in situ*, except in *Xenopus* oocyte germinal vesicles, where they form a lattice of 10 nm filaments in the region of the lamina (Aebi et al 1986). Studies of the ultrastructure of the lamina in other cell types have revealed little more than an amorphous electron-dense meshwork of ill-defined fibrils (Belmont et al 1993).

It has been shown that lamins are involved in establishing and maintaining nuclear shape and mechanical stability. These functions have been revealed in experiments employing dominant negative lamin mutants (Lopez-Soler et al 2001, Moir et al 1994, 2000a, Spann et al 1997, Liu et al 2000), *Lmna* knockout mice (Sullivan et al 1999), and RNA$_i$ in *C. elegans* (Gruenbaum et al 2002). Under these conditions, the normal organization of lamins is disrupted, and is coincident with abnormal nuclear shape and fragility, as well as other nuclear functions (Newport et al 1990, Meier et al 1991). Genetic evidence supports the role of lamins in maintaining nuclear shape and architecture. For example, a mutation in *Drosophila* lamin Dm$_0$, that reduces lamin expression, inhibits nuclear membrane assembly (Lenz-Bohme et al 1997). Another Dm$_0$ mutant has profound effects on nuclear morphology and *Drosophila* development (Guillemin et al 2001). Cells from patients with Emery-Dreifuss Muscular Dystrophy (EDMD, Emery & Dreifuss 1966), Hutchinson-Gilford Progeria Syndrome (HGPS; Gilford 1904, Hutchinson 1886) and numerous other human 'laminopathies' are also characterized by nuclei with abnormal shapes (Favreau et al 2003, Novelli et al 2002, Vigouroux et al 2001).

Nuclear lamins also function in DNA replication as demonstrated by their colocalization with DNA replication factories (Moir et al 1994, Kennedy et al

2000), and the finding that in *Xenopus* nuclei assembled *in vitro*, the reduction of lamin B3 (XLB3 [this lamin is different from the human LB3]) by immunodepletion inhibits DNA synthesis (Newport et al 1990). In addition, dominant negative mutants of XLB3 specifically block the elongation phase of replication in the *Xenopus* system (Spann et al 1997, Moir et al 1994). There is also evidence that lamins are involved in transcription. For example, the retinoblastoma protein (pRb) has been shown to interact with LA either directly or indirectly through interactions with lamin associated proteins (LAPs) (Kennedy et al 2000, Ozaki et al 1994, Mancini et al 1994, Markiewicz et al 2002). In addition, LAP2 (Foisner & Gerace 1993) interacts with mGCL, a vertebrate transcription factor (Nili et al 2001). These studies suggest that complexes containing LA, an LAP and either pRb or mGCL may be involved in transcription.

The functions proposed for lamins are consistent with the observations that they have many binding partners. For example, *in vitro* it has been found that lamins bind to chromatin (Taniura et al 1995). The lamin/chromatin interactions probably involve LAPs, several of which possess the conserved LEM (LAP2, emerin, MAN1) domain (Dechat et al 1998, Lin et al 2000). LEM domains bind BAF (barrier to autointegration factor) (Shumaker et al 2001), which in turn is thought to bind and compact multiple strands of DNA. Based upon these findings, it is likely that nuclear lamins are involved in chromatin organization and consequently gene activity (Goldman et al 2002, Holaska et al 2002, Hutchison et al 2001, Morris 2001, Wilson 2000, Gruenbaum et al 2003). Other lamin binding partners can be divided into integral proteins of the inner nuclear membrane and nuclear proteins without transmembranous domains (Shumaker et al 2003, Gruenbaum et al 2003, Goldman et al 2002). The integral membrane binding proteins include LAP1 (Foisner & Gerace 1993), LAP2 (Foisner & Gerace 1993), emerin (Bione et al 1994, Manilal et al 1996), MAN1(Dechat et al 2000, Lee et al 2000), otefin (Harel et al 1989, Ashery-Padan et al 1997), Nurim (Rolls et al 1999), RING finger binding protein (RFBP) (Gruenbaum et al 2003), lamin B receptor (Courvalin et al 1992), A-kinase anchoring protein 149 (AKAP 149) (Steen & Collas 2001), and Syne/Nesprin/Anc-1 (Shumaker et al 2003, Starr & Han 2003). Other non-membrane associated lamin binding proteins are germ cell-less (GCL) (Jongens et al 1994), PP1 phosphatase (Steen et al 2000), young arrest (YA) (Goldman et al 2002), the transcription factors Oct1 (Imai et al 1997) and pRB (Kennedy et al 2000, Ozaki et al 1994), and the nuclear pore protein NUP 153 (Favreau et al 1996, Smythe et al 2000). Based upon a recent proteomic analysis of the inner nuclear envelope membrane, there are over 60 additional novel components (termed NETs [Nuclear Envelope Targeting proteins]) found in this membrane, many of which are potential lamin binding partners (Schirmer et al 2003).

Based upon the findings that there are numerous lamin binding proteins, with a potentially wide range of functions, we have hypothesized that the lamins form the core of a complex interactive network serving to link various nuclear structures and/or compartments together (Goldman et al 2002, Shumaker et al 2003). This has been dramatically demonstrated in fibroblasts derived from $Lmna^{-/-}$ mice. Nuclei within these cells exhibit abnormal shapes in the form of blebs, lacking emerin and with greatly reduced amounts of LB, LAP2, heterochromatin and the nuclear pore component NUP153 (Sullivan et al 1999). Furthermore, the $Lmna^{-/-}$ mice appear normal at birth, but within 2–3 weeks their growth rate slows, and they exhibit weakness in their forelimbs, an abnormal gait, and a loss of skeletal muscle mass (Sullivan et al 1999). These phenotypes resemble various types of human muscular dystrophies (for reviews see Goldman et al 2002, Mounkes et al 2001, Helbling-Leclerc et al 2002, Gruenbaum et al 2003, Shumaker et al 2003). These diseases have joined a host of other human disorders collectively known as laminopathies (see above). In addition to muscular dystrophies, the laminopathies include cardiomyopathies, lipodystrophies and premature aging diseases such as HGPS and atypical Werner's Syndrome (Chen et al 2003) (for reviews, see Goldman et al 2002, Gruenbaum et al 2003, Shumaker et al 2003, Burke & Stewart 2002, Holaska et al 2002). These diseases, the vast majority caused by point mutations in $LMNA$, have highlighted the importance of the nuclear lamins in the structure and function of the vertebrate cell nucleus. To begin to understand the molecular basis of these diseases, we have developed techniques and assays aimed at determining the specific roles of lamins in the basic housekeeping chores of the nucleus. These include DNA replication, transcription, and the assembly and disassembly of the nucleus during cell division. The following is a brief summary of our findings to date.

Nuclear lamins and DNA replication

Our early studies on nuclear lamins and DNA replication started with the observation that a significant number of cultured cells contained nucleoplasmic, lamin-rich foci (Fig. 2; Goldman et al 1992, Moir et al 1994). Three-dimensional reconstruction of nuclei from these cells demonstrated that the foci did not contain nuclear pore proteins or membranes and were not continuous through the nucleus (Bridger et al 1993, Moir et al 1994). These observations demonstrated that the foci were not simply invaginations of the nuclear envelope. The finding that the foci were present in a subpopulation of cells led us to the hypothesis that they were cell cycle dependent. To test this, mouse 3T3 cells were examined during two phases of the cell cycle using different synchronization procedures. It was shown that lamin B (LB) foci colocalized with sites of DNA synthesis, as shown by BrDU incorporation during S-phase. These same LB foci contained PCNA (proliferating cell nuclear antigen), a DNA polymerase-δ co-factor. When cells

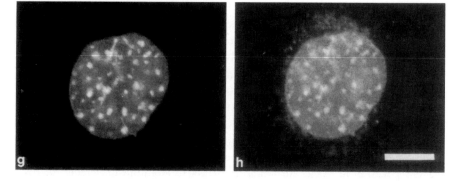

FIG. 2.   During S phase, lamins assemble into intranuclear foci which correlate with sites of BrDU incorporation. Mouse 3T3 cells were exposed to BrDU and then fixed and processed for double indirect immunofluorescence with antibodies directed against lamin B and BrDU. The left image (g) shows the lamin B staining pattern and (h) indicates the location of replication foci (anti-BrDU staining). Note that the majority of lamin B foci colocalize with sites of BrDU incorporation, suggesting that lamins are involved in DNA replication. Bar is $10\,\mu$m. (Reproduced from Moir et al 1994 with permission from The Rockefeller University Press.)

were examined in G1 there was no obvious coincidence between LB and either PCNA or BrDU (Moir et al 1994). These early experiments suggested a role for lamins in DNA replication.

DNA replication can be organized into two distinct stages, initiation and elongation. These stages require different DNA polymerases and cofactors. The initiation stage requires DNA polymerase α, origin recognition complexes (ORC2), mini chromosome maintenance proteins (MCM6) and other factors (Stillman 1994); while the chain elongation phase of DNA replication uses DNA polymerase δ, proliferating cell nuclear antigen (PCNA), replication factor C (RFC), and other factors (Waga et al 1994, Mendez & Stillman 2003). To define the role of lamins in DNA replication we used a mutant lamin lacking its amino-terminal non-α-helical head domain, ΔNLA. This mutant was designed to act in a dominant negative, dose dependent fashion, to inhibit both the assembly of nuclear lamin dimers into polymers and to drive the disassembly of formed lamin polymers. When bacterially expressed ΔNLA was microinjected into mammalian cells it rapidly disrupted normal lamin organization. This resulted in a loss of the peripheral lamin polymer, the nucleoplasmic lamin veil (see above) due to the redistribution of both LA and LB into large foci distributed throughout the nucleoplasm (Spann et al 1997). When ΔNLA or a similar mutant construct of *Xenopus* LB3, ΔNLB3, were added to interphase *Xenopus* egg extracts prior to the addition of chromatin, normal lamin assembly was inhibited and small fragile nuclei were formed (Spann et al 1997). Under these conditions, >95% of DNA replication was inhibited as determined both by biotinylated dUTP and

[^{32}P]-dCTP incorporation (Spann et al 1997). The low level of replication detected was attributable to the synthesis of low molecular weight replication initiation products ($\sim$1500bp). This was confirmed by immunofluorescence observations of these lamin-disrupted nuclei, which revealed that the replication initiation factors required for DNA polymerase-α, XORC2 and XMCM3, were distributed normally. In contrast, the nuclear organization of the replication elongation factors, PCNA and RFC, were disrupted and they colocalized with the large lamin foci that formed in response to ΔNLA (Spann et al 1997). This observation was further supported by the significant inhibition of DNA replication. Using a combination of inhibitors, we were also able to demonstrate that the inhibition of DNA replication in nuclei treated with ΔNLA took place during the transition from the initiation to the elongation phase of replication (Moir et al 2000a). Furthermore, both the effects of ΔNLA on lamin assembly and DNA replication are reversible. This was demonstrated by the rapid recovery of both normal lamin organization and DNA replication when ΔNLA was removed (Moir et al 2000a). These data support the hypothesis that lamins are part of a dynamic nuclear structure that is essential for DNA replication.

Lamins and transcription

The finding that nuclear lamins are distributed throughout the nucleoplasm for much of the cell cycle (Moir et al 2000a), along with their relationship to DNA replication, suggested that lamins might also be involved in other essential functions such as transcription. Initially we tested this hypothesis by injecting ΔNLA into cultured cells. We then assayed for the effects of this dominant-negative lamin mutant on RNA synthesis and the localization of factors involved in RNA synthesis and processing. In order to gain biochemical insights into the role of lamins in transcription we also developed a method for the isolation of transcriptionally active *Xenopus* nuclei from mid-blastula embryos (Spann et al 2002). With this combined approach we showed that the dominant-negative lamin mutant disrupted lamin organization and inhibited RNA polymerase II activity (Spann et al 2002). Specifically, we found that following ΔNLA microinjection and the disruption of the endogenous lamin network in BHK 21 cells, there were significant alterations in the organization of splicing factors (Lamond & Spector 2003). These alterations appeared to be very similar to those found following exposure of cells to the RNA polymerase II inhibitor α-amanitin (Fig. 3). In addition, there was a dramatic reduction in BrUTP incorporation into lamin disrupted nuclei, with only a few clusters of BrUTP incorporation remaining associated with nucleoli. This suggested that lamin disruption did not block RNA polymerase I activity (Wieland 1983), making RNA polymerase II a likely candidate.

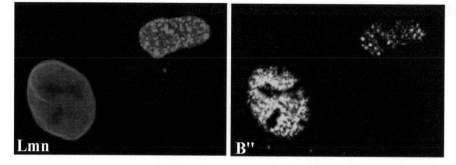

FIG. 3.   Disruption of nuclear lamin organization leads to the altered distribution of splicing factors. The normal distributions of lamin A and the splicing factor B″, a U2-specific binding protein, are shown in the nucleus of the cell at the lower left in both images. Lamins are localized to the nuclear periphery and mainly in a diffuse pattern (the 'veil') throughout the nucleoplasm. The B″ factor is located throughout the nucleus in a 'speckled' pattern. When cells are microinjected with ΔNLA, lamin organization is disrupted as seen by the reorganization of the lamins into discrete foci distributed throughout the nucleoplasm (upper right cell). In this injected cell, B″ is also reorganized into fewer and somewhat larger foci throughout the nucleus. The lamin A and B″ foci do not colocalize in the ΔNLA injected cells. (Reproduced from Spann et al 2002 with permission from The Rockefeller University Press).

In cell-free preparations of *Xenopus* embryonic nuclei, the addition of ΔNLB3 or ΔNLA induced the disruption of the endogenous lamin network. Under these conditions, BrUTP incorporation was greatly reduced as determined by immunofluorescence, thereby revealing that transcriptional activity was inhibited (Spann et al 2002). These preparations of transcriptionally active nuclei were also used to monitor the incorporation of [^{32}P]UTP into RNA. In controls, both tRNA and mRNA were synthesized (rRNA is not synthesized at this stage of gastrulation; Newport & Kirschner 1982). We showed that in these nuclei RNA polymerase II was active as there was a complete inhibition of the synthesis of the higher molecular weight mRNAs following treatment with α-amanitin (Spann et al 2002). Addition of ΔNLA to these embryonic nuclei also blocked mRNA synthesis. No alterations in tRNA-sized products could be detected in lamin disrupted nuclei, providing additional evidence that normal lamin organization is required for RNA polymerase II activity (Spann et al 2002). We also found that the TATA binding protein (TBP), a well characterized co-factor for RNA polymerases, was also associated with the lamin aggregates induced by ΔNLA (Spann et al 2002). Taken together with our observations that lamins are distributed throughout the nucleoplasm, these findings suggest the interesting possibility that lamins form an infrastructure upon which the basal transcription factors required for RNA polymerase II transcription are organized.

Lamins and the assembly of nuclei

In order to begin to understand the role of lamins in nuclear assembly we have observed the behaviour of A- and B-type lamins during the cell cycle using GFP-lamin expression in cultured cells (Moir et al 2000a). During metaphase, GFP-tagged lamins are distributed throughout the cytoplasm in a freely diffusible fashion as determined by their extremely rapid fluorescence recovery after photobleaching (FRAP) (~ 5 s; Moir et al 2000b). During the transition from late anaphase to mid-telophase, GFP-LB1 becomes concentrated on chromosomes near kinetochores. Shortly thereafter, GFP-LB1 was seen to completely surround the decondensing chromosomes as the lamina reformed prior to the completion of nuclear envelope (i.e. membranes and pore complexes) assembly. In contrast, GFP-LA only appeared to accumulate significantly within the nucleus after the nuclear envelope had assembled. In the earliest phases of G1 there was no concentration of LA in the region of the lamina, rather it was distributed uniformly throughout the nucleoplasm. Later in G-1 LA was more concentrated in the lamina. With respect to polymer stability, FRAP experiments carried out ~ 10 min after GFP-LB1 had enclosed the decondensing chromosomes in daughter cells showed complete recovery within 10 min. Within 30 min following enclosure of the decondensing chromosomes, the B1 lamina was more stable as reflected by an increase in the $t_{1/2}$ for fluorescence recovery to ~ 30 min and after ~ 120 min the $t_{1/2}$ was ~ 1.0 h. These studies showed that LB1 assembled into a relatively stable polymer shortly after it became associated with the surface of chromosomes. In contrast, at times equivalent to 60–120 min after LB1 enclosure, GFP-LA recovered its fluorescence within seconds throughout the nucleoplasm. Once GFP-LA became more concentrated in the lamina its recovery rate was slowed dramatically to a $t_{1/2}$ of ~ 90 min. Our findings to date demonstrate that LA assembled into a more stable structure within the lamina after LB1. We also tested for interactions between LB1 and LA using fluorescence resonance energy transfer (FRET). The results demonstrated that LA and LB1 were in very close proximity in the lamina. Interestingly, the results of these FRET analyses also suggested that LA and LB1 interacted throughout the nucleoplasm. These observations regarding interactions between the A- and B-type lamins are further supported by the finding that ΔNLA disrupts both the intranuclear organization of LA and LB (Moir et al 2000a, Spann et al 1997).

These results suggested that B-type lamins played a role in the early stages of nuclear assembly. We tested this further in the *Xenopus* interphase extract nuclear assembly system where we found that the non-α-helical C-terminal domain of LB3 (LB3T) inhibited lamin assembly into the lamina, chromatin decondensation, the fusion of membrane vesicles required for the assembly of the double nuclear envelope membrane and the assembly of nuclear pore complexes. Interestingly,

LB3T inhibited the binding of the 'fusogenic vesicles' required for the formation of the envelope membranes, but had no obvious effect on the ability of 'non-fusogenic vesicles' to bind to chromatin (Lopez-Soler et al 2001).

Acknowledgements

This work was supported by Grants from the NCI and the Muscular Dystrophy Association.

References

Aebi U, Cohn J, Buhle L, Gerace L 1986 The nuclear lamina is a meshwork of intermediate-type filaments. Nature 323:560–564

Ashery-Padan R, Ulitzur N, Arbel A et al 1997 Localization and posttranslational modifications of otefin, a protein required for vesicle attachment to chromatin, during Drosophila melanogaster development. Mol Cell Biol 17:4114–4123

Belmont AS, Zhai Y, Thilenius A 1993 Lamin B distribution and association with peripheral chromatin revealed by optical sectioning and electron microscopy tomography. J Cell Biol 123:1671–1685

Biamonti G, Giacca M, Perini G et al 1992 The gene for a novel human lamin maps at a highly transcribed locus of chromosome 19 which replicates at the onset of S-phase. Mol Cell Biol 12:3499–3506

Bione S, Maestrini E, Rivella S et al 1994 Identification of a novel X-linked gene responsible for Emery-Dreifuss muscular dystrophy. Nat Genet 8:323–327

Bridger JM, Kill IR, O'Farrell M, Hutchison CJ 1993 Internal lamin structures within G1 nuclei of human dermal fibroblasts. J Cell Sci 104:297–306

Burke B, Gerace L 1986 A cell free system to study reassembly of the nuclear envelope at the end of mitosis. Cell 44:639–652

Burke B, Stewart CL 2002 Life at the edge: the nuclear envelope and human disease. Nat Rev Mol Cell Biol 3:575–585

Chen L, Lee L, Kudlow BA et al 2003 LMNA mutations in atypical Werner's syndrome. Lancet 362:440–445

Courvalin JC, Segil N, Blobel G, Worman HJ 1992 The lamin B receptor of the inner nuclear membrane undergoes mitosis- specific phosphorylation and is a substrate for p34cdc2-type protein kinase. J Biol Chem 267:19035–19038

Dechat T, Korbei B, Vaughan OA et al 2000 Lamina-associated polypeptide 2alpha binds intranuclear A-type lamins. J. Cell Sci 113:3473–3484

Dechat T, Gotzmann J, Stockinger A et al 1998 Detergent-salt resistance of LAP2alpha in interphase nuclei and phosphorylation-dependent association with chromosomes early in nuclear assembly implies functions in nuclear structure dynamics. EMBO J 17:4887–4902

Dhe-Paganon S, Werner ED, Chi YI, Shoelson SE 2002 Structure of the globular tail of nuclear lamin. J Biol Chem 277:17381–17384

Emery AE, Dreifuss FE 1966 Unusual type of benign x-linked muscular dystrophy. J Neurol Neurosurg Psychiatry 29:338–342

Erber A, Riemer D, Hofemeister H et al 1999 Characterization of the Hydra lamin and its gene: A molecular phylogeny of metazoan lamins. J Mol Evol 49:260–271

Favreau C, Worman HJ, Wozniak RW, Frappier T, Courvalin JC 1996 Cell cycle-dependent phosphorylation of nucleoporins and nuclear pore membrane protein Gp210. Biochemistry 35:8035–8044

Favreau C, Dubosclard E, Ostlund C et al 2003 Expression of lamin A mutated in the carboxyl-terminal tail generates an aberrant nuclear phenotype similar to that observed in cells from patients with Dunnigan-type partial lipodystrophy and Emery-Dreifuss muscular dystrophy. Exp Cell Res 282:14–23

Fisher DZ, Chaudhary N, Blobel G 1986 cDNA sequencing of nuclear lamins A and C reveals primary and secondary structural homology to intermediate filament proteins. Proc Natl Acad Sci USA 83:6450–6454

Foisner R, Gerace L 1993 Integral membrane proteins of the nuclear envelope interact with lamins and chromosomes, and binding is modulated by mitotic phosphorylation. Cell 73:1267–1279

Furukawa K, Hotta Y 1993 cDNA cloning of a germ cell specific lamin B3 from mouse spermatocytes and analysis of its function by ectopic expression in somatic cells. EMBO J 12:97–106

Furukawa K, Inagaki H, Hotta Y 1994 Identification and cloning of an mRNA coding for a germ cell-specific A- type lamin in mice. Exp Cell Res 212:426–430

Gerace L, Blobel G 1980 The nuclear envelope lamina is reversibly depolymerized during mitosis. Cell 19:277–287

Gilford M 1904 Progeria — a form of senilism. Practitioner 73:188–217

Goldman AE, Moir RD, Montag-Lowy M, Stewart M, Goldman RD 1992 Pathway of incorporation of microinjected lamin A into the nuclear envelope. J Cell Biol 119: 725–735

Goldman RD, Gruenbaum Y, Moir RD, Shumaker DK, Spann TP 2002 Nuclear lamins: building blocks of nuclear architecture. Genes Dev 16:533–547

Gruenbaum Y, Lee KK, Liu J, Cohen M, Wilson KL 2002 The expression, lamin-dependent localization and RNAi depletion phenotype for emerin in C. elegans. J Cell Sci 115:923–929

Gruenbaum Y, Goldman RD, Meyuhas R et al 2003 The nuclear lamina and its functions in the nucleus. Int Rev Cytol 226:1–62

Guillemin K, Williams T, Krasnow MA 2001 A nuclear lamin is required for cytoplasmic organization and egg polarity in Drosophila. Nat Cell Biol 3:848–851

Harel A, Zlotkin E, Nainudel-Epszteyn S et al 1989 Persistence of major nuclear envelope antigens in an envelope-like structure during mitosis in Drosophila melanogaster embryos. J Cell Sci 94:463–470

Heald R, McKeon F 1990 Mutations of phosphorylation sites in lamin A that prevent nuclear lamina disassembly in mitosis. Cell 61:579–589

Helbling-Leclerc A, Bonne G, Schwartz K 2002 Emery-Dreifuss muscular dystrophy. Eur J Hum Genet 10:157–161

Herrmann H, Haner M, Brettel M et al 1996 Structure and assembly properties of the intermediate filament protein vimentin: the role of its head, rod and tail domains. J Mol Biol 264:933–953

Holaska JM, Wilson KL, Mansharamani M 2002 The nuclear envelope, lamins and nuclear assembly. Curr Opin Cell Biol 14:357–364

Hutchinson J 1886 Congenital absence of hair and mammary gland with atrophic condition of skin and its appendages in a boy whose mother had been almost totally bald from alopecia areata from the age of six. Medicochir Trans 69:473–477

Hutchison CJ, Alvarez-Reyes M, Vaughan OA 2001 Lamins in disease: why do ubiquitously expressed nuclear envelope proteins give rise to tissue-specific disease phenotypes? J Cell Sci 114:9–19

Imai S, Nishibayashi S, Takao K et al 1997 Dissociation of Oct-1 from the nuclear peripheral structure induces the cellular aging-associated collagenase gene expression. Mol Biol Cell 8:2407–2419

Jongens TA, Ackerman LD, Swedlow JR, Jan LY, Jan YN 1994 Germ cell-less encodes a cell type-specific nuclear pore-associated protein and functions early in the germ-cell specification pathway of Drosophila. Genes Dev 8:2123–2136

Kennedy BK, Barbie DA, Classon M, Dyson N, Harlow E 2000 Nuclear organization of DNA replication in primary mammalian cells. Genes Dev 14:2855–2868

Krimm I, Ostlund C, Gilquin B et al 2002 The Ig-like structure of the C-terminal domain of lamin A/C, mutated in muscular dystrophies, cardiomyopathy, and partial lipodystrophy. Structure (Camb) 10:811–823

Lamond AI, Spector DL 2003 Nuclear speckles: a model for nuclear organelles. Nat Rev Mol Cell Biol 4:605–612

Lee KK, Gruenbaum Y, Spann P, Liu J, Wilson KL 2000 C. elegans nuclear envelope proteins emerin, MAN1, lamin, and nucleoporins reveal unique timing of nuclear envelope breakdown during mitosis. Mol Biol Cell 11:3089–3099

Lenz-Bohme B, Wismar J, Fuchs S et al 1997 Insertional mutation of the Drosophila nuclear lamin Dm0 gene results in defective nuclear envelopes, clustering of nuclear pore complexes, and accumulation of annulate lamellae. J Cell Biol 137:1001–1016

Lin F, Blake DL, Callebaut I et al 2000 MAN1, an inner nuclear membrane protein that shares the LEM domain with lamina-associated polypeptide 2 and emerin. J Biol Chem 275:4840–4847

Liu J, Ben-Shahar TR, Riemer D et al 2000 Essential roles for Caenorhabditis elegans lamin gene in nuclear organization, cell cycle progression, and spatial organization of nuclear pore complexes. Mol Biol Cell 11:3937–3947

Lopez-Soler RI, Moir RD, Spann TP, Stick R, Goldman RD 2001 A role for nuclear lamins in nuclear envelope assembly. J Cell Biol 154:61–70

Machiels BM, Zorenc AH, Endert JM et al 1996 An alternative splicing product of the lamin A/C gene lacks exon 10. J Biol Chem 271:9249–9253

Mancini MA, Shan B, Nickerson JA, Penman S, Lee WH 1994 The retinoblastoma gene product is a cell cycle-dependent, nuclear matrix-associated protein. Proc Natl Acad Sci USA 91:418–422

Manilal S, Nguyen TM, Sewry CA, Morris GE 1996 The Emery-Dreifuss muscular dystrophy protein, emerin, is a nuclear membrane protein. Hum Mol Genet 5:801–808

Markiewicz E, Dechat T, Foisner R, Quinlan RA, Hutchison CJ 2002 Lamin A/C binding protein LAP2alpha is required for nuclear anchorage of retinoblastoma protein. Mol Biol Cell 13:4401–4413

McKeon FD, Kirschner MW, Caput D 1986 Homologies in both primary and secondary structure between nuclear envelope and intermediate filament proteins. Nature 319:463–468

Meier J, Campbell KH, Ford CC, Stick R, Hutchison CJ 1991 The role of lamin LIII in nuclear assembly and DNA replication, in cell- free extracts of Xenopus eggs. J Cell Sci 98:271–279

Mendez J, Stillman B 2003 Perpetuating the double helix: molecular machines at eukaryotic DNA replication origins. Bioessays 25:1158–1167

Moir RD, Donaldson AD, Stewart M 1991 Expression in Escherichia coli of human lamins A and C: influence of head and tail domains on assembly properties and paracrystal formation. J Cell Sci 99:363–372

Moir RD, Montag-Lowy M, Goldman RD 1994 Dynamic properties of nuclear lamins: lamin B is associated with sites of DNA replication. J Cell Biol 125:1201–1212

Moir RD, Spann TP, Herrmann H, Goldman RD 2000a Disruption of nuclear lamin organization blocks the elongation phase of DNA replication. J Cell Biol 149:1179–1192

Moir RD, Yoon M, Khuon S, Goldman RD 2000b Nuclear lamins A and B1. Different pathways of assembly during nuclear envelope formation in living cells. J Cell Biol 151:1155–1168

Morris GE 2001 The role of the nuclear envelope in Emery-Dreifuss muscular dystrophy. Trends Mol Med 7:572–577

Mounkes LC, Burke B, Stewart CL 2001 The A-type lamins: nuclear structural proteins as a focus for muscular dystrophy and cardiovascular diseases. Trends Cardiovasc Med 11:280–285

Newport J, Kirschner M 1982 A major developmental transition in early Xenopus embryos: I. characterization and timing of cellular changes at the midblastula stage. Cell 30:675–686

Newport JW, Wilson KL, Dunphy WG 1990 A lamin-independent pathway for nuclear envelope assembly. J Cell Biol 111:2247–2259

Nili E, Cojocaru GS, Kalma Y et al 2001 Nuclear membrane protein LAP2beta mediates transcriptional repression alone and together with its binding partner GCL (germ-cell-less). J Cell Sci 114:3297–3307

Novelli G, Muchir A, Sangiuolo F et al 2002 Mandibuloacral dysplasia is caused by a mutation in LMNA-encoding lamin A/C. Am J Hum Genet 71:426–431

Ozaki T, Saijo M, Murakami K et al 1994 Complex formation between lamin A and the retinoblastoma gene product: identification of the domain on lamin A required for its interaction. Oncogene 9:2649–2653

Pollard KM, Chan EK, Grant BJ et al 1990 In vitro posttranslational modification of lamin B cloned from a human T-cell line. Mol Cell Biol 10:2164–2175

Rober RA, Weber K, Osborn M 1989 Differential timing of nuclear lamin A/C expression in the various organs of the mouse embryo and the young animal: a developmental study. Development 105:365–378

Rolls MM, Stein PA, Taylor SS et al 1999 A visual screen of a GFP-fusion library identifies a new type of nuclear envelope membrane protein. J Cell Biol 146:29–44

Schirmer EC, Florens L, Guan T, Yates JR, 3rd, Gerace L 2003 Nuclear membrane proteins with potential disease links found by subtractive proteomics. Science 301:1380–1382

Shumaker DK, Lee KK, Tanhehco YC, Craigie R, Wilson KL 2001 LAP2 binds to BAF.DNA complexes: requirement for the LEM domain and modulation by variable regions. EMBO J 20:1754–1764

Shumaker DK, Kuczmarski ER, Goldman RD 2003 The nucleoskeleton: lamins and actin are major players in essential nuclear functions. Curr Opin Cell Biol 15:358–366

Smythe C, Jenkins HE, Hutchison CJ 2000 Incorporation of the nuclear pore basket protein nup153 into nuclear pore structures is dependent upon lamina assembly: evidence from cell-free extracts of Xenopus eggs. EMBO J 19:3918–3931

Spann TP, Moir RD, Goldman AE, Stick R, Goldman RD 1997 Disruption of nuclear lamin organization alters the distribution of replication factors and inhibits DNA synthesis. J Cell Biol 136:1201–1212

Spann TP, Goldman AE, Wang C, Huang S, Goldman RD 2002 Alteration of nuclear lamin organization inhibits RNA polymerase II- dependent transcription. J Cell Biol 156:603-8.

Starr DA, Han M 2003 ANChors away: an actin based mechanism of nuclear positioning. J Cell Sci 116:211–216

Steen RL, Collas P 2001 Mistargeting of B-type lamins at the end of mitosis: implications on cell survival and regulation of lamins A/C expression. J Cell Biol 153:621–626

Steen RL, Martins SB, Tasken K, Collas P 2000 Recruitment of protein phosphatase 1 to the nuclear envelope by A-kinase anchoring protein AKAP149 is a prerequisite for nuclear lamina assembly. J Cell Biol 150:1251–1262

Stillman B 1994 Smart machines at the DNA replication fork. Cell 78:725–728

Stuurman N, Sasse B, Fisher PA 1996 Intermediate filament protein polymerization: molecular analysis of Drosophila nuclear lamin head-to-tail binding. J Struct Biol 117:1–15

Stuurman N, Heins S, Aebi U 1998 Nuclear lamins: their structure, assembly, and interactions. J Struct Biol 122:42–66

Sullivan T, Escalante-Alcalde D, Bhatt H et al 1999 Loss of A-type lamin expression compromises nuclear envelope integrity leading to muscular dystrophy. J Cell Biol 147:913–920

Taniura H, Glass C, Gerace L 1995 A chromatin binding site in the tail domain of nuclear lamins that interacts with core histones. J Cell Biol 131:33–44

Vigouroux C, Auclair M, Dubosclard E et al 2001 Nuclear envelope disorganization in fibroblasts from lipodystrophic patients with heterozygous R482Q/W mutations in the lamin A/C gene. J Cell Sci 114:4459–4468

Waga S, Hannon G J, Beach D, Stillman B 1994 The p21 inhibitor of cyclin-dependent kinases controls DNA replication by interaction with PCNA. Nature 369:574–578

Wieland T 1983 The toxic peptides from Amanita mushrooms. Int J Pept Protein Res 22: 257–276

Wilson K L 2000 The nuclear envelope, muscular dystrophy and gene expression. Trends Cell Biol 10:125–129

DISCUSSION

Warner: I didn't hear anything about whether these mutations in lamin A generally predispose the mutant cells to apoptosis. Is this true in some laminopathies, but not others? Is there any general statement you can make about this?

Goldman: I can't make a statement about this. We have not noticed any changes in the occurrence of apoptosis, although we haven't been looking for these. I can tell you, however, that the lamins are early nuclear targets cleaved by caspases during programmed cell death.

Capeau: We have performed some studies with fibroblasts from FPLD patients. We have looked at the cell cycle and have seen no difference in the number of cells in the sub-G1 phase corresponding to apoptotic or necrotic cells. This indicates that apoptosis was not increased in these fibroblasts in cell culture. However, depending on the passage, 5–20% of the cells presented with dysmorphic nuclei in this culture.

Hutchison: It is generally incredibly difficult to induce apoptosis in fibroblasts. This may not be the right model. I believe that Bill Earnshaw has shown that under certain circumstances in *in vitro* models, lamin A is a survival factor for apoptosis. (Ruchaud et al 2002). It may be cell-type specific.

Bonne: We have studied a large number of fibroblasts from EDMD patients. We have checked for apoptosis several times and have never seen an increase in apoptotic features in these kinds of fibroblast, but as Chris Hutchison points out, these might not be the best cells to look at.

Gruenbaum: How many passages had these cells had?

Bonne: 15–20. We examined them against age-matched fibroblasts and didn't find any clear differences.

Gruenbaum: This is similar to what people have found for HGPS fibroblasts.

Morris: Cells without emerin actually grow faster than those with it. This suggests that lack of emerin does not induce apoptosis in fibroblasts at least.

Capeau: FPLD fibroblasts seemed to have a decreased proliferation rate when we looked at thymidine incorporation. But we have to be very careful with the experiments, because when the number of passages is increased beyond about 15, there is a decrease in the incorporation of thymidine even in control fibroblasts.

Stewart: We have looked at proliferation and long-term viability in cells from mice that either lack the A-type lamins, or that came from our progeric mouse model. We find that embryonic fibroblasts from the lamin-deficient mice senesce at a much earlier stage than controls, although the cells can be spontaneously transformed and immortalized. In our progeria model we have observed this intriguing phenomenon: there is absolutely no difference in proliferation, long-term viability or senescence between wild-type and mutant embryonic fibroblasts. But if we derive fibroblasts from various adult tissues, such as skin, kidney or muscle, then the progeric cells die within about four passages, whereas the wild-type cells continue to undergo normal proliferation. There seems to be some sort of developmentally regulated phenomenon taking place in terms of the way the cells respond to particular mutations that we've introduced into the lamin A gene.

Goldman: Human progeria cells grow slowly and in our experience will rarely grow much beyond 30 passages.

Stewart: They are also supposedly difficult to establish from biopsies.

Worman: In the HGPS cells it is hard to tell how much of the effect is because of lamin A and how much is because of progerin.

Goldman: In the original paper from the Collins lab there is a blot, showing that the typical HGPS mutant lamin A (progerin) is produced in the cells from patients. If one could do adequate quantitative Western blotting we could probably gain insights into this problem. However, the HGPS cells grow very slowly so obtaining sufficient protein to run gels and to carry out quantitative Westerns is not easy. Furthermore the pictures I showed are samples of the abnormal nuclei. In our experiments to date, there are also cells with apparently normal nuclei. These vary in number as the percentage of abnormal nuclei in cells increases as a function of passage number. We have the progeria construct and we have been looking at the transfection and microinjection of progerin in collaboration with the laboratory of Francis Collins.

Burke: We have done exactly the same experiments, transfecting mutant cDNAs. They are indistinguishable from wild-type and the stability of the protein is normal.

Gerace: These results beg the question of what the actual lamin complement of these cells consists of. This looks and smells like epigenetic changes in the expression of nuclear envelope proteins. How about B-type lamins? Have you examined this?

Goldman: We have not yet had the time to carry out detailed analyses, but to date the B-type lamins look quite normal. There are 162 mutations that have been reported in lamin A, and so far none in lamin B. Therefore it appears that lamin B is essential and that any mutation would be lethal. This has profound implications for lamin A with respect to cell differentiation and differential gene expression.

Young: Along with Dr Karen Reue from UCLA, we have made mice lacking lamin B1, and they are happy until about day 19–20 of embryonic development. You can very easily culture and immortalize fibroblasts lacking lamin B1.

Goldman: What about B2 expression?

Young: Dr Reue has very preliminary data suggesting that lamin B2 expression might be increased in some tissues of the lamin B1 knockout mice.

Gerace: I have one comment relating to the thickened lamina phenotype you described. There are a number of normal human cells that do have thick laminas, including connective tissue cells and smooth muscle cells as shown in histology textbooks such as Bloom and Fawcett.

Goldman: This is true as I remember for adult liver cells in Fawcett's textbook, but there are dramatic differences in relative thickness of the lamina in the cells of normal and HGPS cells.

Gerace: Not in the adult liver, but there are some laminas that are 30–40 nm thick in normal cells, so we can't necessarily say that a thickened lamina will cause disease. I imagine these are terminally differentiated cells. They don't divide, but presumably your fibroblasts are trying to divide.

Goldman: If we take skin fibroblasts from very old people and pass them a number of times, they look normal. The original description of the lamina was based on electron microscopic evidence showing a relatively thick layer of electron-dense material underlying the inner nuclear envelope membrane in tissue cells *in situ*. However, in tissue culture it is very difficult to see this structure using conventional thin section electron microscopy. What we do see is a barely perceptible thin area between peripheral heterochromatin and the inner nuclear envelope membrane. In HGPS cells this becomes very obvious: it is a very thick electron-dense layer.

Wilkins: In the mutant, is that a general feature for all cell types? If not, does it correlate with the cells that show particularly senescent phenotypes?

Goldman: The only cells we have from patients are skin fibroblasts.

Wilson: You mentioned that progeria cells seemed to have fewer pore complexes. Have you looked for lower rates of nucleocytoplasmic transport?

Goldman: There is no detectable difference. I am not sure there is a change in the number of pores, but there is certainly a change in the distribution of pores. We find islands of pores and then large areas devoid of pores.

Wilson: Do you know why the thick lamina phenotype develops so slowly? Perhaps lamins have different rates of turnover at the periphery versus nuclear interior, as suggested by rates of fluorescence recovery after photobleaching (FRAP) experiments.

Goldman: Answering this question from the cultured HGPS cell lines is proving to be difficult because of the time it takes to grow up small numbers of cells. The results of fluorescence recovery experiments in normal cells transfected with GFP-HGPS mutant lamins showed a significant decrease in the rate of fluorescence recovery.

Burke: If one compares the stability of progerin to that of wild-type lamins expressed in HeLa cells there is no obvious difference.

Worman: In cells transfected with a plasmid that expresses lamin A with an epitope tag, the expressed lamin A seems to turn over in about 24 h (Östlund et al 2001). You would think that they are a lot longer lasting.

Gruenbaum: Is there a difference between transfected cells and progeria cells?

Burke: No. The cells transfected with either the progeric lamin or the wild-type lamin look exactly the same.

Morris: We've seen that lamin A with pathogenic mutations can form abnormal aggregates, or 'nuclear foci', when transfected into cultured cells. Similar foci have also been observed in the nuclei of skin fibroblasts from autosomal-dominant Emery-Dreifuss muscular dystrophy (AD-EDMD) patients. In a recent confocal microscopy study (Holt et al 2003), we found some clues as to how nuclear foci are formed. When MEFs (mouse embryonic fibroblasts) from $Lmna^{-/-}$ mice were transfected with lamin A carrying pathogenic mutations in the rod region, nuclear foci appeared to form initially as 'caps' of aggregation under the nuclear membrane (Fig. 1 *[Morris]*). These appeared to grow as invaginations into the nucleoplasm, either from the top or bottom surface of the nucleus and they spanned the entire nucleus in the very largest foci. In $Lmna^{-/-}$ MEFs, emerin is located in the cytoplasmic ER but re-locates to the nuclear rim when the absent lamin A is replaced by transfection. Not unnaturally, emerin also appeared to be attracted to the 'caps' of aggregated lamin A and was found *inside* the invaginations into the nucleoplasm (nuclear foci). Recent studies by Bechert et al (2003) appear superficially to conflict with these data, since nuclear foci were formed by all transfected lamin As (not just pathogenic mutant lamins) in HeLa cells, and emerin was found on the *outside* of these foci. However, closer examination shows that these data can also be consistent with the model shown in Fig. 1. The foci in HeLa cells were also formed at the nuclear rim but they bulged out of the nucleus, instead of invaginating, and, in this case, an outer layer of emerin around the aggregating lamin A was observed (Bechert et al 2003) as would be predicted from the model in Fig. 1 (*Morris*). The formation in HeLa cells of foci by normal and

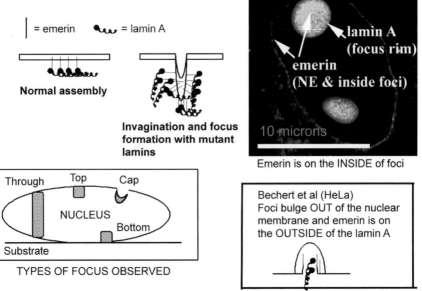

FIG. 1. (*Morris*) See text for details

lipodystrophic lamins, as well as by AD-EDMD lamins, may be partly attributable to the fact that these cells have an endogenous A-type lamina before transfection, whereas transfection of $Lmna^{-/-}$ MEFs creates an A-type lamina for the first time. Focus formation may occur more readily when the normal sites of lamin A accumulation are already occupied. This is not the whole story, however, since transfection of C2C12 mouse myoblasts with lamin A mutants (Ostlund et al 2001) gave results more similar to the MEF data than the HeLa cell data, so the very high levels of transfected lamin A expression obtained in the HeLa cell experiments may be an important factor. Under the right conditions, formation of nuclear foci can be a useful index of defective lamina assembly and might be of use in screening for drug treatments to ameliorate the deleterious effects of lamin mutations.

References

Bechert K, Lagos-Quintana M, Harborth J, Weber K, Osborn M 2003 Effects of expressing lamin A mutant protein causing Emery-Dreifuss muscular dystrophy and familial partial lipodystrophy in HeLa cells. Exp Cell Res 286:75–86

Bloom W, Fawcett DC 1962 Textbook of histology. 1st edn. WB Saunders Co.

Holt I, Ostlund C, Stewart CL, Nguyen thi Man, Worman HJ, Morris GE 2003 Effect of pathogenic mis-sense mutations in lamin A on its interaction with emerin in vivo. J Cell Sci 116:3027–3035

Östlund C, Bonne G, Schwartz K, Worman HJ 2001 Properties of lamin A mutants found in Emery-Dreifuss muscular dystrophy, cardiomyopathy and Dunnigan-type partial lipodystrophy. J Cell Sci 114:4435-4445

Ruchaud S, Korfali N, Villa P et al 2002 Caspase-6 gene disruption reveals a requirement for lamin A cleavage in apoptotic chromatin condensation. EMBO J 21:1967–1977

Aspects of nuclear envelope dynamics in mitotic cells

Brian Burke, Catherine Shanahan*, Davide Salina and Melissa Crisp

*Department of Anatomy and Cell Biology, University of Florida, 1600 SW Archer Rd, Gainesville FL FL 32610-0235, USA and *Department of Medicine, Cambridge University, Cambridge, UK*

Abstract. Major features of the nuclear envelope (NE) are a pair of inner and outer nuclear membranes (INM, ONM) spanned by nuclear pore complexes. While the composition of the ONM resembles that of the endoplasmic reticulum, the INM contains a unique spectrum of proteins. Localization of INM proteins involves a mechanism of selective retention whereby integral proteins are immobilized and concentrated by virtue of interactions with nuclear components. In the case of emerin, INM localization involves interaction with A-type lamins. Interactions between membrane proteins may also play a significant role in INM localization. This conclusion stems from studies on nesprins, a family of membrane proteins that feature a large cytoplasmic domain, a single C-terminal membrane-spanning domain and a small lumenal domain. The nesprin membrane anchor and lumenal (KASH) domains are related to the *Drosophila* Klarsicht protein. Evidence is emerging that this KASH region interacts with other NE proteins and may influence their distributions. Overexpression of GFP-KASH causes loss of emerin and LAP2 from the NE. This is not due to global reorganization of the NE since LAP1 as well as lamins and NPCs remain unaffected. Our results suggest that interactions between NE membrane components are far more extensive and complex than current models suggest.

2005 Nuclear organization in development and disease. Wiley, Chichester (Novartis Foundation Symposium 264) p 22–34

The nuclear envelope in interphase and mitosis

The evolution of compartmental boundaries has provided eukaryotic cells with unique mechanisms with which to regulate multiple activities. Nowhere is this more obvious than in the sequestration of chromosomes within a nuclear envelope (NE), effectively separating the nuclear processes of transcription and replication from translation. A prerequisite for normal cell growth and metabolism, then, is control of molecular traffic across the nuclear envelope. This enhanced level of regulation in eukaryotes, however, comes at the cost of considerably complicating the mechanics of cell division. Progression through mitosis requires that chromosomes within the cell nucleus engage with microtubules (MTs) of the mitotic spindle. In metazoans the mitotic spindle is a cytoplasmic structure, and consequently, in order for chromosomes to align at the spindle equator, the NE must be either partially or completely dispersed.

The most prominent features of the NE are a pair of inner and outer nuclear membranes (INM and ONM) (Gerace & Burke 1988, Gant & Wilson 1997). The ONM displays frequent connections with the endoplasmic reticulum (ER) and features numerous ribosomes. The INM, in contrast, contains a unique set of membrane proteins, is ribosome-free and maintains close contacts with chromatin (Wilson 2000). Despite biochemical differences, the INM and ONM are joined in regions where they are spanned by nuclear pore complexes (NPCs). In this way, the INM, ONM and ER form a single continuous membrane system. Metazoans contain an additional NE structure, the nuclear lamina (Gerace & Burke 1988). In mammalian somatic cells this appears as a thin ($\sim$20 nm) protein meshwork lining the INM and maintaining interactions with both chromatin and INM-specific proteins. The lamina is composed primarily of the A- and B-type lamin family of intermediate filament proteins and plays an essential role in the maintenance of both NE integrity and nuclear organization (Wilson et al 2001).

During mitosis in higher cells, the nuclear lamina and NPCs are disassembled, and at the same time, nuclear membrane components are dispersed throughout the cell (Yang et al 1997, Ostlund et al 1999, Ellenberg et al 1997). These events are regulated by phosphorylation of multiple NE components (Macaulay et al 1995, Gerace & Blobel 1980, Heald & McKeon 1990). While the majority of NE-specific proteins become distributed throughout the mitotic cytoplasm, certain NPC subunits (nucleoporins or Nups) and associated molecules, including Rae1, Nup107 and Nup133 become preferentially associated with kinetochores (Wang et al 2001, Belgareh et al 2001, Babu et al 2003). Another nucleoporin, Nup358, a component of the short (100 nm) filaments that extend from the cytoplasmic face of the NPC during interphase, relocates to both spindle microtubules and kinetochores (Joseph et al 2002). Conversely, certain mitotic checkpoint proteins, such as Mad1, Mad2 and Mps1 that are kinetochore-associated during mitosis, are associated with NPCs during interphase (Campbell et al 2001, Liu et al 2003). In yeast, this localization is mediated by Nup53p, part of a larger complex of NPC proteins (Iouk et al 2002). Remarkably, yeast strains deficient in Mad1p exhibit a reduced rate of nuclear protein import as well as decreased stability of the Nup53p complex (Iouk et al 2002). The implication is that there is a functional relationship between the mitotic apparatus and the nuclear envelope.

Prophase in higher cells is defined by condensation of chromatin, and initiation of events leading to NE breakdown and the dispersal of NE components. The disruption of the nuclear membranes and exposure of chromosomes to the cytoplasm marks the end of prophase. At this time, integral proteins of the INM and NPCs are lost from the nuclear periphery and become distributed throughout the cell (Chaudhary & Courvalin 1993, Ellenberg et al 1997, Yang et al 1997). Subcellular fractionation, and studies on nuclear disassembly and reassembly in *Xenopus* egg extracts suggest that dividing cells contain unique populations of NE-derived vesicles (Newport & Spann 1987, Vigers & Lohka 1991), the basis of models in which nuclear membrane breakdown is accomplished by a process of vesiculation. Other studies, primarily in mammalian systems, suggest that NE breakdown results in the intermingling of ER and INM components (Ellenberg et al 1997, Yang et al 1997). Furthermore, ultrastructural analyses in several mammalian cell types (Zeligs & Wollman 1979, Roos 1973, Robbins & Gonatas 1964) consistently reveal the detachment of membrane cisternae, often described as ER-like, from the nuclear periphery, without extensive vesiculation. Although mechanistically distinct, the notion of intermixing of nuclear membrane components with bulk ER during prophase can nevertheless be reconciled with data supporting the vesicular model (Collas & Courvalin 2000). Were membrane components to enter or to form microdomains within the ER, then subcellular fractionation would be anticipated to yield populations of microsomal vesicles enriched for NE components. Indeed certain nuclear membrane components, including members of the LAP2 family of INM proteins, appear not to be homogeneously distributed within the peripheral ER of cells exiting mitosis. Instead, local cytoplasmic concentrations of these proteins are frequently observed that persist into early G1. The existence of these relatively long-lived foci is consistent with the idea that at least some NE proteins do in fact form microdomains within the mitotic ER.

Cytoplasmic dynein in nuclear membrane dispersal

In the absence of mechanisms inducing vesiculation how is nuclear membrane disruption achieved? There is now a consensus emerging that at least in certain cell types cytoplasmic dynein has an important function in facilitating nuclear membrane dispersal (Burke & Ellenberg 2002). Dynein, a microtubule minus-end-directed motor protein, associates with the cytoplasmic face of the NE during early mitotic prophase. The action of dynein at this time is twofold. Firstly, NE-associated dynein stabilizes the growing ends of astral microtubules in the vicinity of the NE. Secondly, dynein mediates the withdrawal of NE components towards the separating centrosomes. This results in the application of tension across regions of the nuclear membranes that are distal to the centrosomes, and it is in these regions that significant breaches in the nuclear membranes first appear (Beaudouin et al 2002). In this way, cytoplasmic dynein functions to effectively tear open the nuclear membranes (Salina et al 2002). Although yet to be conclusively demonstrated, it is likely that membrane fenestrae, created by the disassembly of nuclear pore complexes, form the epicenter for the initial rupturing of the nuclear membranes.

The mechanisms by which dynein associates with the cytoplasmic face of the NE during mitotic prophase remains obscure. It is known that dynein binding to a variety of organelle membranes, including the NE is mediated by a large multiprotein assembly, the dynactin complex (Schroer 2000). While the dynein/ dynactin binding partner on the NE has yet to be identified it must possess certain atributes. For instance, it must be enriched in the ONM versus the peripheral ER, and furthermore, it must be able to transmit force to the INM and nuclear lamina. Such features initially lead to suggestions that NPCs might actually contain the dynein binding site (Salina et al 2002). Certainly they are concentrated within the NE, are exposed to the cytoplasm and are anchored to the nuclear lamina. However, our immunolocalization studies indicate that dynein is not specifically associated with NPCs during mitotic prophase, rather it is localized to regions of the ONM that lie between NPCs. The implication is that there are ONM specific membrane proteins that are somehow linked to the INM and lamina perhaps via protein complexes that span the perinuclear space. The most promising candidates for dynein/dynactin binding partners are currently represented by members of the nesprin family of nuclear membrane proteins (Zhen et al 2002, Apel et al 2000, Zhang et al 2001, Mislow et al 2002) or possible mammalian homologues of the *C. elegans* Zyg-12 protein (Malone et al 2003).

Outer nuclear membrane proteins

The nesprin protein family in mammalian cells (also known as syne, myne and NUANCE), consists of nesprins 1 and 2 including a variety of differentially spliced isoforms. The nesprins are type II membrane proteins featuring a C-terminal membrane anchor domain and small luminal domain. This C-terminal region of nesprins 1 and 2 shares sequence similarity with the *Drosophila* klarsicht protein (klar), a molecule that is involved in microtubule-dependent nuclear positioning in certain cell types (Starr & Han 2003). A *C. elegans* protein, ANC-1, which appears to be functionally related to the nesprins also shares this klarsicht homology domain (or KASH domain) (Starr & Han 2002). The bulk of each of the nesprin polypeptides, the largest of which exhibit molecular weights in the region of 800 kDa (!), are extra-lumenal and feature multiple spectrin repeats as well as N-terminal calponin homology domains capable of binding actin. At least some members of the nesprin protein family are localized exclusively on the ONM. In this way nesprin proteins may provide a direct link between the NE and the cytoskeleton. Concentration of membrane proteins in the ONM raises some intriguing mechanistic issues. Given the similarity of the ONM and ER how is this preferential localization in the ONM achieved? One possibility is that nesprins could be anchored laterally to NPCs. However, such an association is not borne out in immunocytochemical studies. Another possibility is that nesprin proteins are tethered via their KASH domain, either directly or indirectly, to components of the INM. Such a model predicts the existence of molecular connections spanning the perinuclear space. Such connections should in principle be able to transmit force from the ONM to the INM and nuclear lamina, a prerequisite for nesprin proteins as possible dynein binding partners. This role for the KASH domain makes the prediction that overexpression of the KASH domain alone should displace nesprins from the ONM. This prediction in fact turns out to be the case. Remarkably, cells overexpressing a GFP-KASH fusion protein, lose not only nesprins from the nuclear periphery but also two INM proteins, emerin and LAP2β. This effect is not due to global disruption of NE architecture since another INM protein, LAP1C (Senior & Gerace 1988), as well as A-type lamins and NPCs, are largely unaffected by GFP-KASH overexpression. These results implicate emerin and LAP2β as potential KASH interacting proteins. Other INM proteins that may form a link to the nesprin family are Sun1 and Sun2. These two proteins appear to be mammalian UNC-84 homologues. UNC-84 is known to be required for the appropriate localization of ANC-1, the *C. elegans* nesprin relative (Starr & Han 2002).

NPC proteins in mitotic progression

While NPCs have been ruled out as attachment sites for dynein in nuclear membrane disruption, certain NPC proteins nevertheless have an essential function in mitotic progression. In the case of one of these, Nup358, this mitotic function has been revealed using RNA interference approaches to specifically deplete cells of this protein. Nup358 is also known as Ran binding protein 2 (RanBP2) and is the major component of the 100 nm filaments that extend from the cytoplasmic face of NPCs (Walther et al 2002, Yokoyama et al 1995). Nup358 functions as a docking site for transport substrates in association with their cognate receptors. Nup358 also binds RanGAP1, the activating protein for Ran, a member of the Ras superfamily of small GTPases. Ran functions as a regulator of nucleocytoplasmic transport (Saitoh et al 1997) by defining the directionality of the transport process. Nup358 also has SUMO E3 ligase activity (SUMO being a small ubiquitin-like protein) (Pichler et al 2002). Indeed SUMO modification of RanGAP1 is required for its association with Nup358 (Mahajan et al 1997, Matunis et al 1998). The implication here is that Nup358 functions not just as a docking protein but also as a modulator of the cytoplasmic levels of Ran-GDP versus Ran-GTP. Surprisingly Nup358 is at least partly redundant with respect to the transport process. Nuclei assembled *in vitro* in *Xenopus* egg extracts depleted of Nup358 are transport competent, although the NPCs within these nuclei lack cytoplasmic filaments (Walther et al 2002). Similarly, depletion of Nup358 in HeLa cells by means of RNA interference has little affect on the qualitative import of proteins into the nucleus.

While siRNA-mediated depletion of Nup358 in HeLa cells has little obvious consequence on nuclear protein import, its effects on mitotic progression are dramatic. Depletion of this nucleoporin in HeLa cells results in abnormal spindle assembly associated with defective chromosome congression. The result is missegregation of chromatids. A role for Nup358 in spindle formation and chromatid segregation has also been highlighted in several other recent studies. In *C. elegans* early embryos (Askjaer et al 2002), an RNA interference approach revealed that depletion of Nup358 leads to a phenotype very similar to that observed in HeLa cells. In both experimental systems, while asters are present, chromosome capture is at best inefficient and only aberrant spindles are formed. Identical effects have been observed in *C. elegans* embryos depleted of CENP-A, a centromere-specific histone that is essential for normal kinetochore formation (Oegema et al 2001). These latter observations confirm an essential role for kinetochores in the establishment of normal spindle morphology.

There is now a growing body of evidence that Nup358 is required for normal kinetochore formation and assembly. Electron microscope (EM) studies of Nup358-depleted prometaphase cells reveal highly abnormal kinetochore morphologies that feature partial or complete loss of the characteristic trilaminar plate-like structure as well as incomplete condensation of subjacent centromeric heterochromatin. Significantly, certain kinetochore components, including the checkpoint proteins, Mad1, Mad2 and Zw10, are mislocalized in prometaphase cells depleted of Nup358. Antibody microinjection experiments have demonstrated that interference with the assembly of one of these, Zw10, leads to bypass of the spindle assembly checkpoint, the appearance of lagging chromatids and aneuploidy (Chan et al 2000). Similarly, depletion of the kinetochore protein CENP-E in mammalian cells gives rise to a spectrum of abnormalities that is virtually identical to those observed following Nup358 depletion. Indeed CENP-E is one of the kinetochore proteins that is mislocalized in cells depleted of Nup358. Very similar defects have also been reported in studies involving depletion of several other kinetochore proteins including Hec1, hMis12 and *Drosophila* Mast/Orbit (Martin-Lluesma et al 2002, Maiato et al 2002, Goshima et al 2003).

While depletion of Nup358 clearly perturbs kinetochore structure, preventing microtubule capture and leading to defective chromosome congression, it is not yet clear how Nup358 actually functions at the kinetochore. The most likely role for Nup358 might be related to its ability to attract and bind proteins of the Ran system. Indeed, RanGap1 and SUMO-I have both been found at the kinetochore (Joseph et al 2002). Given that SUMO-modified RanGAP1 binds Nup358 at the

NPC, it is tempting to imagine that Nup358 might perform a similar function at the kinetochore during mitosis. In *C. elegans*, depletion of RCC1, the nucleotide exchange factor for Ran, by RNA interference produces effects similar to, although less severe than, those observed following Nup358 depletion (Askjaer et al 2002). Since Ran has been shown to be essential for kinetochore-microtubule interaction these various Ran system components could well function to modulate the cycling of proteins on and off the kinetochore. A second, although not exclusive possibility is that the Nup358 SUMO ligase activity might be required for proper kinetochore organization. Support for this view is provided by the intriguing finding that SUMO-1 can act as a suppressor of certain CENP-C mutations in vertebrate cells (Fukagawa et al 2001).

Still unclear is why NPC or NE components play any role at all in mitotic progression. Perhaps as suggested by Joseph et al (2002) the cycling of proteins between the NE and mitotic apparatus might provide a fail-safe mechanism that defines the interphase versus mitotic status of the cell (Joseph et al 2002). In any event, this surprising bifunctional relationship may provide novel insights into the evolutionary origins of the NE and mitotic apparatus.

References

Apel ED, Lewis RM, Grady RM, Sanes JR 2000 Syne-1, a dystrophin- and Klarsicht-related protein associated with synaptic nuclei at the neuromuscular junction. J Biol Chem 275:31986–31995

Askjaer P, Galy V, Hannak E, Mattaj IW 2002 Ran GTPase cycle and importins alpha and beta are essential for spindle formation and nuclear envelope assembly in living Caenorhabditis elegans embryos. Mol Biol Cell 13:4355–4370

Babu JR, Jeganathan KB, Baker DJ et al 2003 Rae1 is an essential mitotic checkpoint regulator that cooperates with Bub3 to prevent chromosome missegregation. J Cell Biol 160:341–353

Beaudouin J, Gerlich D, Daigle N, Eils R, Ellenberg J 2002 Nuclear envelope breakdown proceeds by microtubule-induced tearing of the lamina. Cell 108:83–96

Belgareh N, Rabut G, Bai SW et al 2001 An evolutionarily conserved NPC subcomplex, which redistributes in part to kinetochores in mammalian cells. J Cell Biol 154:1147–1160

Burke B, Ellenberg J 2002 Remodelling the walls of the nucleus. Nat Rev Mol Cell Biol 3: 487–497

Campbell MS, Chan GK, Yen TJ 2001 Mitotic checkpoint proteins HsMAD1 and HsMAD2 are associated with nuclear pore complexes in interphase. J Cell Sci 114:953–963

Chan GK, Jablonski SA, Starr DA, Goldberg ML, Yen TJ 2000 Human Zw10 and ROD are mitotic checkpoint proteins that bind to kinetochores. Nat Cell Biol 2:944–947

Chaudhary N, Courvalin JC 1993 Stepwise reassembly of the nuclear envelope at the end of mitosis. J Cell Biol 122:295–306

Collas I, Courvalin JC 2000 Sorting nuclear membrane proteins at mitosis. Trends Cell Biol 10:5–8

Ellenberg J, Siggia ED, Moreira JE et al 1997 Nuclear membrane dynamics and reassembly in living cells: targeting of an inner nuclear membrane protein in interphase and mitosis. J Cell Biol 138:1193–1206

Fukagawa T, Regnier V, Ikemura T 2001 Creation and characterization of temperature-sensitive CENP-C mutants in vertebrate cells. Nucleic Acids Res 29:3796–3803

Gant TM, Wilson KL 1997 Nuclear assembly. Annu Rev Cell Dev Biol 13:669–695

Gerace L, Blobel G 1980 The nuclear envelope lamina is reversibly depolymerized during mitosis. Cell 19:277–287

Gerace L, Burke B 1988 Functional organization of the nuclear envelope. Ann Rev Cell Biol 4:335–374

Goshima G, Kiyomitsu T, Yoda K, Yanagida M 2003 Human centromere chromatin protein hMis12, essential for equal segregation, is independent of CENP-A loading pathway. J Cell Biol 160:25–39

Heald R, McKeon F 1990 Mutations of phosphorylation sites in lamin A that prevent nuclear lamina disassembly in mitosis. Cell 61:579–589

Iouk T, Kerscher O, Scott RJ, Basrai MA, Wozniak RW 2002 The yeast nuclear pore complex functionally interacts with components of the spindle assembly checkpoint. J Cell Biol 159:807–819

Joseph J, Tan SH, Karpova TS, McNally JG, Dasso M 2002 SUMO-1 targets RanGAP1 to kinetochores and mitotic spindles. J Cell Biol 156:595–602

Liu ST, Chan GK, Hittle JC et al 2003 Human MPS1 kinase is required for mitotic arrest induced by the loss of CENP-E from kinetochores. Mol Biol Cell 14:1638–1651

Macaulay C, Meier E, Forbes DJ 1995 Differential mitotic phosphorylation of proteins of the nuclear pore complex. J Biol Chem 270:254–262

Mahajan R, Delphin C, Guan T, Gerace L, Melchior F 1997 A small ubiquitin-related polypeptide involved in targeting RanGAP1 to nuclear pore complex protein RanBP2. Cell 88:97–107

Maiato H, Sampaio P, Lemos CL et al 2002 MAST/Orbit has a role in microtubule-kinetochore attachment and is essential for chromosome alignment and maintenance of spindle bipolarity. J Cell Biol 157:749–760

Malone CJ, Misner L, Le Bot N et al 2003 The C. elegans hook protein, ZYG-12, mediates the essential attachment between the centrosome and nucleus. Cell 115:825–836

Martin-Lluesma S, Stucke VM, Nigg EA 2002 Role of Hec1 in spindle checkpoint signaling and kinetochore recruitment of Mad1/Mad2. Science 297:2267–2270

Matunis MJ, Wu J, Blobel G 1998 SUMO-1 modification and its role in targeting the Ran GTPase-activating protein, RanGAP1, to the nuclear pore complex. J Cell Biol 140:499–509

Mislow JM, Kim MS, Davis DB, McNally EM 2002 Myne-1, a spectrin repeat transmembrane protein of the myocyte inner nuclear membrane, interacts with lamin A/C. J Cell Sci 115:61–70

Newport J, Spann T 1987 Disassembly of the nucleus in mitotic extracts: membrane vesicularization, lamin disassembly, and chromosome condensation are independent processes. Cell 48:219–230

Oegema K, Desai A, Rybina S, Kirkham M, Hyman AA 2001 Functional analysis of kinetochore assembly in Caenorhabditis elegans. J Cell Biol 153:1209–1226

Ostlund C, Ellenberg J, Hallberg E, Lippincott-Schwartz J, Worman HJ 1999 Intracellular trafficking of emerin, the Emery-Dreifuss muscular dystrophy protein. J Cell Sci 112:1709–1719

Pichler A, Gast A, Seeler JS, Dejean A, Melchior F 2002 The nucleoporin RanBP2 has SUMO1 E3 ligase activity. Cell 108:109–120

Robbins E, Gonatas NK 1964 The ultrastructure of a mammalian cell during the mitotic cycle. J Cell Biol 21:429–463

Roos UP 1973 Light and electron microscopy of rat kangaroo cells in mitosis. II. Kinetochore structure and function. Chromosoma 41:195–220

Saitoh H, Pu R, Cavenagh M, Dasso M 1997 RanBP2 associates with Ubc9p and a modified form of RanGAP1. Proc Natl Acad Sci USA 94:3736–3741

Salina D, Bodoor K, Eckley DM et al 2002 Cytoplasmic dynein as a facilitator of nuclear envelope breakdown. Cell 108:97–107

Schroer TA 2000 Motors, clutches and brakes for membrane traffic: a commemorative review in honor of Thomas Kreis. Traffic 1:3–10

Senior A, Gerace L 1988 Integral membrane proteins specific to the inner nuclear membrane and associated with the nuclear lamina. J Cell Biol 107:2029–2036

Starr DA, Han M 2002 Role of ANC-1 in tethering nuclei to the actin cytoskeleton. Science 298:406–409

Starr DA, Han M 2003 ANChors away: an actin based mechanism of nuclear positioning. J Cell Sci 116:211–216

Walther TC, Pickersgill HS, Cordes VC et al 2002 The cytoplasmic filaments of the nuclear pore complex are dispensable for selective nuclear protein import. J Cell Biol 158:63–77

Wang X, Babu JR, Harden JM et al 2001 The mitotic checkpoint protein hBUB3 and the mRNA export factor hRAE1 interact with GLE2p-binding sequence (GLEBS)-containing proteins. J Biol Chem 276:26559–26567

Wilson KL 2000 The nuclear envelope, muscular dystrophy and gene expression. Trends Cell Biol 10:125–129

Wilson KL, Zastrow MS, Lee KK 2001 Lamins and disease: insights into nuclear infrastructure. Cell 104:647–650

Vigers GP, Lohka MJ 1991 A distinct vesicle population targets membranes and pore complexes to the nuclear envelope in Xenopus eggs. J Cell Biol 112:545–556

Yang L, Guan T, Gerace L 1997 Integral membrane proteins of the nuclear envelope are dispersed throughout the endoplasmic reticulum during mitosis. J Cell Biol 137:1199–1210

Yokoyama N, Hayashi N, Seki T et al 1995 A giant nucleopore protein that binds Ran/TC4. Nature 376:184–188

Zeligs JD, Wollman SH 1979 Mitosis in thyroid follicular epithelial cells in vivo. III. Cytokinesis J Ultrastruct Res 66:288–303

Zhang Q, Skepper JN, Yang F et al 2001 Nesprins: a novel family of spectrin-repeat-containing proteins that localize to the nuclear membrane in multiple tissues. J Cell Sci 114:4485–4498

Zhen YY, Libotte T, Munck M, Noegel AA, Korenbaum E 2002 NUANCE, a giant protein connecting the nucleus and actin cytoskeleton. J Cell Sci 115:3207–3222

DISCUSSION

Wilson: Might the localization of nesprin 1α at the inner nuclear membrane also be disrupted by overexpressing the mutant you described?

Burke: Yes, it probably does. I didn't mention that we have done exactly the same experiments using the nesprin 1 KASH domain and we see the same effects: the two are interchangeable. I wouldn't like to leave you with the impression that we are doing something exclusive to the nesprin 2 family members.

Wilson: Nesprin 1α is the tightest binding partner that we know of so far for emerin. The affinity is 4 nM. When emerin became dislocalized away from the nuclear inner membrane, did it move to centrosomes? Which membranes are these blobs associated with?

Burke: They appear to be associated with the ER. They are not associated with centrosomes. We have looked at various other membrane compartments and endosomal compartments, and these aggregates appear to be restricted to the ER.

Starr: Is the part of the KASH domain that Brian Burke overexpresses the same as the domain of Nesprin protein that Kathy Wilson sees interacting with emerin?

Wilson: No. The KASH domain is behind the membrane, and the part that binds emerin is exposed.

Gruenbaum: What are the exact stages of nuclear envelope breakdown with respect to lamins, nuclear pore complexes and COP-1-dependent disassembly?

Burke: Starting with the role of dynein, at least in mammalian systems this appears to be facilitating nuclear membrane dispersal. It is obviously not essential for it because you can treat cells with nocodazole or interfere with dynein function in various ways, and still get nuclear membrane breakdown eventually. However, it is significantly delayed. In terms of the other NE components, the NPCs begin to disassemble quite quickly. The nuclear lamina also begins to disassemble during prophase. I would suggest that the disassembly of the lamina is a pre-requisite for this tearing open of the nuclear membranes during prometaphase. Jan Ellenberg, who had a companion paper to ours, was initially suggesting that the action of dynein was not only to tear open the nuclear membranes but also to facilitate ripping apart of the nuclear lamina. I don't buy that, and I don't think he buys that now.

Wilson: Mark Terasaki reported that pore complex disassembly leaves empty holes in the nuclear envelope (like those in a donut). With the NPC complex gone, the hole is no longer constrained, and can enlarge. This is a tenable way to explain nuclear membrane disassembly. I don't like the word 'tearing' because it implies a loss of integrity and spilling of the lumenal contents of the nuclear envelope, which would be unprecedented in cell biology.

Burke: That is a reasonable point, and the idea that I have always had in my mind is that vacated pore complexes, or membrane fenestrae left behind, form the epicentre for nuclear membrane breakdown.

Goldman: The nuclear pores have dozens of different components. It is not clear how these are sequentially dismantled during the nuclear disassembly process. I think Kathy Wilson is right, though: the use of the terms 'tearing' and 'ripping' bothers a lot of cell biologists. You showed very nicely that there is a controlled sequence of events that is controlled by a motor, dynein. If this is true then it is more of a dismantling than a tearing process. In addition, there has been no mention of the important and essential role of phosphorylation of nuclear envelope proteins by MPF which probably triggers many of the activities related to nuclear envelope breakdown. We know that at least lamins are an early substrate for $p34^{cdc2}$ kinase and that this phosphorylation event is essential for triggering nuclear envelope breakdown.

Burke: There is no question of that. If you treat cells with nocodazole and thus delay nuclear envelope breakdown, and then look at the A type lamins you see that they are clearly disassembled.

Goldman: When you say that it is slower, do you mean that the process changes? Cargo transport driven by dynein along microtubules is undoubtedly a highly regulated process in light of the remarkable complexity of the large number of proteins involved in normal dynein function. Therefore you would expect to see some differences in the distribution of one of the many components of the disassembled nuclear envelope in mitotic cells. When we compare a normal mitotic and a nocodazole-arrested mitotic cell, we find that the degree of disassembly and the distribution of lamins throughout the cytoplasm looks about the same.

Burke: I know what you are getting at. All I can say is that it is taking more time to get from A to B, but it isn't clear whether a different road has been taken.

Goldman: Timing of the process of nuclear breakdown during mitosis is obviously critical, and it may be the essential feature here.

Hutchison: I'd like to ask about your Nup358 distribution and the RNAi experiment. When you do this, do you generate a genuine anaphase checkpoint arrest? Is this a mechanism for making sure that the cell knows that each daughter is going to get approximately half of what the mother has provided?

Burke: With Nup358 RNAi we see a metaphase arrest. The spindle assembly checkpoint appears to be substantially intact. These experiments take days to do. Over a period of hours we see a gradual escape from the metaphase arrest and the accumulation of cells within the culture that contain multiple micronuclei and pairs of cells connected by mis-segregated chromosomes. Everything we see in these RNAi-treated cultures can be accounted for simply by the spindle assembly defect and then slow escape from the checkpoint.

Julien: There are now some mouse models of motor neuron disease with dynein mutations. There are also human beings with dynactin mutations. It might be interesting to look at the distribution of the nuclear components in these.

Burke: It would.

Davies: These mice are available from the repository at Harwell.

Malone: We have shown that ZYG-12 in *C. elegans* is a nuclear envelope protein that is required for the localization of dynein to the nuclear envelope. There is a two-hybrid interaction between a conserved domain of ZYG-12 and a dynein component (Malone et al 2003).

Burke: This is something I have been seriously thinking about. The only complication in terms of our work is that you see the interaction directly with one of the intermediate chains of dynein, whereas we can eliminate dynein binding to the nuclear envelope by messing with dynactin. The molecular details of the interaction seem to be different.

Malone: ZYG-12 is an integral membrane protein that might be in the outer nuclear envelope. In *C. elegans* we have identified several such proteins that have genetically defined cytoplasmic functions, and they all depend on the Sun family for their localization.

Shackleton: I wanted to mention some work that we have been doing on Sun1, the mouse homologue of UNC-84. We identified it as a lamin-binding protein in a two-hybrid screen. We found that it is the N-terminal domain of Sun1 that interacts with lamins and the C-terminal domain appears to be on the exterior of the nuclear envelope. This suggests that it spans both membranes and would therefore fulfil the criteria you are suggesting for being a lamin-binding protein. In fact, we have also shown that if Sun1 is depleted by RNAi, dynein binding to the nuclear envelope is reduced. I have not been able to show any direct interaction with any dynein component, though.

Goldman: Surrounding the nuclear surface is a cargo for dynein and kinesin, which for 90% of the cells that divide doesn't change during mitosis. These are the intermediate filaments. They are close relatives of lamins and they form a tight cage around the surface, and there have been reports that they anchor to the nuclear envelope. I wouldn't ignore these because they turn out to be major cargoes for molecular motors. We have found that $\sim 30\%$ of dynein in cultured fibroblasts is associated with intermediate filaments, and they form a tight cage around the nucleus.

Gerace: Do dynein inhibitors in interphase affect the positioning of the centrosome? This thought is directed at the notion that nuclear-associated dynein might have a function in interphase cells aside from the mitotic disassembly.

Burke: I don't know.

Malone: In *C. elegans* dynein isn't needed for establishing and maintaining the association of the centrosome with the envelope.

Gruenbaum: I would like to make a general comment. The bridging and interaction between the nucleoskeleton and cytoskeleton are probably dynamic, since vertebrate nuclei rotate independently of the cytoplasm. We should keep it in mind.

Goldman: Yosef Gruenbaum is referring to some old observations taken with time lapse movies showing that nuclei do rotate and spin. Just before mitosis nuclei frequently rotate for some unknown reason.

Burke: Exactly what is moving?

Gruenbaum: According to the literature, it is the whole nucleus.

Burke: If you look at a lot of these photobleaching experiments the lamina remains static.

Goldman: It recovers its fluorescence, which means that the spot doesn't move. The movements Yosef is talking about occur intermittently. The problem is that the nucleolus is usually used as a marker and we can't be sure that there isn't slippage inside the nucleus.

Starr: There are cell types where the nuclei are just moving around apparently randomly. Root hairs of *Arabidopsis* and myotubes are examples (Chytilova et al 2000, Englander & Rubin 1987).

Goldman: David Soll has beautiful data on *Dictyostelium* showing that the nuclei move coincident with cell locomotion, and that the position of the nucleus is critical for engagement in normal locomotory activity.

References

Chytilova E, Macas J, Sliwinska E, Rafelski SM, Lambert GM, Galbraith DW 2000 Nuclear dynamics in Arabidopsis thaliana. Mol Biol Cell 11:2733–2741

Englander LL, Rubin LL 1987 Acetylcholine receptor clustering and nuclear movement in muscle fibers in culture. J Cell Biol 104:87–95

Malone CJ, Misner L, Le Bot N et al 2003 The C. elegans Hook protein, ZYG-12, mediates the essential attachment between the centrosome and nucleus. Cell 115:825–836

Components of the nuclear envelope and their role in human disease

Howard J. Worman

Departments of Medicine and of Anatomy and Cell Biology, College of Physicians and Surgeons, Columbia University, 630 West 168th Street, 10th Floor, Room 508, New York, NY 10032, USA

Abstract. The nuclear envelope is composed of the nuclear lamina, nuclear pore complexes and nuclear membranes. The outer nuclear membrane is very similar to the rough endoplasmic reticulum. The pore membranes contain unique integral proteins and are associated with nuclear pore complexes. The inner nuclear membrane is associated with heterochromatin and the nuclear lamina, a meshwork of intermediate filament proteins called lamins. In humans, lamins are encoded by three genetic loci, *LMNA*, *LMNB1* and *LMNB2*. Mutations in *LMNA* cause a spectrum of inherited diseases, including autosomal dominant Emery-Dreifuss muscular dystrophy and related striated muscle disorders, partial lipodystrophies, a peripheral neuropathy and progeria syndromes. Eighty or more transmembrane proteins may reside primarily in the inner nuclear membrane but only several have been fairly well characterized. These include emerin, which is mutated in X-linked Emery-Dreifuss muscular dystrophy, LAP2, MAN1 and LBR. LBR binds to B-type lamins and chromatin proteins and shares sequence similarities with sterol reductases. Heterozygous mutations in LBR cause Pelger-Huët anomaly, characterized by morphologically abnormal neutrophil nuclei, and homozygous mutations cause HEM/Greenberg skeletal dysplasia, characterized by developmental abnormalities and 3 β-hydroxysterol-δ-14-reductase deficiency. Further studies of nuclear envelope proteins may uncover additional unsuspected relationships to human disease.

2005 Nuclear organization in development and disease. Wiley, Chichester (Novartis Foundation Symposium 264) p 35–50

The nuclear envelope is a membranous structure that surrounds the nucleus of eukaryotic cells. Its membrane can be divided into three morphologically distinct but continuous domains: the outer nuclear membrane, pore membranes and inner nuclear membrane (Worman & Courvalin 2000). The outer nuclear membrane is directly continuous with the rough endoplasmic reticulum, similarly containing ribosomes on its cytoplasmic surface. The pore membranes connect the inner and outer nuclear membranes at numerous points and are associated with the nuclear pore complexes. In mammals, the pore complexes are composed of approximately

50 proteins, most of which have been identified (Cronshaw et al 2002). The inner nuclear membrane is associated with heterochromatin and the nuclear lamina, a meshwork of intermediate filament proteins called lamins. Recently, mutations in nuclear lamins and integral proteins of the inner nuclear membrane have been shown to cause a wide range of inherited diseases. This paper will summarize some of our work on nuclear envelope components and their relationships to human diseases.

Human nuclear lamins

In humans, three genetic loci encode nuclear lamins. *LMNA*, which is localized to chromosome 1q21.2, encodes the 'A-type' lamins (Lin & Worman 1993, Wydner et al 1996) (Fig. 1). Lamins A and C, the major 'A-type' lamins, are identical for their first 566 amino acids with differences in their C-terminal tails (Fisher et al 1986, Lin & Worman 1993). Lamins A and C are expressed in most differentiated somatic cells but are lacking from some undifferentiated and cancer cells (see references in Worman & Courvalin 2000, 2002). *LMNA* also encodes lamin C2, a germ cell-specific isoform (Furukawa et al 1994), and lamin AΔ10, a minor somatic cell isoform (Machiels et al 1996). Lamin A is synthesized as a precursor protein, prelamin A, from which the last 18 amino acids removed by endoproteolytic cleavages after C-terminal farnesylation and carboxymethylation (Sinensky et al 1994).

Two human genes encode 'B-type' lamins. *LMNB1* on chromosome 5q23.3-q31.1 encodes lamin B1 (Lin & Worman 1995, Wydner et al 1996), which

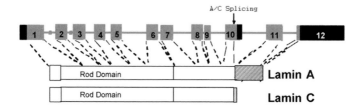

FIG. 1.   Structural organization of *LMNA*. The lamin A/C mRNA coding region is 12 exons (numbered 1 to 12; protein coding regions grey and regions coding untranslated RNA black). Alternative splicing of RNA encoded by exon 10 leads to the generation of lamins A and C. Lamins A and C are identical for their first 566 amino acids, including the head domains (unlabelled rectangular areas at left of each protein), central rod domains (labelled) and the first portions of their tail domains (unlabelled rectangular areas at right of each protein). Lamin A (actually prelamin A) has 98 unique amino acids (diagonal stripes) at its carboxyl-terminus and lamin C has six unique amino acids (grey shading). Splicing of RNA encoded by exon 10 may lead to the synthesis of mRNA for the minor isoform lamin AΔ10.

appears to be expressed in all somatic cells. *LMNB2* on chromosome 19p13.3 encodes lamin B2 (Biamonti et al 1992) and lamin B3, a germ cell-specific isoform (Furukawa & Hotta 1993). B-type lamins are farnesylated and carboxymethylated at their C-termini. Both B-type and A-type lamins bind to several integral proteins of the inner nuclear membrane (Worman & Courvalin 2000).

Integral proteins of the inner nuclear membrane

Topologies of integral membrane proteins synthesized on the rough endoplasmic reticulum, including those destined to reach the inner nuclear membrane, are determined by signal and stop-transfer sequences (Blobel 1980). Newly synthesized integral proteins of the inner nuclear membrane first appear in the rough endoplasmic reticulum (and essentially identical outer nuclear membrane) and reach the inner nuclear membrane by lateral diffusion (Ellenberg et al 1997). While most integral proteins of the endoplasmic reticulum membrane are generally also in the outer nuclear membrane (and perhaps some are continuously diffusing in and out of the inner nuclear membrane via the pore membranes), some are preferentially concentrated in the pore or inner nuclear membranes during interphase.

Several studies have demonstrated that integral proteins are targeted to and concentrated in the inner nuclear membrane by a 'diffusion-retention' mechanism (Soullam & Worman 1993, 1995, Ellenberg et al 1997, Östlund et al 1999, Wu et al 2002). All integral proteins synthesized on the rough endoplasmic reticulum could potentially reach the inner nuclear membrane by lateral diffusion in the interconnected membranes of the nuclear envelope, unless their diffusion is restrained by binding to cytoplasmic structures or resident endoplasmic reticulum proteins. If the size of the cytoplasmic domain of a protein is greater than that of the lateral channels of the nuclear pore complexes, it may also be unable to reach the inner nuclear membrane by diffusion (Soullam & Worman 1995, Wu et al 2002). Binding to components of the lamina and chromatin retains resident integral proteins in the inner nuclear membrane. Oligomerization of transmembrane segments may also contribute to retention.

A recent proteomics analysis suggests that approximately 80 transmembrane proteins are concentrated in or diffusing through the inner nuclear membrane during interphase (Schirmer et al 2003). Some of the better-characterized integral proteins of the inner nuclear membrane have been recently reviewed elsewhere (Worman & Courvalin 2000, Holaska et al 2002). Two inner nuclear membrane proteins, LBR and MAN1, are a major focus of research in our laboratory.

LBR is a polytopic protein with an amino-terminal, nucleoplasmic domain of approximately 200 amino acids followed by a hydrophobic portion with eight transmembrane segments (Worman et al 1990, Ye & Worman 1994) (Fig. 2).

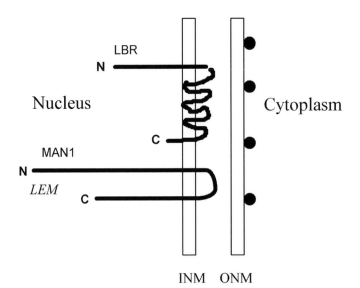

FIG. 2.   Schematic diagrams showing topologies of LBR and MAN1 in the inner nuclear membrane. LBR has an N-terminal nucleoplasmic domain followed by a hydrophobic domain with eight putative transmembrane segments in the inner nuclear membrane (INM). MAN1 has an N-terminal nucleoplasmic domain that contains a LEM domain at the N-terminus, two transmembrane segments and a second C-terminal nucleoplasmic domain. ONM indicates outer nuclear membrane with ribosomes (black ovals) on its outer surface. N and C respectively indicate N-termini and C-termini of proteins.

LBR was identified by its binding to B-type lamins (Worman et al 1988). The human *LBR* gene is on chromosome 1q42.1 (Wydner et al 1996); its first four protein coding exons encode the nucleoplasmic domain while the remainder encodes the hydrophobic portion (Schuler et al 1994). The hydrophobic portion of LBR is similar in sequence to sterol reductases of yeasts, plants, humans and other animals, which are diffusely localized throughout the endoplasmic reticulum (Holmer et al 1998). The nucleoplasmic domain of LBR, which binds to both B-type lamins and chromatin proteins (Ye & Worman 1994, 1996), can mediate inner nuclear membrane targeting (Soullam & Worman 1993).

MAN1, encoded by a gene on chromosome 12q14, is a member of the 'group' of proteins that contains the LEM domain (Lin et al 2000). The LEM domain is 50 amino acids and mainly composed of two large parallel alpha helices (Laguri et al 2001). It is found in LAP2, emerin and MAN1 (hence its name) as well as several other proteins, some of which are also in the inner nuclear membrane. MAN1 contains two transmembrane segments and two domains that face the nucleoplasm (Fig. 2). MAN1 and related proteins may regulate transcription. In

TABLE 1   Inherited diseases caused by mutations in lamins A and C

Striated Muscle Diseases (variable skeletal muscle involvement and cardiomyopathy)

Autosomal Dominant Emery-Dreifuss Muscular Dystrophy

Dilated Cardiomyopathy with Conduction Defect 1

Limb Girdle Muscular Dystrophy Type 1B

Partial Lipodystrophy Syndromes

Dunnigan-type Partial Lipodystrophy

Mandibuloacral Dysplasia (with developmental anomalies)

Peripheral Neuropathy

Charcot-Marie-Tooth Disorder Type 2B1

'Premature Ageing' Syndromes

Hutchinson-Gilford Progeria

Atypical Werner Syndrome

The diseases can be grouped into four categories: (1) disorders affecting striated (skeletal and cardiac) muscle, (2) partial lipodystrophy syndromes with or without developmental abnormalities, (3) peripheral neuropathy and (4) syndromes with features of 'premature ageing.'

Xenopus, the second nucleoplasmic domain of MAN1 binds Smad1 and antagonizes bone morphogenetic protein signalling (Osada et al 2003). A cytoplasmic protein called SANE also regulates bone morphogenetic protein signalling (Raju et al 2003). SANE contains a LEM domain and region similar to the second nucleoplasmic domain of MAN1 but lacks the portion identified by Wu et al (2002) responsible for inner nuclear membrane targeting.

Inner nuclear membrane proteins and inherited diseases

Emerin, an integral protein of the inner nuclear membrane, was initially identified by positional cloners as mutated in X-linked Emery-Dreifuss muscular dystrophy (Bione et al 1994). Subsequently, mutations in lamins A and C were shown to cause several human inherited diseases (Table 1). These include autosomal dominant Emery-Dreifuss muscular dystrophy, which is clinically indistinguishable from the X-linked form caused by emerin mutations, related diseases of skeletal and cardiac muscle, Dunnigan-type partial lipodystrophy, mandibuloacral dysplasia and a Charcot-Marie-Tooth type II peripheral neuropathy. Since our most recent previous review (Worman & Courvalin 2002), a dominant *de novo* mutation in lamin A has also been shown to cause Hutchinson-Gilford progeria syndrome, a condition with features of premature aging (Eriksson et al 2003, De Sandre-Giovannoli et al 2003). This mutation creates an aberrant RNA splice donor site, leading to expression of a truncated prelamin A and apparently reduced concentrations of wild-type lamins A and C. Mutations in lamins A and C have also recently been shown to cause atypical Werner syndrome, another disorder with features of premature aging (Chen et al 2003).

Positional cloning has shown that mutations in LBR cause Pelger-Huët anomaly, an autosomal dominant condition characterized only by abnormal nuclear shape and chromatin organization in blood neutrophils (Hoffmann et al 2002). Recently, homozygous LBR mutations have been identified in a subject with HEM/Greenberg skeletal dysplasia, a usually lethal condition with 3 β-hydroxysterol-δ-14-reductase deficiency (Waterham et al 2003). These findings are consistent with LBR being a 'bi-functional' protein, with a nucleoplasmic domain involved in chromatin and lamin binding and a hydrophobic domain with sterol reductase activity.

Summary of conclusions

Work by several cell biology laboratories has provided us with a rudimentary picture of the nuclear envelope and its components. However, there are many more components that remain to be characterized, especially with regards to their functions. Recently, the nuclear envelope has attracted the interests of a wide range of investigators because of the increasing number of inherited diseases caused by mutations in its protein components. Some of the most pressing questions arise from its role in these diseases. How do mutations in lamins A and C cause a wide variety of clinical phenotypes? How do mutations in these proteins, which are expressed in virtually all differentiated somatic cells, lead to tissue-specific disorders? How do mutations in emerin and lamins cause the same disease? How can Pelger-Huët anomaly and HEM/Greenberg skeletal dysplasia, caused respectively by heterozygous and homozygous mutations in LBR, be explained by what is known about the protein's functions? These questions have led to a coming together of cell biologists, geneticists and clinicians at this Novartis Foundation Symposium.

Acknowledgements

I apologize to the authors of papers that could not be cited because of space limitations. The National Institutes of Health, Muscular Dystrophy Association, American Diabetes Association, American Cancer Society and Human Frontiers Science Program have supported most of our work discussed in this review.

References

Biamonti G, Giacca M, Perini G et al 1992 The gene for a novel human lamin maps at a highly transcribed locus of chromosome 19 which replicates at the onset of S-phase. Mol Cell Biol 12:3499–3506

Bione S, Maestrini E, Rivella S et al 1994 Identification of a novel X-linked gene responsible for Emery-Dreifuss muscular dystrophy. Nat Genet 8:323–327

Blobel G 1980 Intracellular protein topogenesis. Proc Natl Acad Sci USA 77:1496–1500

Chen L, Lee L, Kudlow BA et al 2003 *LMNA* mutations in atypical Werner's syndrome. Lancet 362:440–445

Cronshaw JM, Krutchinsky AN, Zhang W, Chait BT, Matunis M 2002 Proteomic analysis of the mammalian nuclear pore complex. J Cell Biol 158:915–927

De Sandre-Giovannoli A, Bernard R, Cau P et al 2003 Lamin A truncation in Hutchinson-Gilford progeria. Science 300:2055

Ellenberg J, Siggia ED, Moreira JE et al 1997 Nuclear membrane dynamics and reassembly in living cells: targeting of an inner nuclear membrane protein in interphase and mitosis. J Cell Biol 138:1193–1206

Eriksson M, Brown WT, Gordon LB et al 2003 Recurrent *de novo* point mutations in lamin A cause Hutchinson-Gilford progeria syndrome. Nature 423:293–298

Fisher DZ, Chaudhary N, Blobel G 1986 cDNA sequencing of nuclear lamins A and C reveals primary and secondary structural homology to intermediate filament proteins. Proc Natl Acad Sci USA 83:6450–6454

Furukawa K, Hotta Y 1993 cDNA cloning of a germ cell specific lamin B3 from mouse spermatocytes and analysis of its function by ectopic expression in somatic cells. EMBO J 12:97–106

Furukawa K, Inagaki H, Hotta Y 1994 Identification and cloning of an mRNA coding for a germ cell-specific A-type lamin in mice. Exp Cell Res 212:426–430

Hoffmann K, Dreger CK, Olins AL et al 2002 Mutations in the gene encoding the lamin B receptor produce an altered nuclear morphology in granulocytes (Pelger-Huët anomaly). Nat Genet 31:410–414

Holaska JM, Wilson KL, Mansharamani M 2002 The nuclear envelope, lamins and nuclear assembly. Curr Opin Cell Biol 14:357–364

Holmer L, Pezhman A, Worman HJ 1998 The human lamin B receptor/sterol reductase multigene family. Genomics 54:469–476

Laguri C, Gilquin B, Wolff N et al 2001 Structural characterization of the LEM motif common to three human inner nuclear membrane proteins. Structure 9:503–511

Lin F, Worman HJ 1993 Structural organization of the human gene encoding nuclear lamin A and nuclear lamin C. J Biol Chem 268:16321–16326

Lin F, Worman HJ 1995 Structural organization of the human gene (*LMNB1*) encoding nuclear lamin B1. Genomics 27:230–236

Lin F, Blake DL, Callebaut I et al 2000 MAN1, an inner nuclear membrane protein that shares the LEM domain with lamina-associated polypeptide 2 and emerin. J Biol Chem 275:4840–4847

Machiels BM, Zorenc AH, Endert JM et al 1996 An alternative splicing product of the lamin A/C gene lacks exon 10. J Biol Chem 271:9249–9253

Osada S, Ohmori SY, Taira M 2003 XMAN1, an inner nuclear membrane protein, antagonizes BMP signaling by interacting with Smad1 in *Xenopus* embryos. Development 130:1783–1794

Östlund C, Ellenberg J, Hallberg E, Lippincott-Schwartz J, Worman HJ 1999 Intracellular trafficking of emerin, the Emery-Dreifuss muscular dystrophy protein. J Cell Sci 112:1709–1719

Raju GP, Dimova N, Klein PS, Huang HC 2003 SANE, a novel LEM domain protein, regulates bone morphogenetic protein signaling through interaction with Smad1. J Biol Chem 278:428–437

Schirmer EC, Florens L, Guan T, Yates JR 3rd, Gerace L 2003 Nuclear membrane proteins with potential disease links found by subtractive proteomics. Science 301:1380–1382

Schuler E, Lin F, Worman HJ 1994 Characterization of the human gene encoding LBR, an integral protein of the nuclear envelope inner membrane. J Biol Chem 269:11312–11317

Sinensky M, Fantle K, Trujillo M, McLain T, Kupfer A, Dalton M 1994 The processing pathway of prelamin A. J Cell Sci 107:61–67

Soullam B, Worman HJ 1993 The amino-terminal domain of the lamin B receptor is a nuclear envelope targeting signal. J Cell Biol 120:1093–1100

Soullam B, Worman HJ 1995 Signals and structural features involved in integral membrane protein targeting to the inner nuclear membrane. J Cell Biol 130:15–27

Waterham HR, Koster J, Mooyer P et al 2003 Autosomal recessive HEM/Greenberg skeletal dysplasia is caused by 3 beta-hydroxysterol delta 14-reductase deficiency due to mutations in the lamin B receptor gene. Am J Hum Genet 72:1013–1017

Worman HJ, Courvalin JC 2000 The inner nuclear membrane. J Membr Biol 177:1–11

Worman HJ, Courvalin JC 2002 The nuclear lamina and inherited disease. Trends Cell Biol 12:591–598

Worman HJ, Yuan J, Blobel G, Georgatos SD 1988 A lamin B receptor in the nuclear envelope. Proc Natl Acad Sci USA 85:8531–8534

Worman HJ, Evans CD, Blobel G 1990 The lamin B receptor of the nuclear envelope inner membrane: a polytopic protein with eight potential transmembrane domains. J Cell Biol 111:1535–1542

Wu W, Lin F, Worman HJ 2002 Intracellular trafficking of MAN1, an integral protein of the nuclear envelope inner membrane. J Cell Sci 115:1361–1371

Wydner KL, McNeil JA, Lin F, Worman HJ, Lawrence JB 1996 Chromosomal assignment of human nuclear envelope protein genes *LMNA*, *LMNB1*, and *LBR* by fluorescence *in situ* hybridization. Genomics 32:474–478

Ye Q, Worman HJ 1994 Primary structure analysis and lamin B and DNA binding of human LBR, an integral protein of the nuclear envelope inner membrane. J Biol Chem 269:11306–11311

Ye Q, Worman HJ 1996 Interaction between an integral protein of the nuclear envelope inner membrane and human chromodomain proteins homologous to *Drosophila* HP1. J Biol Chem 271:14653–14656

DISCUSSION

Wilson: MAN1 is an interesting protein. Yosef Gruenbaum and I have evidence that MAN1 is essential for cell viability in *C. elegans*. Furthermore reducing (but not eliminating) MAN1 in emerin-null cells caused synthetic lethality (Liu et al 2003). MAN1 is clearly doing something important. Can you comment on how your result relates to evidence that MAN1 binds Smad1, to mediate bone morphogenetic protein (BMP) signalling in *Xenopus* (Raju et al 2003, Osada et al 2003)? MAN1 now appears to mediate signal transduction for the TGFβ family, as you've just shown, in addition to BMPs.

Worman: I'd say that the difference between Smad1, 2 and 3 may be at a molecular level. Perhaps in one organism MAN1 is binding more strongly to Smad1 and in another organism it is binding more strongly to Smad2. We weren't able to show binding to Smad1. We didn't look at anything in that pathway. In the *Xenopus* paper you refer to, Osada et al (2003) said there was weak binding to Smad2 and 3, but then they didn't look at consequences of TGFβ signalling. They got this through a genetic screen looking at BMP signalling. Can MAN1 somehow be involved in regulating signalling mediated by Smad1, 2 or 3 in different tissues and organisms? This is all I can guess.

Young: We have made *Man1* knockout mice and the heterozygotes are normal. The homozygotes die at embryonic day 8.5–9.5. *Man1* is expressed in many different tissues in adult mice. In embryos, *Man1* is expressed in a lot of tissues, including prominent expression in the neural tube.

Wilson: Is *Man1* truly absent from certain tissues, or is it just lower-level expression?

Young: I can't really say. β-galactosidase staining is less sensitive than a Northern blot in our hands.

Stewart: A few years ago we characterized embryos deficient in the TGFβ-regulated transcription factor Smad2 (Heyer et al 1999). Embryos lacking Smad2 die due to gastrulation defects, as Smad2 is important for mediating interactions between the embryonic endoderm and the embryonic ectoderm during gastrulation. It is possible that loss of MAN1 is affecting critical developmental interactions during gastrulation in the embryo.

Gerace: Howard Worman, do you propose that MAN1 is involved in positive stimulation of the signalling pathway, or attenuation of it?

Worman: From the overexpression studies we think it is attenuation. When we do this, the TGFβ response goes down. Can it do the opposite in real cells in real animals? I don't know. I think work that could potentially be done with knockout mice would be very interesting. There are a lot of physiological responses to TGFβ and things that signal through the same pathway that you could look at in these mice to see whether they have an attenuated or enhanced response. The knockout mice give us somewhere to look.

Capeau: I wonder whether something could be done by acting on TGFβ signalling pathways in patients with lamin mutations. Normally TGFβ is inhibiting cell growth. If TGFβ signalling was enhanced in patients' cells, this would result in altered growth. In that situation, inhibition of TGFβ could possibly reverse the abnormal growth seen in premature ageing. Has anyone done experiments on TGFβ signalling in cells from patients?

Worman: We have tried to do such experiments, but they aren't easy. These cells are difficult to culture and they don't grow much. We have tried with cells from patients with Emery-Dreifuss muscular dystrophy and in Colin Stewart's knockout mouse cells, but we have not obtained good responses to TGFβ.

Stewart: In the mouse there are some spontaneous mutations in the lamin B receptor (LBR)(Shultz et al 2003). These mice are known as the Ichthyosis line, and they are viable but they are bald. Is there a connection between LBR and alopecia in progeria? Is there any idea how LBR could be involved in a failure of the hair to regenerate?

Oshima: In progeria it is the hair follicle cells that are affected and which fail to divide.

Goldman: Patients lose their hair, but how do people know that it is a basic defect in the hair follicle that is responsible?

Oshima: I don't know.

Gruenbaum: It is an interesting issue; whether the diseases caused by LBR are due to loss of LBR enzymatic activity. At least in *Drosophila*, Georg Krohne has recently shown that while *LBR* is an essential gene, the mutants don't have this sterol C-14 reductase activity (Wagner et al 2004). It is an essential gene

Worman: Do *Drosophila* make cholesterol?

Gruenbaum: This is another interesting point: although the structure of the *LBR* gene looks the same, *Drosophila* cells do not have a sterol C-14 reductase activity. In the mouse lamin A knockout, what happens to MAN1? Does it stay in the nuclear periphery or is it displaced like emerin?

Stewart: We don't know yet. We are still characterizing an antibody we raised to the N-terminus of MAN1, which looks promising.

Goldman: The fact that it is termed the LBR implies that it binds lamin B in some way.

Worman: That is how it was discovered, as a protein that binds to B-type lamins.

Goldman: Is LBR required to get lamin B in the nuclear lamina?

Worman: Probably not. There's an old experiment in which Eric Nigg expressed lamin B2 in *Schizosaccharomyces pombe* and got a lamina (Enoch et al 1991).

Goldman: People argue about whether it actually got in or was stuck on the outside. It may be transported to the nucleus but it might not get through the nuclear pore and form a proper lamina.

Worman: There's also a paper from Hoffmann et al (2002) where they had LBR knockout cells from a subject with homozygous Pelger-Huët anomaly. I think they did label with anti-lamin B antibodies and the lamin B was primarily localized to the nuclear envelope. So there are other things that bind lamin B apart from the LBR.

Goldman: Are these specific for lamin B? Does LBR bind lamin A?

Worman: From experiments we did a long time ago, binding of A-type lamins to lamin-depleted nuclear envelopes seemed weak (Worman et al 1988).

Burke: You can turn the whole issue around. Lamin B binding might reflect the fact that for some reason you need to localize the LBR itself to the inner nuclear membrane. There may be some aspect of sterol metabolism that requires this protein specifically localized to the nuclear envelope. There are odd connections between the nuclear envelope and sterol metabolism.

Worman: There are also oxysterol receptors that are nuclear receptors (Janowski et al 1996). One wonders whether the cell may want to reduce sterols to enhance their binding to these receptors to activate a signalling pathway.

Burke: Those lamins might be receptors for LBR rather than the other way round.

Worman: If the function is reducing sterols, then it wouldn't matter where LBR is found, unless it is something special bound to something in the nucleus.

Goldman: During nuclear assembly following mitosis, is LBR associated with chromatin prior to lamin B?

Worman: LBR is one of the first things targeted in reassembly isn't it?

Courvalin: Yes.

Goldman: At that time, is it associated with membranes?

Courvalin: Yes.

Goldman: So before it begins to decondense, chromatin is associated with LBR and membranes.

Courvalin: Yes, and there are no lamins associated with chromosomes at that time.

Gruenbaum: We should keep in mind that during mitosis there are always lamins associated with the membranes and come with the membranes when the nuclear envelope re-assembles.

Goldman: It is true that there is a membrane fraction in mitotic cells that contains lamin. I would warn everyone that while light microscopy is a great tool, you must take such results with a grain of salt. The limit of the resolution is significant. The most recent revelation of how many components are in the inner nuclear envelope membrane must make us all think about what really is interacting with what.

Gerace: Do you think these results with MAN1 are relevant to the phenotype where you get inhibition of transcription when you release lamins from their normal site in the nuclear periphery? Could this be a release of transcriptional repressors?

Worman: We have tried this in some cells with lamin mutations, but we couldn't get it to work.

Goldman: There is a literature on RB binding to lamin, which dates back about 10 or more years.

Bonne: I am not a cell biologist and so this may be a naïve observation, but in one of your pictures when you have cells with lamin A mutations some of the cells show disturbed nuclear membranes with blebbing. The opposite is seen with LBR mutations: neutrophils lose their blebbing. Is there something underlying this? On one side, lamin mutation is increasing the blebs and on the other side, depletion of LBR decreases them.

Goldman: Nuclear blebs have been seen for many years. Now it is fashionable to talk about them. They certainly have functional implications, but no one really knows what is going on. Pathologists have noticed for years that transformation from a premalignant to a malignant state is frequently accompanied by nuclear shape changes. Nuclear morphology and positioning within cells are frequently used diagnostically. Furthermore it appears that many tumours have lobulated nuclei as well as an abnormal distribution of lamins.

Bonne: When we observe the very first cells of patients we have only a few aberrant nuclei. Specialists said that this was normal. We had difficulty in working out whether this was specific or not.

Hutchison: Loss of expression of A-type lamins is very common at the point at which tumours progress. So why don't people with laminopathies more commonly get tumours? They live long enough.

Oshima: There is one case of Hutchinson-Gilford progeria where the patient died of a mesenchymal tumour in the chest wall. This is rare.

Goldman: For all these laminopathies, there doesn't seem to be a connection with tumours.

Stewart: Harry Dietz has described a new mutation in the lamin A gene that causes cardiocutaneous progeria (personal communication). This mutation seems to be associated with the development of skin cancers, within one family carrying the mutation.

Worman: Most of the tumours we have looked at still do have the normal amounts of lamins A and C. It is rare tumours that don't.

Stewart: We have introduced our lamin deficient mutation into a background where the mice are deficient for the cycle regulators p16/p19^{arf}. The p16/p19^{arf} homozygous null mice tend to develop tumours starting at around six months. However in mice that are p16/p19^{arf} null, but heterozygous for lamin A deficiency, the onset of detectable tumours starts at around two months.

Fatkin: Dr Worman's comments open up the very interesting question of the tissue-specific phenotypes associated with lamin mutations. One thing that hasn't been determined yet is the relative levels of lamin A/C expression in different tissues. For example, in cardiomyopathies associated with *LMNA* mutations, there is generally no clinical evidence of skeletal muscle involvement. Whether this reflects differences in the level of expression of lamin A/C in the heart and skeletal muscle is not known. We also don't know whether there is associated up-regulation or down-regulation of lamin B. A major limitation of human studies is the difficulty in getting tissue samples for these types of comparisons.

Capeau: We have looked at lamin levels in tissues from patients and there is no change from normal.

Goldman: Do you do this by quantitative Western blotting of total protein derived from tissues?

Capeau: Yes. However, when we looked at the nuclear morphology in cells from patients with laminopathies, we observed a decreased labelling of lamin B in the blebs, while lamins A/C were present. This is a general finding. I don't know why. In cancer cells with abnormal shaped nuclei, is there something wrong with the partition of lamins?

Goldman: We don't know whether A and B interact *in vivo*. I don't know of anyone that has done an A/B binding experiment.

Gerace: In vitro they avidly bind to each other.

Goldman: Do they actually interact in the cell, though? We know cells will survive quite nicely with just B type lamins. A is only turned on during gastrulation when cells begin to differentiate. Yet when A is mutated, many people are now finding that there are alterations in B, which is probably the best evidence that they interact in some way *in vivo*.

Morris: If you see local reductions in lamin B where lamin A is present using antibodies, it might be due to masking rather than a reduction of lamin B.

Goldman: That is an important point. Masking of epitopes is quite common, especially in the world of intermediate filaments, which form extremely tight, complex polymers, making epitopes very difficult to get to.

Bonne: In various cells of the laminopathy patients the nuclei have abnormal shapes, and we have repeatedly observed the absence of lamin B in nuclear blebs. Do you think this is due to epitope masking?

Goldman: No, but you have to be careful using just immunofluorescence microscopy, and it is hard to get other forms of evidence. Ultrastructure would help. I am assuming this is correct, because we see the same thing in a lot of cell lines that we have been growing from patients. One has to be cautious in interpreting whether or not there is a B-rich region without A and vice versa.

Wilson: The herniations seen in lamin-disrupted nuclei appear to lack many nuclear envelope structural proteins. This might be expected if herniations are places where nuclear lamina structure is (by definition) disrupted or absent. There may be different mechanisms of herniation, which could be informative. For example lamin A-null mouse cells have a 'softened' bulging region of the nuclear envelope, which might be mechanistically different from a 'Mickey Mouse ear' herniation.

Goldman: It can be very dynamic. Werner Greene and colleagues have reported that in cells transfected with VPR1, the HIV protein that is involved in disrupting cell cycle progression, there are transient herniations and the nucleus actually breaks open. Nucleoplasmic contents come out of the nucleus and cytoplasmic contents go into the nucleus, but then it reseals and the cell looks perfectly normal.

Burke: We have seen exactly the same thing in the lamin A knockout fibroblasts.

Goldman: Another interesting feature in the VPR story is that there seems as if there is a weak point at one pole of an elliptically shaped nucleus.

Burke: That is probably true. But it is impossible to define which pole happens to be weak with respect to any other landmark in the cell. There is no obvious correlation. I would add that we have preliminary data showing that we can make these herniations go away by treating the cells with cytochalasin B.

Disrupting the actin cytoskeleton causes them to vanish, and when the cytochalasin is washed out they come back again.

Stewart: Do these herniations or blebs always correspond to the electron microscopy (EM) pictures where the outer nuclear membrane is separated from the inner nuclear membrane?

Burke: They must do, because this is the only region in the EM pictures which don't have nuclear pore complexes. By definition you can't have a nuclear pore complex in something blown out like that. These herniations are devoid of pores when viewed at the light microscope level.

Goldman: They seem to be accumulated in the neck region of blebs.

Burke: If you look at the distribution of pores in the lamin A null fibroblasts, they are more or less uniform across most of the nucleus, except for one herniated region. There are other lamin mutant cells with dramatic pore complex clustering and far worse looking herniations.

Capeau: I am not sure these blebs are so transient. At the beginning we made videos and in one cell we saw the blebs remaining for 24 h.

Goldman: In some cases blebbing is transient. Also, some of these transient blebbing structures could be related to cell cycle stages. We find that cells derived from Hutchinson-Gilford progeria patients grow slowly, and this may also be a relevant factor.

Wilson: Is there evidence for interactions between lamin A and B *in vivo*? When we expressed a BAF mutant in living cells, the assembly of A-type lamins was completely blocked as cells exited mitosis with no detectable effect on the assembly of B type lamins (Haraguchi et al 2001). The A and B lamin networks might be completely independent of each other.

Goldman: I don't think they are completely independent. When we put dominant negative mutants into cultured mammalian cells, A and B type lamins are altered.

Wilson: There are a handful of proteins capable of binding both A and B type lamins. You could have a domino effect on B-type lamins because you messed up proteins that bind both types.

Gruenbaum: One of the most important areas in our field is lamin structure. Our current knowledge only includes the structure of the lamina in *Xenopus* oocytes *in vivo* or the head-to-tail polymerization, the IgG fold and lamin dimerization *in vitro*. There is currently only one lamin that makes stable intermediate filaments *in vitro*, the *C. elegans* lamin, as shown by Klaus Weber and Ueli Aebi (Karabinos et al 2003).

Young: I have a comment about EM. With the lamin B1 knockout fibroblasts the blebbing observable by light-level immunocytochemical studies is stunning: it is worse than any blebbing that we have ever seen with any lamin A mutations. But when we look by EM, we really don't see any abnormalities. Part of this could be that we are not experts at this kind of thing, but another explanation could be that thin-section EMs are only about 1/500th of the thickness of a nucleus. Despite the fact that we know that blebbing is present in 50% of cells — when we look at hundreds of EMs — we really don't see much difference between the knockout cells and the wild-type cells, whether it be in the nuclear lamina itself, the inner or outer nuclear membranes, or the distribution of the chromatin. I think this has to do with the sampling of very thin sections.

Goldman: So you see abnormally blebbed nuclei in the B1 knockout.

Young: Yes, it is stunning, but when we look at EMs we don't see anything.

Wilson: Have you done serial sections to 3D reconstruct?

Young: No. We have done a sampling of thin sections.

Goldman: You did say before that you had some compensation by lamin B2.

Young: In terms of gene expression, Dr Reue has obtained preliminary data indicating that lamin B2 expression may be increased in some tissues. In the lamin B1-deficient fibroblasts, there is a lot of blebbing. We have done a lot of EM studies with *Zmpste24* knockout cells (which have an accumulation of prelamin A). We don't see anything wrong with these cells by EM, even though we see blebbing using light-level immunocytochemical techniques. We have looked at hundreds of high quality EMs. I have concluded that blebs are difficult to observe by EM of thin sections.

Gerace: One other possibility needs to be excluded. Presumably these blebs are dynamic, and when cells are chemically fixed, potential fixation artefacts could cause blebs to disappear.

Young: That is a potential explanation. However, we added fixative directly to the cells that were growing on the plastic plates.

Goldman: It would be interesting to see whether any of these structural changes make a difference in terms of nuclear function.

Young: The nuclei of these cells are leaky.

Goldman: Leakiness could cause enormous problems, especially if it is bidirectional between the nucleus and the cytoplasm.

Collas: What about the localization of inner nuclear membrane proteins in the blebs. Are there any?

Goldman: This would be easy to see by EM.

Bonne: Emerin is still present in the blebs.

Goldman: It is quite possible that lobulation or blebbing has something to do with nuclear mechanics. When the shape of the nucleus changes, the positions of nucleoli and chromatin are also altered. These positional changes could cause malfunctions of various nuclear components.

References

Enoch T, Peter M, Nurse P, Nigg EA 1991 p34cdc2 acts as a lamin kinase in fission yeast. J Cell Biol 112:797–807

Haraguchi T, Koujin T, Segura-Totten M et al 2001 BAF is required for emerin assembly into the reforming nuclear envelope. J Cell Sci 114:4575–4585

Heyer J, Escalante-Alcalde D, Lia M et al 1999 Postgastrulation Smad2-deficient embryos show defects in embryo turning and anterior morphogenesis. Proc Natl Acad Sci USA 96: 12595–12600

Hoffmann K, Dreger CK, Olins AL et al 2002 Mutations in the gene encoding the lamin B receptor produce an altered nuclear morphology in granulocytes (Pelger-Huet anomaly). Nat Genet 31:410–414

Janowski BA, Willy PJ, Devi TR, Falck JR, Mangelsdorf DJ 1996 An oxysterol signalling pathway mediated by the nuclear receptor LXR alpha. Nature 383:728–731

Karabinos A, Schunemann J, Meyer M, Aebi U, Weber K 2003 The single nuclear lamin of Caenorhabditis elegans forms in vitro stable intermediate filaments and paracrystals with a reduced axial periodicity. J Mol Biol 325:241–247

Liu J, Lee KK, Segura-Totten M, Neufeld E, Wilson KL, Gruenbaum Y 2003 MAN1 and emerin have overlapping function(s) essential for chromosome segregation and cell division in Caenorhabditis elegans. Proc Natl Acad Sci USA 100:4598–4603

Osada S, Ohmori SY, Taira M 2003 XMAN1, an inner nuclear membrane protein, antagonizes BMP signaling by interacting with Smad1 in Xenopus embryos. Development 130:1783–1794

Raju GP, Dimova N, Klein PS, Huang HC 2003 SANE, a novel LEM domain protein, regulates bone morphogenetic protein signaling through interaction with Smad1. J Biol Chem 278:428–437

Shultz LD, Lyons BL, Burzenski LM et al 2003 Mutations at the mouse ichthyosis locus are within the lamin B receptor gene: a single gene model for human Pelger-Huët anomaly. Hum Mol Genet 12:61–69

Wagner N, Weber D, Seitz S, Krohne G 2004 The lamin B receptor of Drosophila melanogaster. J Cell Sci 117:2015–2028

Worman HJ, Yuan J, Blobel G, Georgatos SD 1988 A lamin B receptor in the nuclear envelope. Proc Natl Acad Sci USA 85:8531–8534

Nuclear membrane protein emerin: roles in gene regulation, actin dynamics and human disease

Katherine L. Wilson, James M. Holaska, Rocio Montes de Oca, Kathryn Tifft, Michael Zastrow, Miriam Segura-Totten*, Malini Mansharamani and Luiza Bengtsson

*Department of Cell Biology, Johns Hopkins University School of Medicine, 725 N. Wolfe Street, Baltimore, MD 21205, USA, and *Department of Science and Technology, Universidad Metropolitana, Puerto Rico*

Abstract. Loss of emerin, a nuclear membrane protein, causes Emery-Dreifuss muscular dystrophy (EDMD), characterized by muscle weakening, contractures of major tendons and potentially lethal cardiac conduction system defects. Emerin has a LEM-domain and therefore binds barrier-to-autointegration factor (BAF), a conserved chromatin protein essential for cell division. BAF recruits emerin to chromatin and regulates higher-order chromatin structure during nuclear assembly. Emerin also binds filaments formed by A-type lamins, mutations in which also cause EDMD. Other partners for emerin include nesprin-1α and transcriptional regulators such as germ cell-less (GCL). The binding affinities of these partners range from 4 nM (nesprin-1α) to 200 nM (BAF), and are physiologically significant. Biochemical studies therefore provide a valid means to predict the properties of emerin–lamin complexes *in vivo*. Emerin and lamin A together form stable complexes with either BAF or GCL *in vitro*. BAF, however, competes with GCL for binding to emerin *in vitro*. These and additional partners, notably actin and nuclear myosin I, suggest disease-relevant roles for emerin in gene regulation and the mechanical integrity of the nucleus.

2005 Nuclear organization in development and disease. Wiley, Chichester (Novartis Foundation Symposium 264) p 51–62

Emerin is an inner membrane protein of the nuclear envelope

The nuclear envelope has two parallel membranes, which merge at nuclear pore complexes. The outer membrane is continuous with the endoplasmic reticulum (ER), contains ER resident proteins and functions as part of the ER network (Burke & Stewart 2002). However, the inner membrane contains a special set of integral membrane proteins, which are thought to diffuse from the ER to the inner membrane and to be retained by binding to stable intranuclear structures (Östlund & Worman 2003). The most important 'anchoring' structures are polymers of nuclear intermediate filament proteins named lamins, which lie immediately

51

adjacent to the inner membrane (Cohen et al 2001). Lamins, a key architectural element of metazoan nuclei, are also found throughout the interior of the nucleus. The thickness, lengths and configurations of lamin polymers in the nucleus are important unsolved questions in cell biology (Goldman et al 2002). Most nuclear membrane proteins bind directly to either A-type or B-type lamins (or both); this attachment is important for their localization and function (Burke & Stewart 2002).

Emerin is an integral membrane protein found specifically at the nuclear inner membrane (Bengtsson & Wilson 2004). Emerin has a transmembrane domain near its C-terminus which anchors it to the membrane. The bulk of emerin (~ 220 residues) is exposed and interacts with lamin filaments and soluble partners inside the nucleus. The N-terminus of emerin consists of a folded 'LEM domain', which mediates direct binding to one known partner: barrier-to-autointegration factor (BAF; Segura-Totten & Wilson 2004). Interestingly, many other nuclear proteins (LAP2α, LAP2β, MAN1, Lem3, Lem2/NET-25) also have this domain and collectively comprise the 'LEM domain' protein family (Foisner 2003, Bengtsson & Wilson 2004, Lee & Wilson 2004). The structures of BAF and several LEM domains were determined in nuclear magnetic resonance (NMR) or X-ray crystallography studies (see Segura-Totten & Wilson 2004). Aside from its LEM domain, the structure of emerin is unknown, with only scant predicted secondary structure (Bengtsson & Wilson 2004). Nevertheless a 'functional' map of emerin has emerged from biochemical studies of 27 missense mutations (Lee et al 2001, Holaska et al 2003, 2004, Wilkinson et al 2003, Haraguchi et al 2004). Each mutation disrupts binding to one or a few partners, but not other partners. Collectively these studies suggest three functional regions in emerin: the LEM-domain (which binds BAF), a central lamin-binding domain, and two 'repressor binding' domains (RBD-1, RBD-2) that flank and overlap partially with the lamin-binding domain. RBD-2, located near emerin's transmembrane domain, is also critical for binding and stabilizing actin filaments (Holaska et al 2004).

A special partnership between emerin and A-type lamins: disruption of either protein can cause Emery-Dreifuss muscular dystrophy

Emerin binds both A- and B-type lamins (Bengtsson & Wilson 2004). However, most investigators focus on its binding to A-type lamins for two reasons. First, emerin is mislocalized throughout the ER network in cells that lack A-type lamins (Sullivan et al 1999), meaning that A-type lamin filaments (or other proteins that depend on A-type lamins) are needed to anchor emerin at the inner membrane. Second, a disease named Emery-Dreifuss muscular dystrophy (EDMD) is caused either by loss of emerin or dominant missense mutations in A-type lamins (Bonne et al 2003). Thus, the interaction between emerin and

A-type lamins is particularly important for the three tissues affected by this disease: major tendons (Achilles, neck, elbow), certain skeletal muscles and the conduction system of the heart (Bonne et al 2003). However many additional tissues that normally express both proteins are unaffected by disease. How do we explain this tissue-specific phenotype? One possibility is that unaffected tissues express other LEM-domain proteins that overlap with emerin in function, and 'cover' its loss in emerin-deficient patients. This model is supported by biochemical evidence that emerin binds many of the same partners as LAP2β and MAN1 (Nili et al 2001, M. Mansharamani. and K.L.Wilson, unpublished observations), and by *in vivo* evidence that emerin and MAN1 have essential overlapping functions in *Caenorhabditis elegans* embryos (Liu et al 2003). Future tests of this 'overlapping function' model will require characterizing each LEM-domain protein and determining which functions and partners are unique versus shared. A second model, not mutually exclusive, is that disease-affected tissues express specific proteins that depend uniquely on emerin for their function in the nucleus. Future tests of this 'tissue-specific interactor' model will require the identification of tissue-specific partners and pathways regulated by each LEM-domain protein.

Emerin is a general binding partner for many proteins that regulate gene expression

One major hypothesis for the disease mechanisms of EDMD and other laminopathies is that emerin and other lamin-anchored proteins regulate tissue-specific gene expression (Wilson 2000, Cohen et al 2001, Östlund & Worman 2003). While direct evidence is still lacking, indirect evidence abounds. For example, emerin binds directly to transcription regulator germ cell-less (GCL) both *in vitro* and *in vivo*, and BAF competes for the binding site *in vitro* (Holaska et al 2003). The selection of splice sites on mRNA is influenced by an emerin-binding protein named YT521-B (Wilkinson et al 2003). Emerin interacts with an anti-death transcription regulator named Btf *in vitro*, which localizes near the nuclear envelope in cells triggered to undergo apoptosis (Haraguchi et al 2004). Emerin also binds a putative transcriptional regulator named Lmo7 (Holaska et al 2004). Both Btf and Lmo7 are potentially disease-relevant: Btf binding to emerin is specifically sensitive to the S54F disease-causing missense mutation in emerin (Haraguchi et al 2004), and loss of Lmo7 in mice causes muscle defects (Semenova et al 2003). These findings suggest that a variety of gene regulatory complexes interact with emerin at the inner nuclear membrane. Loss of interaction due to mutations in either emerin or A-type lamins may contribute directly to tissue-specific disease. Further indirect evidence that emerin can influence gene expression comes from a DNA microarray analysis of mRNA expression in fibroblasts from an emerin-null patient (Tsukuhara et al 2002). The expression of about 60 genes increased or decreased in emerin-null cells, and these effects were reversed for 28 genes by expression of exogenous emerin. Interestingly, emerin-null cells upregulated the genes encoding A-type lamins, αII-spectrin, a fibroblast growth factor (FGF) receptor, signalling protein Wnt-13 and gene regulator α-catenin (Tsukuhara et al 2002).

The idea that emerin and other nuclear membrane proteins are involved in tissue-specific gene regulation is further supported by studies of LAP2β and MAN1. Both LAP2β and MAN1 directly repress transcription *in vivo* through direct binding to GCL or Smad, respectively (Nili et al 2001, Raju et al 2003, Osada et al 2003). The functional overlap between MAN1 and emerin in *C. elegans* (Liu et al 2003) further supports roles for emerin in gene expression. Similar roles, direct or indirect, are suggested for LAP2α, a LEM-domain protein that associates with intranuclear A-type lamins and appears to provide a scaffold for retinoblastoma protein (Foisner 2003, Markiewicz et al 2002). Finally, we note that the shared binding partner for all LEM-domain proteins, BAF, is itself a tissue-specific transcriptional repressor *in vivo* (Wang et al 2002).

The involvement of lamin-binding nuclear membrane proteins in transcriptional regulation is a new frontier. Do these proteins regulate transcription by imposing 'higher-order' chromatin structure on genes targeted by proximity to the nuclear envelope? Further work is needed to determine (a) which signalling pathways converge on nuclear membrane proteins, (b) the nature and extent of functional overlap between emerin and other LEM-domain proteins, and (c) the role of BAF (Liu et al 2003, Bengtsson & Wilson 2004).

Emerin's partner BAF binds DNA, represses gene expression and has essential structural roles in nuclear assembly

BAF is found only in metazoans, and interacts with all LEM-domain proteins (reviewed by Bengtsson & Wilson 2004). The binding site on BAF for the LEM-domain is located centrally on the BAF dimer structure (Segura-Totten et al 2002). Missense mutations in BAF have been tested biochemically for their effects on DNA-binding, emerin-binding and nuclear assembly in *Xenopus* extracts (Segura-Totten et al 2002). Interestingly, a mutation at BAF residue 53 specifically disrupts binding to emerin (but not DNA), and this mutant form of BAF arrests nuclear assembly very early, with highly compressed chromatin and failure to recruit nuclear membranes (Segura-Totten et al 2002). BAF also has structural roles during nuclear assembly in somatic cells (Haraguchi et al 2001). In mammalian cells BAF concentrates at the surface of telophase chromosomes during mitosis, and is essential to recruit emerin to the chromatin surface during nuclear assembly (Haraguchi et al 2001). When nuclear assembly was examined in detail in *Xenopus* egg extracts, BAF had two opposite effects depending on its concentration: at low concentrations of exogenous BAF, chromatin decondensation and nuclear growth were enhanced suggesting that BAF has positive roles in 'opening' chromatin structure and facilitating the growth phase of nuclear assembly (Segura-Totten et al 2002). However at higher concentrations, exogenous BAF potently compressed chromatin structure and blocked assembly. RNAi-studies in *C. elegans* show that BAF is essential for cell viability during mitosis (Zheng et al 2000, Y. Gruenbaum and K. L. Wilson, unpublished observations). Understanding emerin's role in human disease will depend on understanding its essential partner, BAF.

Emerin might contribute to an actin-based cortical network at the nuclear membrane

The second major hypothesis for the disease mechanism of EDMD and other laminopathies is that mutations in A-type lamins cause the nucleus to become mechanically unstable leading to non-specific defects in nuclear integrity or nuclear function that might lead to cell death (Burke & Stewart 2002, Bonne et al 2003). At the cell surface, the plasma membrane is reinforced by a network of filaments formed by actin and spectrin; these filaments bind to each other and to integral membrane proteins (Bennett & Baines 2001). This network, the 'cortex', anchors and structurally reinforces protein complexes including dystrophin and dystroglycan at the cell surface (Östlund & Worman 2003). There is now evidence for an actin-dependent network at the nuclear inner membrane. The first hint came from the discovery that emerin binds tightly (4 nM affinity) to a nuclear membrane protein named nesprin-1α, which has seven 'spectrin repeat' (SR) domains and is related to much larger nesprin isoforms that bind actin (Mislow et al 2002, Zhen et al 2002). The second hint comes from a study in which emerin was used to affinity purify proteins from HeLa cell nuclear extracts. Eight proteins were identified by mass spectrometry and proteomic fingerprinting including actin, αII-spectrin and the nuclear-specific isoform of a molecular motor, myosin I (Holaska et al 2004). Both F-actin and nuclear myosin I each bind directly to emerin, and myosin remains bound even in the presence of 10 mM ATP (Holaska et al 2004). Interestingly emerin specifically binds the pointed ('minus') end of actin filaments, and stabilizes F-actin *in vitro*. A closely related LEM-domain protein, LAP2β, was 19-fold less active, suggesting that emerin might specifically contribute to the formation or stability of an actin cortical network at the nuclear inner membrane.

   Given that the nuclear envelope must anchor, organize and protect the human genome, its reinforcement by an actin-based network is not surprising. Actin is constantly entering the nucleus, and levels of actin appear to be regulated by a dedicated export receptor (exportin 6) which binds actin in a complex with profilin (Stuven et al 2003). Activities including mRNA transcription, export of certain mRNAs and proteins, and chromatin remodeling all depend on actin (Pederson & Aebi 2002). Furthermore nuclear assembly after mitosis involves both actin and nuclear isoforms of actin-scaffolding protein 4.1 (Krauss et al 2003). Despite long-held skepticism, the existence of nuclear actin is now indisputable and our challenge is to understand its forms and functions (Pederson & Aebi 2002, Olave et al 2002). Keeping an open mind on nuclear actin is important for understanding the fundamental properties of the nucleus, and for determining if mechanical defects contribute to the tissue pathology of EDMD or other laminopathies.

Final comments

Emerin and A-type lamins are hypothesized to form numerous 'scaffolds' at the inner nuclear membrane, upon which many other proteins depend (Zastrow et al 2004). Considering emerin alone, these 'dependent' proteins include transcription regulators, signalling proteins and architectural proteins such as actin and myosin I (Bengtsson & Wilson 2004). Imagine the combinatorial possibilities for structure and function in human cells, which have six genes encoding LEM-domain proteins (Lee & Wilson 2004) and also have B-type lamin filaments! This field is poised to make rapid progress in understanding the biochemistry of protein–protein interactions that depend on nuclear lamina infrastructure. We are optimistic that interactions responsible for the mechanical integrity of the nucleus, and the relationship between lamins and human disease can be unravelled.

Acknowledgements

We apologize to authors whose work was not cited directly due to space constraints. We gratefully acknowledge support from the American Heart Association to Katherine Wilson (Scott B. Deutschman Memorial Research Award). This work was also supported by grants from the National Institutes of Health (T32 HL07227 and F32 GM067397 to J.M.H. and RO1 GM48646 to K.L.W) and the Deutsche Forschungsgemeinschaft (to L.B.).

References

Bengtsson L, Wilson KL 2004 Multiple and surprising new functions for emerin, a nuclear membrane protein. Curr Opin Cell Biol 16:73–79
Bennett V, Baines AJ 2001 Spectrin and ankyrin-based pathways: metazoan inventions for integrating cells into tissues. Physiol Rev 81:1353–1392
Bonne G, Yaou RB, Beroud C et al 2003 108th ENMC International Workshop, 3rd Workshop of the MYO-CLUSTER project: EUROMEN, 7th International Emery-Dreifuss Muscular Dystrophy (EDMD) Workshop, 13–15 September 2002, Naarden, The Netherlands. Neuromuscul Disord 13:508–515
Burke B, Stewart CL 2002 Life at the edge: the nuclear envelope and human disease. Nat Rev Mol Cell Biol 3:575–585

Cohen M, Lee KK, Wilson KL, Gruenbaum Y 2001 Transcriptional repression, apoptosis, human disease and the functional evolution of the nuclear lamina. Trends Biochem Sci 26:41–47

Foisner R 2003 Cell cycle dynamics of the nuclear envelope. ScientificWorldJournal 3:1–20

Goldman RD, Gruenbaum Y, Moir RD, Shumaker DK, Spann TP 2002 Nuclear lamins: building blocks of nuclear architecture. Genes Dev 16:533–547

Haraguchi T, Koujin T, Segura-Totten M et al 2001 BAF is required for emerin assembly into the reforming nuclear envelope. J Cell Sci 114:4575–4585

Haraguchi T, Holaska JM, Yamane M et al 2004 Emerin binding to Btf, a death-promoting transcriptional repressor, is disrupted by a missense mutation that causes Emery-Dreifuss muscular dystrophy. Eur J Biochem 271:1035–1045

Holaska J, Lee KK, Kowalski A, Wilson KL 2003 Transcriptional repressor germ cell-less (GCL) and barrier-to-autointegration factor (BAF) compete for binding to emerin in vitro. J Biol Chem 278:6969–6975

Holaska J, Kowalski AK, Wilson KL 2004 Emerin caps the pointed end of actin filaments: evidence for an actin cortical network at the nuclear inner membrane. PLoS Biol 2:1354–1362

Krauss SW, Chen C, Penman S, Heald R 2003 Nuclear actin and protein 4.1: essential interactions during nuclear assembly in vitro. Proc Natl Acad Sci USA 100:10752–10757

Lee KK, Wilson KL 2004 All in the family: evidence for four new LEM-domain proteins Lem2 (NET-25), Lem3, Lem4 and Lem5, in the human genome. In: Hutchinson C, Bryant J, Evans D (ed) Communication and gene regulation at the nuclear envelope. BIOS Scientific Publishers, Oxford, in press

Lee KK, Haraguchi T, Lee RS, Koujin T, Hiraoka Y, Wilson KL 2001 Distinct functional domains in emerin bind lamin A and DNA-bridging protein BAF. J Cell Sci 114:4567–4573

Liu J, Lee KK, Segura-Totten M, Neufeld E, Wilson KL, Gruenbaum Y 2003 MAN1 and emerin have overlapping function(s) essential for chromosome segregation and cell division in *Caenorhabditis elegans*. Proc Natl Acad Sci USA 100:4598–4603

Markiewicz E, Dechat T, Foisner R, Quinlan RA, Hutchison CJ 2002 Lamin A/C binding protein LAP2α is required for nuclear anchorage of retinoblastoma protein. Mol Biol Cell 13:4401–4413

Mislow JMK, Holaska JM, Kim MS et al 2002 Nesprin-1α self-associates and binds directly to emerin and lamin A *in vitro*. FEBS Lett 525:135–140

Nili E, Cojocaru GS, Kalma Y et al 2001 Nuclear membrane protein LAP2β mediates transcriptional repression alone and together with its binding partner GCL (germ-cell-less). J Cell Sci 114:3297–3307

Olave IA, Reck-Peterson SL, Crabtree GR 2002 Nuclear actin and actin-related proteins in chromatin remodeling. Annu Rev Biochem 71:755–781

Osada S, Ohmori SY, Taira M 2003 XMAN1, an inner nuclear membrane protein, antagonizes BMP signaling by interacting with Smad1 in *Xenopus* embryos. Development 130:1783–1794

Östlund C, Worman H 2003 Nuclear envelope proteins and neuromuscular diseases. Muscle Nerve 27:393–406

Pederson T, Aebi U 2002 Actin in the nucleus: what form and what for? J Struct Biol 140:3–9

Raju GP, Dimova N, Klein PS, Huang HC 2003 SANE, a novel LEM domain protein, regulates BMP signaling through interaction with Smad1. J Biol Chem 278:428–437

Segura-Totten M, Kowalski AM, Craigie R, Wilson KL 2002 Barrier-to-autointegration factor: major roles in chromatin decondensation and nuclear assembly. J Cell Biol 158:475–485

Segura-Totten M, Wilson KL 2004 BAF: roles in chromatin, nuclear structure and retrovirus integration. Trends Cell Biol 14:261–266

Semenova E, Wang X, Jablonski MM, Levorse J, Tilghman SM 2003 An engineered 800 kilobase deletion of Uchl3 and Lmo7 on mouse chromosome 14 causes defects in viability, postnatal growth and degeneration of muscle and retina. Hum Mol Genet 12:1301–1312

Stuven T, Hartmann E, Gorlich D 2003 Exportin 6: a novel nuclear export receptor that is specific for profilin-actin complexes. EMBO J 22:5928–5940

Sullivan T, Escalante-Alcalde D, Bhatt H et al 1999 Loss of A-type lamin expression compromises nuclear envelope integrity leading to muscular dystrophy. J Cell Biol 147:913–920

Tsukuhara T, Tsujino S, Arahata K 2002 cDNA microarray analysis of gene expression in fibroblasts of patients with X-linked Emery-Dreifuss muscular dystrophy. Muscle Nerve 25:898–901

Wang X, Xu S, Rivolta C et al 2002 Barrier to autointegration factor interacts with the cone-rod homeobox and represses its transactivation function. J Biol Chem 277:43288–43300

Wilkinson FL, Holaska JM, Sharma A et al 2003 Emerin interacts *in vitro* with the splicing-associated factor, YT521-B. Eur J Biochem 270:2459–2466

Wilson KL 2000 The nuclear envelope, muscular dystrophy and gene expression. Trends Cell Biol 10:125–129

Zastrow MS, Vlcek S, Wilson KL 2004 Proteins that bind A-type lamins: integrating isolated clues. J Cell Sci 117:979–987

Zhen YY, Libotte T, Munck M, Noegel AA, Korenbaum E 2002 NUANCE, a giant protein connecting the nucleus and actin cytoskeleton. J Cell Sci 115:3207–3222

Zheng R, Ghirlando R, Lee MS, Mizuuchi K, Krause M, Craigie R 2000 Barrier-to-autointegration factor (BAF) bridges DNA in a discrete, higher-order nucleoprotein complex. Proc Natl Acad Sci USA 97:8997–9002

DISCUSSION

Fatkin: I was interested to hear that emerin binds to F-actin. I know this is a controversial area, but I thought most of the nuclear actin was supposed to be G-actin or oligomeric actin.

Wilson: This is why we did such extensive biochemical studies. Phalloidin does not stain the nucleus much, so for years it was assumed that nuclei lack F-actin. However, there is a thoughtful review by Thoru Pederson and Ueli Aebi (Pederson & Aebi 2002) saying that actin has multiple forms, not just G-actin or long F-actin filaments, even in the cytoplasm. In fact there are many interesting short filament forms, which may be directly relevant in the nucleus. We hypothesize that actin is not allowed to form long filaments inside the nucleus, because they might interfere with chromatin. Dirk Gorlich showed that actin constantly enters the nucleus and is continually exported by a dedicated exportin, in a complex with profilin. If actin export is blocked, actin forms huge filaments in the nucleus that look like they would clearly disrupt nuclear function if not also nuclear structure. Our view is that emerin should be explored and considered in the same way as cytoplasmic actin-binding proteins, to understand its properties. Emerin's ability to cap the pointed end of actin filaments is very rare. Only three other proteins are known to do this; two are cytoplasmic and the third is a chromatin-remodelling protein in the nucleus. Of these three, emerin behaves most like tropomodulin, which stabilizes short actin filaments in junctional complexes at the cell surface. There is definitely actin in the nucleus. We speculate that the most primitive functions of actin might be nuclear, and might involve short forms of F-actin.

Gruenbaum: An interesting experiment would be to overexpress this fragment of emerin in the cytoplasm, and see whether it affects cytoplasmic actin.

Ellis: Given that mutations in cardiac actin actually cause a dilated cardiomyopathy or hypertrophic cardiomyopathy, and we don't know how emerin causes a cardiac defect, would it be worth looking at patients who don't have muscular dystrophy but have other heart diseases to see whether emerin is mislocalized? Would it be worth making mutations in actin and seeing whether these can affect emerin binding to it?

Wilson: Those are excellent ideas, although we worry that actin mutations might affect 50 other functions in the cell. We haven't tried to address actin yet, because the emerin experiments have taken some time.

Julien: In patients with muscular dystrophy, are there any differences in the assembly of sarcomeric actin?

Wilson: I have no idea.

Ellis: Not as far as I know. All the dystrophin-associated proteins are normal and targeted to the cell surface.

Julien: There is always a possibility that a small amount of the non-nuclear emerin stabilizes the cytoplasmic actin.

Ellis: There is so much actin in the cell that it may be that if a small amount were affected you wouldn't see it.

Wilson: There are at least 80 distinct actin-binding proteins in the cytoplasm. They each have defined functions. I think emerin is the tip of the iceberg in terms of nuclear actin-binding proteins. We also tested LAP2β because it is the protein most homologous to emerin. LAP2β also enhances actin polymerisation rates, but is 20-fold less active than emerin. We're excited about this because it's the first function for which emerin might actually be specialized. Every other partner we tested for emerin, so far, overlaps with LAP2β. Until now we haven't known of any particular function for emerin that can explain why its loss matters.

Shumaker: Do you have evidence that emerin binds to both actin and myosin at the same time?

Wilson: No, we hope to test this soon. It requires rigorous tri-molecular biochemistry. You can't just do pull-downs. You have to know the affinity and binding curve for each pair of interactors, and use appropriate concentrations of each protein. We used this approach previously to show that emerin and lamin A form tertiary complexes with either BAF or GCL (Holaska et al 2003).

Ellis: The last four amino acids of human emerin are GNPF. I know you did not look at the transmembrane tail, but three of these amino acids are conserved with the LAPs. Have you done any alanine mutagenesis to find a function for them?

Wilson: That's a good point. The transmembrane domain and lumenal domain residues of emerin are also highly conserved, suggesting they have an important function. This is an unexplored area.

Ellis: Some of the ER proteins have these four-letter motifs. There is KDEL for soluble ER proteins and KKXX at the C-terminus of ER membrane proteins. Emerin's motif is not that rigorous a motif so it is possible that it has something to do with its targeting to the ER.

Wilson: KDEL and KKXX signals are used mainly to retrieve proteins that escape the ER into the secretory pathway. This might make sense for emerin, but it lacks both signals. Do you know of any disease-causing mutations at the C-terminus of emerin?

Ellis: Unfortunately there aren't any mutations that are viable at the C-terminus.

Goldman: If it is correct that there are nuclear specific forms of profilin and that most of the actin in the nucleus is in the G form, it would be hard to imagine how a contractile event would take place. You implied that there was tension being produced, which means that bridges between filamentous actin, not G-actin, and myosin must be made and broken. There have been preliminary reports on specific forms of nuclear myosin, although there has been no follow-up to demonstrate function.

Wilson: I think there is follow-up coming.

Goldman: I was very excited by this nuclear form of myosin. But it hasn't gone much further than the initial report (Pestic-Dragovich et al 2000).

Wilson: We are on the edge of ignorance. The de Lanerolle lab is studying myosin's role in transcription (Pestic-Dragovich et al 2000). We don't know whether emerin has anything to do with transcription when bound to actin and myosin. There may be many distinct emerin-based complexes. We imagine that emerin might be anchored by some partners, while simultaneously binding other more dynamic partners. When bound to actin and nuclear myosin 1 it might not be involved in transcription, but might instead be contributing to the mechanical stiffness or stability of the nuclear envelope. It is also formally possible that emerin–actin complexes interconnect with the lamin network, because lamins also bind actin.

Goldman: Let's assume that you have a form of myosin just below the nuclear envelope, and actin is present. You say that emerin binds both actin and myosin.

Wilson: Yes, in pair-wise binding assays. We don't yet know if emerin binds myosin and actin simultaneously.

Goldman: We should be able to begin to find out where the myosin binds. The myosin head would have to bind to actin. Therefore, it is likely that emerin would bind somewhere else on myosin.

Wilson: Yes, however there's no evidence that transcription complexes concentrate near the envelope. If anything, they are in the interior. That's why actin and myosin are both likely to have multiple roles in the nucleus.

Goldman: You show overlap of the binding regions for lamin and actin. Can you compete for this binding using excess lamin A or actin?

Wilson: We hope to do this study very soon.

Goldman: You might expect that this level of interaction might be regulated in such a way that when one is binding the other is not. Or they may not be overlapping binding sites.

Wilson: I should point out that missense mutations only tell us which residues are likely to be important for the interaction, because mutations at these sites disrupt binding. These experiments don't show which regions of emerin are sufficient for binding, or prove direct interaction at the mutated site. A mutation in one place might disturb the folded structure elsewhere, and disrupt binding indirectly. The only known folded region in emerin is the LEM domain. Our working model is that the rest of emerin is loosely extended, and undergoes some kind of induced fit when binding to specific partners.

Morris: Is that based on structural analyses of your molecule without the C-terminus?

Wilson: This working model comes from unpublished subtilisin digestions of LAP2β, which is homologous to emerin. Subtilisin digests accessible, extended, non-folded regions of proteins, and it chewed up LAP2 outside the LEM-domain. I would be very happy to hear that someone has the crystal or NMR structure of any other part of emerin. Our mutations are mapping functional regions but they aren't showing us its structure.

Gerace: What end of the actin filaments does the myosin walk to?

Wilson: It is a barbed end-directed motor. It could therefore either pull F-actin toward emerin, or walk away from emerin.

Gerace: It would want to move away from the nuclear envelope end, then.

Wilson: Yes, if myosin I could let go of emerin, it could move towards the nuclear interior, or along the inner membrane.

Goldman: Provided that there was a long enough filament.

Gruenbaum: I find it interesting that there is possible competition for binding emerin between the transcription factors and the structural factors. Would GCL or BAF actually compete for the binding of emerin to actin?

Wilson: We predict that BAF can bind independently of all partners, since they interact with emerin outside the LEM-domain. We haven't done the key competition experiments yet. Deciphering what co-binds and what competes will be key to understanding how many different higher-order complexes might form.

Gruenbaum: In *C. elegans* GCL can bind directly to lamins. Can the vertebrate GCL bind other proteins such as lamins? Also, can it be displaced once it is bound?

Wilson: We tested human GCL and lamin A, and they do not bind to each other directly. However, GCL and lamin can co-bind to emerin.

Gruenbaum: What about binding to actin?

Wilson: We haven't yet tested if GCL binds actin.

Goldman: Your finding that emerin binds to actin is very interesting. The interaction between lamin A and actin brings to mind an old observation by Gerhard Wiche showing that plectin binds to lamin B. Plectin has an actin and an intermediate filament binding domain, making it a perfect candidate for linking lamins with nuclear actin. However, I haven't seen any follow-up on this story.

Wilson: The nuclear actin literature is quite beautiful. There are nuclear-specific isoforms of many different actin-binding proteins. I think the people who published these nice papers were faced with so much negativity that they fled back to the cytoplasm. I am fascinated by these long proteins (nesprins) that connect to the nucleus, the fact that exterior nesprins bind actin, and the potential cross-talk between plectin, intermediate filaments, lamins and actin. We recently identified titin as a binding partner for lamin A. The era of long proteins inside the nucleus is about to begin.

Goldman: There is a way to induce rapid polymerization of actin in the nucleus using low concentrations of DMSO. If you did this, where would emerin be?

Wilson: Emerin would probably remain at the nuclear inner membrane, but we haven't tried.

Wilkins: You sketched a somewhat daunting picture of what you are facing in terms of figuring out all these varying networks in different cells. There are two approaches to addressing this. One is the genetic approach that makes use of model systems, finding suppressors and enhancers. The other is by the informatics approach that is being developed with the aim of making theoretical predictions. Are you thinking along these lines?

Wilson: Yes, but I must confess that I am a little scared. The evidence that MAN1 and LAP2β regulate gene expression, and Rb binds directly to lamins, suggests that we have to be ready for domino or indirect effects. I am very excited about the mouse and *C. elegans in vivo* work, for obvious reasons. I think the tissue phenotypes are best studied in real tissues.

References

Holaska J, Lee KK, Kowalski A, Wilson KL 2003 Transcriptional repressor germ cell-less (GCL) and barrier-to-autointegration factor (BAF) compete for binding to emerin in vitro. J Biol Chem 278:6969–6975

Pederson T, Aebi U 2002 Actin in the nucleus: what form and what for? J Struct Biol 140:3–9

Pestic-Dragovich L, Stojiljkovic L, Philimonento AA et al 2000 A myosin I isoform in the nucleus. Science 290:337–341

Identification of novel integral membrane proteins of the nuclear envelope with potential disease links using subtractive proteomics

Eric C. Schirmer, Laurence Florens, Tinglu Guan, John R. Yates III and Larry Gerace[1]

Department of Cell Biology, The Scripps Research Institute, La Jolla CA, USA 92037, USA

Abstract. Lamin A and some integral membrane proteins of the nuclear envelope (NE) have been linked to human diseases, mostly dystrophies. To comprehensively identify integral membrane proteins specific to the nuclear envelope, we have carried out a subtractive proteomics analysis of NEs isolated from rodent liver using Multidimensional Protein Identification Technology (MudPIT). An NE fraction and a nucleus-depleted membrane fraction were separately analyzed by MudPIT and proteins appearing in both fractions were 'subtracted' from the NE fraction. This identified 67 novel putative NE transmembrane proteins in addition to the 13 that had been previously characterized. Most or all of the new proteins we identified are likely to be *bona fide* NE Transmembrane proteins (NETs), since all eight of the first group of proteins we tested in a cell transfection assay target to the NE. Moreover, five of the eight NETs remained associated with the nuclear periphery after extraction with Triton-X100, suggesting an association with the nuclear lamin polymer. 27 of the proteins occur in chromosomal regions where 18 different human dystrophies have been mapped, making these proteins disease candidates. We have analysed the expression of these proteins using transcriptome databases, providing direction for future functional analysis of these novel proteins.

2005 Nuclear organization in development and disease. Wiley, Chichester (Novartis Foundation Symposium 264) p 63–80

The nuclear envelope (NE), which forms the boundary of the eukaryotic nucleus, is important both for segregation of nuclear metabolism from the cytosol, and for organization of nuclear infrastructure. The NE is formed by inner and outer nuclear membranes that are joined at the nuclear pore complexes, and is lined by

---

[1]This paper was presented at the symposium by Larry Gerace to whom correspondence should be addressed.

the nuclear lamina (reviewed by Gerace & Burke 1988). The outer nuclear membrane is continuous with the peripheral endoplasmic reticulum (ER) and is functionally similar to the latter, whereas the inner nuclear membrane has a number of specific protein constituents, most of which are associated with the nuclear lamina. The lamina consists of a meshwork of intermediate type-filament proteins (lamins), as well as a number of more minor components, most notably integral membrane proteins of the inner membrane. Studies in vertebrate cells have described four major lamin isotypes and 11 integral proteins (plus their splice variants) specific to the inner nuclear membrane as well as two integral proteins of the pore complex (reviewed by Stuurman et al 1998, Goldman et al 2002, Hutchison 2002). Recently the NE has received widespread attention because a number of inherited human diseases ('laminopathies') have been linked to mutations in lamin A and in certain lamina-associated integral proteins of the inner nuclear membrane (reviewed by Mounkes et al 2003, Burke & Stewart 2002, Worman & Courvalin 2002, Wilson et al 2001). Since the actual number of integral proteins specific to the NE is unknown, we decided to carry out a comprehensive proteomic analysis of NEs isolated from rodent liver, as a way of expediting an understanding of NE functions and relationship to disease (Schirmer et al 2003).

Results and discussion

For this work we took advantage of a recently developed proteomic method, termed MultiDimensional Protein Identification Technology (MudPIT), which allows identification of a large number of different proteins in a complex mixture (Washburn et al 2001, Wolters et al 2001, Tabb et al 2002, Sadygov & Yates 2003). In this procedure, proteolytic digests of protein mixtures are subjected to multiple liquid chromatography steps, and isolated peptides are first fractionated in an ion trap mass spectrometer. Individual ionic species from this step are then fragmented by collision-induced dissociation with the helium bath gas, and the resulting ions are analyzed in a second mass spectrometry step. The fragmentation pattern from the last step often yields amino acid sequences of individual peptides. By this procedure it is possible to identify up to several thousand proteins in a single fraction without the need for one or two-dimensional polyacrylamide gel electrophoresis.

To identify NE-specific proteins, we used a 'subtractive proteomics' approach (Schirmer et al 2003). For this we isolated a microsomal membrane (MM) fraction from the post-nuclear supernatant of liver homogenates. This fraction contains cytoplasmic membranes that contaminate isolated NEs, as well as proteins shared with the peripheral ER. However, the MM fraction is essentially devoid of NE proteins, since the rapidly-sedimenting nuclei are removed by the initial centrifugation step. Therefore, by comparing the proteins that appear in the MM fraction to those that appear in isolated NEs, we were able to 'subtract' proteins common to both fractions and thereby define a set of NE-specific membrane proteins (illustrated in Fig. 1A).

To focus our study on membrane-integrated proteins, we analyzed NEs and MMs that had been pre-extracted with 0.1 M NaOH to enrich for integral membrane proteins (Fig. 1B). We also analysed NEs that had been pre-extracted with high salt/nonionic detergent, to help identify proteins tightly associated with the detergent- and salt-insoluble lamin polymer. In thin section electron microscopy, the isolated NEs showed the typical double membrane with NPCs, and the MM appeared mainly as vesicles (Fig. 1C). NaOH extraction caused the release of virtually all electron-dense-nonmembranous structures from both samples, and the NEs became vesiculated (Fig. 1C). The three fractions were proteolytically cleaved and analyzed by MudPIT. Over 30 000 peptides were obtained, allowing nearly 2400 separate protein identifications between the three fractions. All previously characterized integral membrane proteins of the NE were detected in NaOH-extracted NEs, but were absent from NaOH-extracted MMs. Furthermore, no lamins were recovered in the NaOH-extracted MMs. Together, these data support the premises underlying the use of the subtractive approach to determine NE-specific proteins. Moreover, all 31 proteins of the mammalian nuclear pore complex that have been previously identified (Cronshaw et al 2002) were represented in the salt/detergent-extracted fraction. This indicates that our analysis is likely to be comprehensive.

The broad dynamic range of the MudPIT analysis enabled the identification of 1830 different proteins in salt/detergent extracted NEs, 566 proteins in NaOH-extracted NEs, and 652 proteins in NaOH-extracted MMs (Schirmer et al 2003). The overlap between these groups is shown in the schematic diagram in Fig. 2A. 41% of the proteins in NaOH-extracted NEs also appeared in NaOH-extracted MM, immediately removing these common proteins from consideration. We then narrowed our search to proteins of salt/detergent extracted NEs and NaOH-extracted NEs. After eliminating proteins with previously defined functions in the nucleus (e.g. histones and HP1) or in the cytoplasm (e.g. mitochondrial proteins), we were left with 337 uncharacterized open reading frames (ORFs) (Fig. 2B). The transmembrane prediction algorithm, TMPred, predicted that 21 novel ORFs in the NaOH-extracted NEs and 53 novel ORFs in the salt/detergent-extracted NEs have at least one transmembrane stretch. These two sets partially overlapped (Fig. 2B), and together they identified 67 novel putative integral proteins of the NE (Schirmer et al 2003).

The determination of such a large number of new putative NE proteins from this analysis was unexpected, since only 13 integral proteins of the NE, plus their splice variants, had been identified previously. To obtain an insight into whether

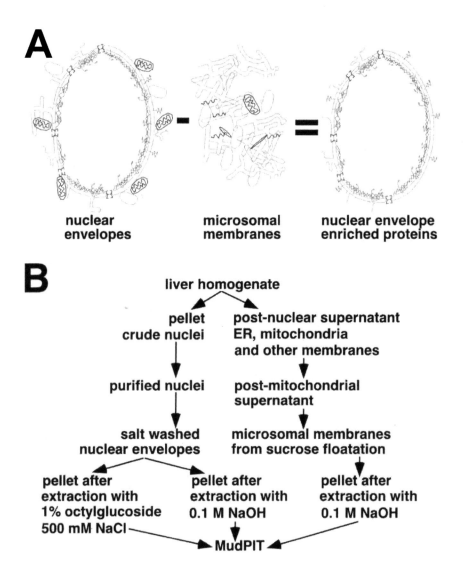

such a large set of novel proteins might exist in the NE, we selected a representative set of the proteins to examine their ability to be targeted to the NE in transiently transfected cells after engineering them with N-terminal epitope tags. Eight full-length cDNAs were recovered for this analysis, reflecting a range of protein sizes (112 to 674 residues), numbers of predicted transmembrane sequences (one to five) and numbers of peptide hits obtained from the MudPIT analysis (one to 19 hits, roughly representing protein abundance). All eight proteins that we tested were

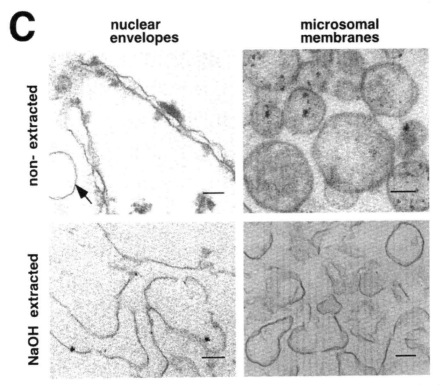

FIG. 1. Strategy for subtractive proteomics of the NE. (A) Cartoon illustrating that isolated NEs, in addition to containing the lamina, NPCs and NE-specific proteins, also contain ER-like membranes (such as the outer nuclear membrane) and contaminating cytoplasmic membranes, such as mitochondrial membranes. Determination separately of the proteins in a microsomal membrane fraction allows 'subtraction' of the proteins common to both the NE and microsomal membrane fraction, thereby identifying NE-specific proteins. (B) Flow diagram illustrating the methods used to isolate the three fractions used for MudPIT analysis. NEs and MMs from rat and mouse liver were extracted with 0.1 M NaOH to enrich for transmembrane proteins. Separately, rat NEs were extracted with 1% octylglucoside + 500 mM NaCl to enrich for proteins tightly bound to the nuclear lamina. (C) Thin section electron micrographs of salt-washed rat NEs and isolated rat MMs, either without any further extraction, or after extraction with 0.1 M NaOH.

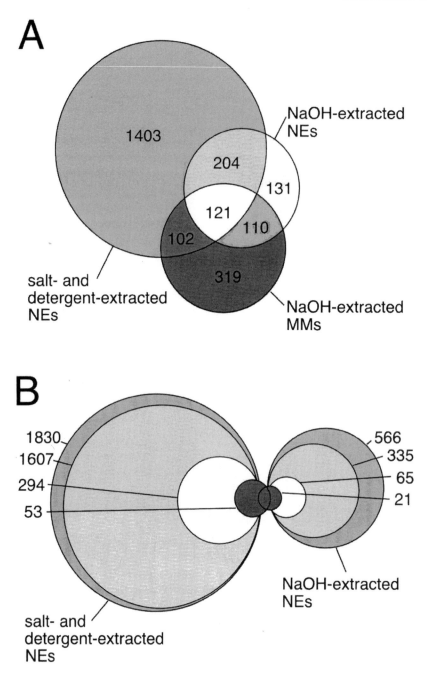

concentrated to a significant extent at the NE in the transfected cells, coinciding with the localization of endogenous lamins (Fig. 3; see also Schirmer et al 2003). A partial localization in peripheral ER-like structures also was observed, which varied in extent with the protein examined and the level of expression. The pattern that was seen for the transfected proteins-concentrated NE localization together with variable cytoplasmic staining-was similar to that obtained when cells were transfected with the well-characterized inner nuclear membrane protein LAP2β (Fig. 3A). However the pattern was different from the pattern revealed with an ER-selective dye, which shows no preferential concentration at the NE. For five of the proteins we examined (Fig. 3B), a concentrated localization at the NE was retained after cells were treated with Triton-X100, whereas most labelling in the peripheral ER was lost. This result was similar to that obtained for LAP2β (Fig. 3B), which is tethered to the lamina in a Triton-resistant linkage. These results suggest that these novel NE proteins may have a direct or indirect linkage to nuclear lamins. Their concentration at the nuclear rim strongly argues for an NE function, although it remains possible that some of the proteins have functions in the peripheral ER as well.

Considered together, the first eight proteins that we examined from the proteomic analysis appear to be *bona fide* nuclear envelope transmembrane proteins (NETs), suggesting that many of the remaining 59 proteins are likely to be NETs. We speculate that we identified such a large number of novel NE transmembrane proteins, as compared to an earlier proteomic analysis on mammalian NEs (Dreger et al 2001), because MudPIT is extremely sensitive; it avoids loss or poor resolution of integral proteins that can occur with the 2D gel electrophoresis used previously (e.g. many membrane proteins migrate together outside the range of the ampholines), and we analysed nuclei from rodent liver,

---

FIG. 2.   Schematic diagrams of proteins identified in the various fractions. (A) Venn diagrams of proteins identified separately in the three fractions by MudPIT, with numbers indicating the proteins unique to each fraction and overlapping with the other fractions. Protein identifications were generated by search spectra against a database of 106 360 human, rat and mouse sequences, including ~25 000 sequences recently added to the rat databases (Schirmer et al 2003). (B) Strategy for determination of the focus group of putative transmembrane proteins of the NE. In the salt and detergent extracted NEs, of the 1830 total hits, 1607 remained after subtracting MM proteins, 294 were previously uncharacterized ORFs, and 53 were predicted to be transmembrane proteins. In NaOH-extracted NEs, of the 566 total hits, 335 remained after subtracting MM proteins, 65 were previously uncharacterized ORFs, and 21 were predicted to be transmembrane proteins. ~50% of the putative transmembrane proteins in the NaOH-extracted NEs also were found in the corresponding fraction of the salt and detergent-extracted NEs.

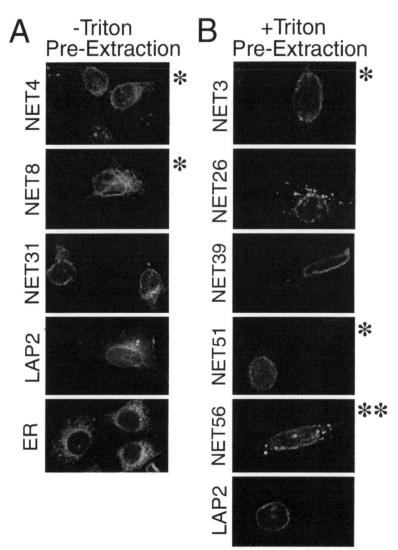

FIG. 3.   Localization of nuclear envelope transmembrane proteins (NETs) in transfected cells. Full-length cDNAs for the indicated NETs and for LAP2β were N-terminally tagged with an HA-epitope inserted into an expression vector driven by a cytomegalovirus promoter and transiently transfected into COS7 or HeLa cells. (A) Cells were fixed and stained with antibodies to the HA-epitope tag for immunofluorescence microscopy. To label the total ER, cells were treated with the ER stain DiOC6. (B) Cells were pre-extracted by three washes with 1% Triton X-100 in PBS, fixed, and labelled with antibodies to the HA tag for immunofluorescence microscopy. * indicates proteins with putative links to disease (see Fig. 4). ** This protein was recently identified as 'dullard', although its localization was not reported.

as opposed to the single cultured cell line analysed before (Dreger et al 2001). Liver contains a wide range of differentiated cell types, including hepatocytes, Kupffer cells, endothelial cells, lipocytes, bile duct cells and smooth muscle cells, although most cell types are minor in terms of their relative mass as compared to hepatocytes. Indeed, we identified two integral membrane proteins of the NE, Syne-1 and Syne-2 (nesprins), which are preferentially expressed in muscle (reviewed by Burke & Stewart 2002) and which were not detected in the previous study. The group of proteins we identified includes two (numbers 25 and 66) with the LEM domain, a motif that occurs in three integral proteins of the inner nuclear membrane (LAP2, emerin, MAN1), and one (# 9) that appears to be related to LAP1 through a gene duplication. Twelve of the proteins contained functional domains similar to those of enzymes such as phosphatases, acetyltransferases, and glycosyltransferases.

To investigate whether the confirmed and putative NETs might be expressed selectively in certain tissue types, we examined the GNF (Genomics Institute of Novartis Research Foundation, La Jolla) transcriptome database, which provides information on the relative levels of mRNA expression in different human and mouse tissues as determined by microarray analysis (Su et al 2002) (Fig. 4). 27 of the corresponding genes were represented on the microarray chips used for screening the human tissues, and 35 were found on the chips used for screening mouse tissues. In some cases (indicated by §), duplicate chips were analysed and gave similar results. Only in the case of putative NET 33 were the results significantly different in two different experiments (Fig. 4, right panel). The expression patterns varied widely: some proteins were expressed at moderate to high levels in many or most tissues shown, whereas others were at low or undetectable levels. A number of the proteins were expressed at substantial levels in skeletal muscle, cardiac myocytes and in adipocytes, representing some of the major tissues affected in laminopathies (Mounkes et al 2003, Worman & Courvalin 2002). The finding that a number of the putative NETs are not present at substantial levels in liver does not contradict our proteomic analysis, since MudPIT has a much higher dynamic range than microarray analysis, and some of the proteins may be expressed in only a minor cell population of liver (see above).

Up to now, ~13 laminopathies, the majority of which are dystrophies and neuropathies, have been linked to mutations in either lamin A or in either of two lamina-associated integral proteins of the inner nuclear membrane (emerin and LBR) (reviewed by Mounkes et al 2003, Burke & Stewart 2002, Worman & Courvalin 2002, Wilson et al 2001). There remain about 300 human dystrophies and neuropathies that have not yet been linked to a specific gene(s), although over 30 of these dystrophies and neuropathies have been mapped to discrete regions of the human genome. We examined the chromosome locations of the proteins revealed by our proteomic analysis, with respect to the regions containing the partially mapped dystrophies. Five of the 67 proteins we identified

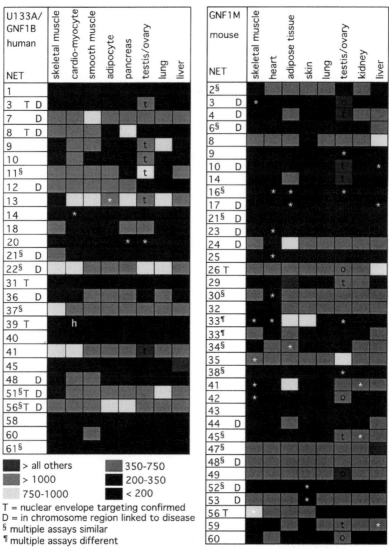

FIG. 4. Possible association of putative and confirmed NETs with genetic diseases. Chromosome locations of the genes encoding the putative NETs were determined with the use of Genbank's human genome resources. Potential dystrophy links were obtained from dystrophy databases as in Schirmer et al (2003). The list shown here was updated to accommodate more recent publications on linkage analyses of the pertinent dystrophies, as compared to that presented in Schirmer et al (2003). Proteins designated as NETs have confirmed NE localization.

did not have apparent human homologues. Of the remaining proteins, 44% (27 genes) mapped within chromosomal regions linked to 18 of these dystrophies (Fig. 5). Although most of these dystrophy-linked regions contain 100–400 genes, two lines of reasoning indicate that our list of proteins contains good candidates for disease genes. First, the candidate dystrophy genes we have identified occur at a significantly increased frequency in chromosomal regions linked to dystrophies (34%) as compared to the random probability that a gene would reside in these areas (21%). Moreover, if the dystrophies mapped to very large chromosome territories are excluded (i.e. dystrophies in chromosome 7q, 8q and 14), the remaining genes are linked to dystrophies at an even higher relative frequency (33% as compared to 15%). Second, nuclear lamina proteins, including integral membrane proteins, have been linked to eight dystrophies and neuropathies, and five of the eight NETs we have analysed so far are suggested to be lamina-associated integral proteins. It is important to note that mutations in either lamin A or in a lamin A-associated integral protein (emerin) can cause the same disease, i.e. Emery-Dreifuss muscular dystrophy (EDMD). Thus, it appears that the functions of the network formed by the lamin polymer together with its associated integral membrane proteins can be disrupted by mutating multiple distinct interacting partners. In this regard, it is noteworthy that two of the dystrophies associated with chromosome regions containing putative NETs (i.e. limb-girdle muscular dystrophy and Charcot-Marie tooth disease) have variants that have been linked to mutations in lamin A.

Several non-exclusive models have been proposed to account for the link between human diseases and mutations in nuclear lamina components (discussed in Mounkes et al 2003, Burke & Stewart 2002, Worman & Courvalin 2002, Wilson et al 2001). One model is based on the proposed role of the lamina in higher order chromosome organization, which is thought to help control the distinctive patterns of gene silencing and activity in different cell types. In this hypothesis, mutations in lamina components could affect chromatin tethering to the lamina, and thereby alter epigenetically determined patterns of gene expression. A second model involves the widely accepted role of the lamina in mechanically stabilizing the nucleus and NE. In this view, disruption or weakening of certain key interactions in the lamina could affect NE integrity and cell viability. A third class of model involves the suggested interactions of the NE with the cytoplasmic cytoskeleton. In this case, mutations in lamina components could affect nuclear positioning and/or signalling between the nucleus and the cytoplasm.

Underlying all of these models is the notion that the lamina comprises a complex network of interconnected proteins, which in turn associate with chromatin, nuclear membranes, and perhaps cytoskeletal elements. One of the major conundrums associated with laminopathies is that all diseases described to date involve lamina proteins that are expressed in a widespread fashion in most

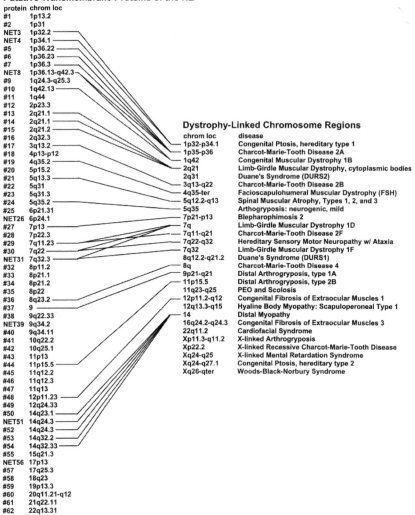

FIG. 5.   Tissue expression of putative NETs by examination of transcriptome database. Data were obtained from the U122A/GNF1B human (left) and GNF1M mouse (right) transcriptome databases assembled by Su et al (2002) for the indicated tissues, and for a human cardiomyocyte cell line (left panel). T, nuclear envelope targeting confirmed in transfected cells; D, in chromosome region linked to dystrophy. Relative level of mRNA expression is indicated by numerical values.

differentiated cells, yet the diseases are manifest only in a small number of tissues (Mounkes et al 2003). This supports the view that the many components of the lamina have overlapping functions in lamina organization and/or its association with other structures. Thus, mutations in a particular lamina component become limiting only in situations where the ensemble of other lamina components in that particular cell type cannot functionally accommodate the loss or gain of function imposed by the mutant protein. This underscores the need for a systems approach to analyse the nuclear lamina in different specialized cells. Our identification of a group of novel NETs is expected to significantly advance this goal.

Summary and conclusions

We have carried out a near-comprehensive analysis of the complement of integral membrane proteins in NEs isolated from rodent liver, using MudPIT proteomics combined with subtractive and comparative approaches. We have identified 67 novel putative NETs with these methods. Since all eight of the first group of proteins we analysed targeted to the NE in transfected cells, we believe that most of the remainder will also turn out to be NETs. Similar to previously described integral proteins of the NE, many of the genes are expressed in different tissues, although the expression levels of some vary widely between different tissue types. Our results provide the basis for thoroughly understanding the network of molecular interactions at the nuclear lamina that are involved in its functions and which, when altered, lead to human diseases.

Acknowledgements

This work was supported by the NIH through grants F32GM19085 to ECS, RO1GM28521 to LG, and RO1RR11823 to JRY.

References

Burke B, Stewart C 2002 Life at the edge: the nuclear envelope and human disease. Nat Rev Mol Cell Biol 3:575–585

Cronshaw J, Krutchinsky A, Zhang W, Chait B, Matunis M 2002 Proteomic analysis of the mammalian nuclear pore complex. J Cell Biol 158:915–927

Dreger M, Bengtsson L, Schoneberg T, Otto H, Hucho F 2001 Nuclear envelope proteomics: novel integral membrane proteins of the inner nuclear membrane. Proc Natl Acad Sci USA 98:11943–11948

Gerace L, Burke B 1988 Functional organization of the nuclear envelope. Annu Rev Cell Biol 4:335–374

Goldman R, Gruenbaum Y, Moir R, Shumaker D, Spann T 2002 Nuclear lamins: building blocks of nuclear architecture. Genes Dev 16:533–547

Hutchison C 2002 Lamins: building blocks or regulators of gene expression? Nat Rev Mol Cell Biol 3:848–858

Mounkes L, Kozlov S, Burke B, Stewart CL 2003 The laminopathies: nuclear structure meets disease. Curr Opin Genet Dev 13:223–230

Sadygov RG, Yates JR 3rd 2003 A hypergeometric probability model for protein identification and validation using tandem mass spectral data and protein sequence databases. Anal Chem 75:3792–3798

Schirmer EC, Florens L, Guan T, Yates JR 3rd, Gerace L 2003 Nuclear membrane proteins with potential disease links found by subtractive proteomics. Science 301:1380–1382

Stuurman N, Heins S, Aebi U 1998 Nuclear lamins: their structure, assembly, and interactions. J Struct Biol 122:42–66

Su AI, Cooke MP, Ching KA et al 2002 Large-scale analysis of the human and mouse transcriptomes. Proc Natl Acad Sci USA 99:4465–4470

Tabb DL, McDonald WH, Yates JR 3rd 2002 DTASelect and Contrast: tools for assembling and comparing protein identifications from shotgun proteomics. J Proteome Res 1:21–26

Washburn MP, Wolters D, Yates JR 3rd 2001 Large-scale analysis of the yeast proteome by multidimensional protein identification technology. Nat Biotechnol 19:242–247

Wilson KL, Zastrow MS, Lee KK 2001 Lamins and disease: insights into nuclear infrastructure. Cell 104:647–650

Wolters DA, Washburn MP, Yates JR 3rd 2001 An automated multidimensional protein identification technology for shotgun proteomics. Anal Chem 73:5683–5690

Worman HJ, Courvalin JC 2002 The nuclear lamina and inherited disease. Trends Cell Biol 12:591–598

DISCUSSION

Young: In your mass spectrometry analysis, did you see an enrichment of isoprenylcysteine carboxyl methyltransferase or Zmpste24 — two enzymes involved in the posttranslational modification of lamins?

Gerace: That's a good question. I need to look at the original data.

Levy: In your transcriptome analysis, among the genes that you mapped in the potential muscular dystrophy region, did you confirm each time that they were expressed either in the muscle or peripheral nerves?

Gerace: In terms of the cell type-specific expression patterns, there was not any discrete differentiation signature readily apparent.

Levy: I'm not a cell biologist, but a geneticist. I looked at your list of syndromes, and some of these disorders have already been mapped and the genes have been identified. Some of these genes have nothing to do with the nuclear envelope.

Gerace: We looked at the literature extensively because there were concerns that some of these disease genes have already been mapped. Our conclusion was that the literature is not that decisive. There have been good candidate disease genes identified for some of these dystrophies, but the literature is less than 100% conclusive.

Levy: The only point that is not conclusive is the CMT type 2a, because there is only one patient, and it is probably not the right gene. It is a point of debate.

Gerace: I think there will only be a relatively small number of these genes linked to disease in the final analysis. Nonetheless, this work has revealed some candidates that the human geneticists should keep their eyes on.

Levy: Of course. Facioscapulohumeral muscular dystrophy (FSH) is very interesting.

Davies: With regards to FSH, haven't the microarray analyses been done by several groups?

Gerace: This is a controversial area. One group argues that there is overexpression of these genes adjacent to the subtelomeric repeat, but another group doing microarray analysis says that there is no overexpression. There was raging controversy between these two groups at a recent FSH meeting. There is no way I can see of resolving this dispute. In our case, by making antibodies and looking we can at least for one protein determine whether it is overexpressed or not. It may not be.

Davies: It could be down-regulated. You still don't know what the effect of that repeat is going to be. They are assuming it is overexpression. Have they looked at your protein 19 in their microarray?

Gerace: No, but they looked at three genes in the area. *FRG1* and *2*, and *ANT1*. All three were substantially overexpressed in the FSH patients that had the deletion repeats.

Davies: Why did they miss your protein 19? Is the genome annotated with this gene?

Gerace: I don't know but it has become evident that there are a lot of annotation errors in the published genome sequences.

Worman: The proteins identified in the proteomics analysis may be mostly in the inner nuclear membrane (INM), but there may also be some that diffuse freely between the ER and INM in interphase. With your subtractive effort can you say roughly how many of the proteins may have been subtracted out because they are also in the ER.

Gruenbaum: The number of proteins in the INM that you described is probably correct for mammalian liver cells. There are probably other INM proteins with tissue- or developmental-specific expression. In that case, 67+11 would probably be a minimal number and we should expect this number to grow.

Gerace: I would allow for this. What is interesting is that many of these proteins are widely expressed. Perhaps the ones we are not getting are the ones that have a highly restricted expression pattern.

Goldman: Have you tried any other nuclei? Have you tried brain?

Gerace: No, because the procedures for making nuclei are not nearly as good for other tissues.

Goldman: Liver is the classical material because it is a soft tissue. What about kidney?

Gerace: Brain would be better.

Goldman: I think brain would be next on my list.

Gruenbaum: What about total embryo?

Goldman: The idea would be to show that there is a difference in expression of nuclear envelope proteins in different cell types.

Davies: What happens to protein modifications in your separation techniques?

Gerace: They are lost. We are looking at matches to the predicted sequence of the protein from the genome. If you get something that doesn't match it is uninterpretable.

Davies: This is therefore the tip of the iceberg.

Gerace: The hope is that we would be getting extensive coverage of the proteins.

Goldman: You only see the proteins that are stable under these extraction conditions. Those that might be more dynamic and labile with respect to their associations might be equally interesting, but they are likely to be lost during extraction.

Gerace: We are looking at proteins that are integral to the nuclear envelope, and the operational definition for integral proteins is resistance to sodium hydroxide extraction. I think it more likely that we would be losing them because they aren't expressed in one of the cell types in liver.

Goldman: But you could be losing peripherally associated proteins.

Gerace: We are only looking at integral proteins. There are a lot of data that remain to be mined, but I don't see any efficient way to do this. There are many peripheral membrane proteins that we find in this fraction, and we don't know whether these are contaminants from chromatin or cytoplasmic intermediate filaments, for example. We get a lot of cytosolic contamination in the fraction. Fortunately we are able to eliminate this by focusing on the integral proteins. This allows us to do the subtractive technique better.

Wilson: One binding partner for emerin that came from a two-hybrid search was an RNA splice choice factor. Several of your NET proteins have predicted RNA binding motifs, or helicase-like regions. Do these proteins have any logical connections to each other, such as membership in the same complexes?

Gerace: This is something that an informatics analysis might answer. We haven't made any connections yet. It is an interesting idea.

Morris: Splicing may seem a far-fetched concept to introduce into EDMD, but if you compare this with what is happening in myotonic dystrophy, a lot of clinical features of myotonic dystrophy have been shown to be due to altered splicing (discussed in Wilkinson et al 2003). Amazingly, myotonic dystrophy shares some of the clinical features of EDMD, such as cardiac conduction problems and muscle wasting.

Gerace: I am trying to think of this in terms of the earlier discussion on transcriptional repressors binding to lamins. Is this conceptually a different type of function we are thinking about here? Could this be related as a way of sequestering splicing factors?

Morris: So many things seem to be binding to emerin. It is very difficult to work it out.

Wilson: So far we are finding direct binding partners for lamins, and missing all information about higher-order complexes that each partner might represent. RNA splicing complexes have many components. We must begin to think about how protein complexes and machines interact with lamin filaments.

Worman: We have photobleached emerin in the nuclear envelope. The bleached area gradually recovers (Östlund et al 1999).

Gerace: We may well have been subtracting things that are somewhat enriched in the nuclear envelope, and therefore have lost proteins that are of interest. The thing about mass spectrometry, including MudPIT, is that it is non-quantitative. The number of peptide hits roughly corresponds to protein abundance, but the people who do proteomics analysis use the term 'abundance' very loosely. It is likely that some of the proteins we have subtracted will turn out to be proteins with a function in the nuclear envelope and that are somewhat enriched in the nuclear envelope.

Worman: I can imagine a cell where emerin is expressed at such a high level that perhaps 20% is in the ER, and it might get subtracted out.

Gerace: All I can say is that the 11 proteins that have been characterized so far were in the nuclear envelope fraction but were not detectably in the microsomal membrane fraction.

Wilson: I noticed that at least one protein, when it was expressed, was also in the nuclear interior.

Gerace: That was LAP2. We get that with some of the proteins. It is either proteolysis of the transfected protein, or it could reflect a phenomenon like invagination of membrane sheets or some other mechanism for large-scale relocalization of integral proteins to the nuclear interior when they are overexpressed. These transfection studies should be examined carefully because they can be ambiguous, and there is often noise. The best information comes when you get a positive signal in the nuclear envelope that is significantly above the background level.

Goldman: If you consider the growth of the nucleus after a cell division, once you enclose chromatin with a nuclear envelope, the nucleus gets larger. How does this increase in surface area take place? How is new inner and outer nuclear envelope membrane being added continually?

Gerace: Brian Burke touched on this earlier.

Goldman: Was it emerin photobleaching you were doing, Brian?

Burke: No, I've never photobleached anything.

Worman: We have photobleached emerin. The bleached area just stays there.

Goldman: Suppose you photobleach the nuclear envelope after you have expressed emerin-GFP. Is there a rapid turnover of these proteins?

Worman: For LBR it almost never turns over. It is slower than the lamins. The photobleached area in the nuclear envelope almost never fills in during interphase (Ellenberg et al 1997). The other ones we looked at, MAN1 (Wu et al 2002) and emerin (Östlund et al 1999), do recover reasonably fast. They diffuse within the plane of the INM, but not back out into the ER.

Goldman: This is important: it is a highly mobile fraction versus an immobilized fraction. Maybe a lot of this has to do with what binds to lamins and what doesn't bind to lamins.

Worman: If the numbers are right, the diffusion within the inner nuclear membrane for MAN1 and emerin seems to be about one-third of what it is in the ER. Maybe they bind to something that slows their diffusion and they can't get out because they are too big to fit through the lateral channels of the pore complex.

Young: I was going to comment about these 67 new proteins. One of the things we have done in our lab is gene trapping in mouse ES cells. Currently we have frozen down about 9000 ES cells corresponding to about 3000 knocked out genes in mice (about 10% of all mouse genes). We can be fairly confident that for 10% of your 67 genes, it would be possible to order ES cells and make knockouts. For example, Kathy Wilson made a comment about the RNA splicing protein that interacts with lamin A (YT521): this is one of the knockouts that is on our list.

Stewart: We have looked at both the BayGenomics database and the German gene trap consortium lists. We found that for about 15 of the novel nuclear envelope genes identified by Larry, there are ES cell clones carrying genetrap insertions.

Worman: There are also secret databases, such as the one by Lexicon Genetics.

Stewart: You might not want to touch them, as they have rather restrictive licensing agreements!

References

Ellenberg J, Siggia ED, Moreira JE et al 1997 Nuclear membrane dynamics and reassembly in living cells: targeting of an inner nuclear membrane protein in interphase and mitosis. J Cell Biol 138:1193–1206

Östlund C, Ellenberg J, Hallberg E, Lippincott-Schwartz J, Worman HJ 1999 Intracellular trafficking of emerin, the Emery-Dreifuss muscular dystrophy protein. J Cell Sci 112:1709–1719

Wilkinson FL, Holaska JM, Zhang Z et al 2003 Emerin interacts in vitro with the splicing-associated factor, YT521-B. Eur J Biochem 270:2459–2466

Wu W, Lin F, Worman HJ 2002 Intracellular trafficking of MAN1, an integral protein of the nuclear envelope inner membrane. J Cell Sci 115:1361–1371

Genetics of laminopathies

Rabah Ben Yaou, Antoine Muchir, Takuro Arimura, Catherine Massart, Laurence Demay*, Pascale Richard* and Gisèle Bonne[1]

*Inserm U582, Institut de Myologie, Batiment Babinski, Groupe Hospitalier Pitié-Salpétrière, 47, Boulevard de l'Hôpital, Paris and *UF Cardiogénétique et Myogénétique, Groupe Hospitalier Pitié-Salpétrière, Assistance Publique-Hôpitaux de Paris, France*

Abstract. Laminopathies are now recognized as a group of disorders due to mutations of the *LMNA* gene, which encodes A-type lamins. Primarily, mutations in *LMNA* have been associated to the autosomal forms of Emery-Dreifuss muscular dystrophy, a rare slowly progressive humero-peroneal muscular dystrophy accompanied by early contractures and dilated cardiomyopathy with conduction defects. *LMNA* mutations have been reported to be responsible for up to 10 distinct phenotypes that affect specifically either the skeletal and/or cardiac muscle, the adipose tissue, the peripheral nervous tissue, the bone tissue or more recently premature ageing. So far more than 180 different *LMNA* mutations have been identified in 903 individuals. The first studies of phenotype/genotype relationships revealed no clear relation between the phenotype and the type and/or the localization of the mutation, except perhaps for the globular tail domain of lamins A/C. Studies of the consequences of *LMNA* mutations in the skin cultured fibroblasts from the patients reveal abnormal nuclei in variable proportions, with dysmorphic nuclei exhibiting abnormal patterns of expression of B-type lamins and emerin. Finally, the development of KO and KI *LMNA* mice, will certainly give further insight into the pathophysiological mechanisms associated with *LMNA* mutations. For example, $Lmna^{H222P/H222P}$ mice harbour phenotypes reminiscent of Emery-Dreifuss muscular dystrophy.

2005 Nuclear organization in development and disease. Wiley, Chichester (Novartis Foundation Symposium 264) p 81–97

The laminopathy story began in 1999 with the identification of the first mutation of the *LMNA* gene encoding the A-type lamins, proteins of the nuclear envelope, in patients affected with autosomal dominant Emery-Dreifuss muscular dystrophy (EDMD) (Bonne et al 1999). This inherited disorder is characterized by a triad of symptoms:

* early contractures of the Achilles tendons, elbows and post cervical muscles

---

[1]This paper was presented at the symposium by Gisèle Bonne to whom correspondence should be addressed.

- slow progressive muscle wasting and weakness with a distinctive humero-peroneal distribution early in the course of the disease, and
- a dilated cardiomyopathy that invariably develops by adulthood, usually presenting as cardiac conduction defects requiring pacing.

Thus affected individuals may die suddenly from heart block, or develop progressive heart failure. Skeletal muscle biopsies from patients with EDMD show dystrophic changes with a few necrotic and regenerating fibres, but this is not specific to this muscular disease (Emery 2000). Two major modes of inheritance exist, X-linked and autosomal dominant (XL- and AD-EDMD). Rare cases of autosomal recessive transmission have also been reported. Mutations of the *EMD* (formerly named *STA*) gene are responsible for the X-linked forms (Bione et al 1994). It encodes emerin, an integral membrane protein of the nuclear envelope (Manilal et al 1996). In 1999, we localized the locus to chromosome 1q21-q22 and identified the disease gene of the AD-EDMD: *LMNA*, which encodes two other proteins of the nuclear envelope (Bonne et al 1999). Since this first identification, the role of nuclear envelope components has become more widely recognized with the *LMNA* gene implicated in up to nine other diseases that specifically affect either the skeletal and/or cardiac muscle, the adipose tissue, or the peripheral nervous tissue and more recently *LMNA* has been implicated in premature ageing syndromes, thus leading to the new concept of 'laminopathies'.

The clinical and genetic spectrum of laminopathies

Since the identification of the first *LMNA* mutations, collaborative networks have expanded. We have formally established two specific networks. The first one is the French network on 'Emery-Dreifuss Muscular Dystrophy and other Nuclear Envelopathies' (coordinators: Gisèle Bonne & Dominique Recan) and the second is the European Consortium 'MYOCluster-EUROMEN' (coordinators: Gisèle Bonne & Ketty Schwartz).

The wide screening of *LMNA* mutation allowed identification of other *LMNA* mutations not only in AD-EDMD patients (Bonne et al 2003) but also in a patient with an autosomal recessive form of EDMD (AR-EDMD) (di Barletta et al 2000), as well as in two other autosomal dominant diseases: a form of limb girdle muscular dystrophy associated with cardiac conduction defects (LGMD1B) (Muchir et al 2000), and a form of dilated cardiomyopathy with conduction defect disease (DCM-CD) (Bécane et al 2000, Fatkin et al 1999). Besides these laminopathies affecting striated muscles, *LMNA* mutations have been identified in familial partial lipodystrophy of Dunningan (FPLD), characterized by an abnormal distribution of the adipose tissue associated with insulin resistance and diabetes (Shackleton et al 2000), and in an autosomal recessive form of axonal neuropathy or Charcot-Marie-Tooth disease, type 2 (AR-CMT2B1) characterized by the loss of motor and/or sensory axons (De Sandre-Giovannoli et al 2002). Features of partial lipodystrophy present as clinical symptoms in Mandibuloacral Dysplasia (MAD) led us to screen the *LMNA* gene as a potential candidate for this rare autosomal recessive disorder characterized by mandibular and clavicular hypoplasia, acro-osteolysis, delayed cranial suture closure, joint contractures and aspects of lipodystrophy. We identified a homozygous *LMNA* mutation in five different families with MAD (Novelli et al 2002). More recently, two teams reported at the same time, *de novo* mutations in a rare Hutchinson-Gilford progeria syndrome (HGPS), a syndrome of segmental premature ageing (De Sandre-Giovannoli et al 2003, Eriksson et al 2003). In addition, several cases were reported with overlaps of various types of laminopathies, i.e. association of partial lipodystrophy with muscular dystrophy and/or dilated cardiomyopathy (Garg et al 2002, van der Kooi et al 2002), association of autosomal dominant axonal neuropathy with cardiomyopathy and leukonychia (Goizet et al 2004) and association of generalized lipoatrophy, insulin-resistant diabetes, disseminated leukomelanodermic papules, liver steatosis and cardiomyopathy (LDHCP) (Caux et al 2003). Very recently, *LMNA* mutations were also reported in atypical forms of Werner syndrome (Chen et al 2003), a clinical condition that is very close to that of a patient reported with LDHCP (Bonne & Levy 2003, Vigouroux et al 2003). These results highlight the extreme variability of the phenotypes associated with the large spectrum of mutations identified in the lamin A/C gene and point towards a continuum of clinical disorders.

If we look closer at striated muscle laminopathies, we may find some variations in the various clinical presentations. As for the skeletal symptoms, the analysis of a large panel of patients with muscular dystrophy and dilated cardiomyopathy identified a few cases with variable presentation of the muscular wasting and weakness and/or of the contractures. For example, humero-peroneal wasting and weakness (typical of EDMD) but without any contractures or proximal limb-girdle involvement (typical of LGMD), with early and/or severe contractures, were observed (EUROMEN*, unpublished data). Myopathy specifically affecting the quadriceps was also reported (Charniot et al 2003). All these observations point towards a continuum of skeletal muscle involvement. But still, all patients with these various forms of muscular dystrophy, invariably develop a cardiac disease, i.e. dilated cardiomyopathy with conduction defect

---

*EUROMEN (European Muscle Envelope Nucleopathies). http://www.myocluster.org/euro.htm

disease. And the diagnosis of the laminopathies of the striated muscles is particularly important because of the extreme severity of their cardiac symptoms (more than 40% of the patients died suddenly, the majority of whom never had any cardiac symptoms). A careful cardiac follow-up, implantation of a permanent pacemaker or better, cardiac defibrillators, can considerably reduce the risk of sudden death (Bonne et al 2003, Taylor et al 2003). It has to be noted that the screening of *LMNA* mutations in patients with 'isolated' dilated cardiomyopathy (without clinical symptoms of muscular dystrophy), also allows the identification of several variants of DCM, such as DCM with early atrial fibrillation (Sebillon et al 2003) or DCM with left apical aneurysm without conduction defects (Forissier et al 2003), highlighting again the wide variability of laminopathies.

Not so surprisingly in view of this large clinical spectrum, there is also a wide spectrum of *LMNA* mutations. Up to now, more than 180 different *LMNA* mutations, have been identified, most of them being missense mutations (Fig. 1) (Bonne et al 2003, Brown et al 2001, Vytopil et al 2003). We are setting up a mutation database UMD-*LMNA* that will be available via the internet (*http://www.udm.be*). The database collects not only genetics but also clinical details of each individual carrying an *LMNA* mutation, and the specific tools of this database will allow further and more complete analyses of phenotype/genotype relations (Bonne et al 2003).

The first analysis performed on phenotype/genotype relationships for laminopathies of the striated muscles did not reveal any relation between the phenotype, the type and/or the localization of the mutation. A large clinical variability exists both at intra- and inter-familial levels for these laminopathies (Brown et al 2001, Bonne et al 2003, Vytopil et al 2003). The same mutation can lead within the same family to EDMD, LGMD1B or isolated DCM-CD, i.e. laminopathies of the striated muscle (Bécane et al 2000, Brodsky et al 2000).

In contrast, regarding *LMNA* mutations leading to other types of laminopathies, some 'hotspot' or founder mutations have been reported. The majority of FPLD cases are due to a mutation affecting codon R482, leading to several amino acid exchanges (Bonne et al 2003) and mutations at codon 608 are associated with HGPS (De Sandre-Giovannoli et al 2003, Eriksson et al 2003). Founder mutations were reported in AR-CMT2B1 (R298C) and in MAD (R527H) (De Sandre-Giovannoli et al 2002, Novelli et al 2002).

Pathophysiological mechanisms of lamin A/C gene mutations

Lamins are intermediate filaments, components of the nuclear lamina that underline the inner face of the inner nuclear membrane, the real 'skeleton' of the nuclear envelope. In mammals, there are up to seven different lamins: the A-type

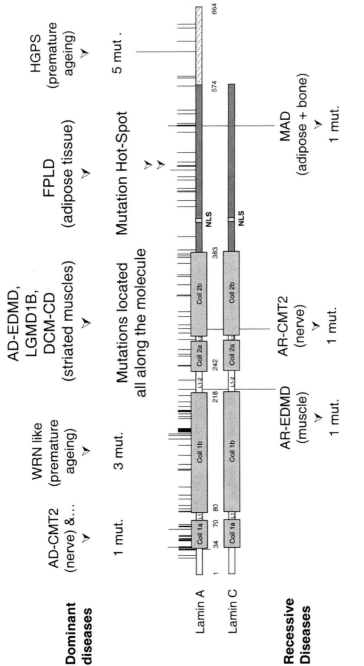

FIG. 1. Schematic representation of *LMNA* mutations identified in various types of laminopathies. Dominant disorders due to heterozygous *LMNA* mutations are depicted on the top of the protein scheme, whereas recessive disorders due to homozygous mutations are presented below.

lamins (A, ΔA10, C and C2 encoded by *LMNA* via alternative splicing) and the B-type lamins (B1, B2 and B3). Undifferentiated cells express only B-type lamins; A-type lamin appears later during embryogenesis. Numerous studies suggest that lamins A and C are implicated in DNA replication, chromatin organization, the spatial organization of the nuclear pore and the correct anchorage of the integral proteins of the inner nuclear membrane. However, their precise role remains unclear (Östlund & Worman 2003).

Studies of the consequence of *LMNA* mutations using the various tissues that were available from patients have been carried out. The first revealed a great variability in the abnormalities observed, not only in muscular biopsies (Sewry et al 2001) but also in skin-cultured fibroblasts. Nuclei with dysmorphic shape and abnormal expression patterns of B-type lamins and emerin were observed in various proportions. The relationship between the abnormalities observed within the tissues and cells of the patients and their genotype or their phenotype are still unclear (Vigouroux et al 2001, Caux et al 2003, Novelli et al 2002, Muchir et al 2004). Indeed, an average of 10 to 15% of FPLD, LDHCP or MAD patient fibroblasts with missense *LMNA* mutations, exhibit dysmorphic nuclei (Vigouroux et al 2001, Caux et al 2003, Novelli et al 2002). We obtained similar results with fibroblasts from EDMD, LGMD1B or DCM-CD patients, although fibroblasts with *LMNA* missense mutations affecting the central rod domain did not exhibit any abnormal nuclei (Muchir et al 2004). Whether this 'domain specific' behaviour reflects truly different and specific pathophysiological mechanisms of mutation is currently unclear. Regarding nonsense *LMNA* mutations, the analysis of fibroblasts of a premature child that died at birth and carried a homozygous *LMNA* nonsense mutation reveals that the complete absence of lamins A and C, due to this mutation, promotes a high proportion of abnormally shaped nuclei with lobules in which B-type lamins, emerin, nesprin-1α, LAP2β and Nup153 were also absent. In addition, nesprin-1α, like emerin, exhibited an aberrant localization in the ER, that was rescued by transfection of either exogenous lamin A or lamin C (Muchir et al 2003).

In order to gain further insights in the pathophysiological mechanisms of *LMNA* mutation, and to understand how mutations in a gene encoding ubiquitously expressed proteins could lead to tissue-specific diseases, we developed two mouse models into which we introduced *LMNA* mutations identified in families presenting a typical EDMD phenotype. The first analysis of phenotype and histopathological character in the first model, KI *Lmna*[H222P], showed that homozygous KI *Lmna*[H222P] mice develop muscular dystrophy and dilated cardiomyopathy reminiscent of human EDMD. We hope to obtain important data from the analysis of this mouse model, and in the long-term to develop valuable tools to investigate possible therapeutic approaches for laminopathies.

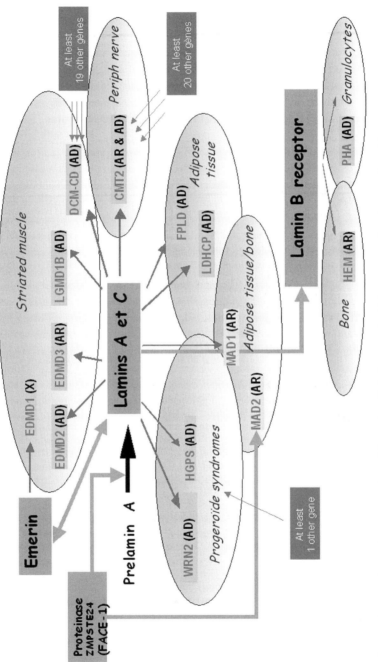

FIG. 2.   Current schematic representation of the various links between the different types of nuclear envelopathies.

In conclusion, the field of laminopathies has dramatically expanded since the identification of the first *LMNA* mutation. Mutations in other nuclear envelope components, i.e. emerin gene mutations in XL-EDMD and LBR gene mutations are responsible for defects in Pelger-Huët anomaly and HEM/Greenberg skeletal dysplasia (Hoffmann et al 2002, Waterham et al 2003). Finally, ZMPSTE24 gene mutations are associated with MAD (Agarwal et al 2003). The spectrum of 'nuclear envelopathies' is certainly still incomplete (Fig. 2) and future analysis will most probably dissect further the various defects of the nucleolemma.

Acknowledgements

We thank all partners from the French Network on 'EDMD and other nuclear envelopathies', the European EUROMEN Network and the international Human Frontier Program Network for the contribution to the wild development of the laminopathic field. A particular thankyou to Professor J. C. Kaplan for fruitful discussions and elaboration of Fig. 2. These studies were supported by grants from European Union Fifth Framework (MYO-CLUSTER/EUROMEN contract #QLG1-1999-00870), 'Association Française contre les Myopathies' (AFM, grant #8185 & #9278) and Human Frontiers Science Program (grant #RGP0057/2001-M101).

References

Agarwal AK, Fryns JP, Auchus RJ, Garg A 2003 Zinc metalloproteinase, ZMPSTE24, is mutated in mandibuloacral dysplasia. Hum Mol Genet 12:1995–2001

Bécane H-M, Bonne G, Varnous S et al 2000 High incidence of sudden death with conduction system and myocardial disease due to lamins A and C gene mutation. Pacing Clin Electrophysiol 23:1661–1666

Bione S, Maestrini E, Rivella S et al 1994 Identification of a novel X-linked gene responsible for Emery-Dreifuss muscular dystrophy. Nat Genet 8:323–327

Bonne G, Levy N 2003 LMNA mutations in atypical Werner's syndrome. Lancet 362:1585–1586

Bonne G, Di Barletta MR, Varnous S et al 1999 Mutations in the gene encoding lamin A/C cause autosomal dominant Emery-Dreifuss muscular dystrophy. Nat Genet 21:285–288

Bonne G, Ben Yaou R, Beroud C et al 2003 108th ENMC International Workshop, 3rd Workshop of the MYO-CLUSTER project: EUROMEN, 7th International Emery-Dreifuss Muscular Dystrophy (EDMD) Workshop, 13–15 September 2002, Naarden, The Netherlands. Neuromusc Disord 13:508–515

Brodsky GL, Muntoni F, Miocic S et al 2000 Lamin A/C gene mutation associated with dilated cardiomyopathy with variable skeletal muscle involvement. Circulation 101:473–476

Brown CA, Lanning RW, McKinney KQ et al 2001 Novel and recurrent mutations in lamin A/C in patients with Emery-Dreifuss muscular dystrophy. Am J Med Genet 102:359–367

Caux F, Dubosclard E, Lascols O et al 2003 A new clinical condition linked to a novel mutation in lamins A and C with generalized lipoatrophy, insulin-resistant diabetes, disseminated leukomelanodermic papules, liver steatosis and cardiomyopathy. J Clin Endocrinol Metab 88:1006–1013

Charniot JC, Pascal C, Bouchier C et al 2003 Functional consequences of an LMNA mutation associated with a new cardiac and non-cardiac phenotype. Hum Mutat 21:473–481

Chen L, Lee L, Kudlow B et al 2003 LMNA mutations in atypical Werner's syndrome. Lancet 362:440–445

De Sandre-Giovannoli A, Bernard R, Cau P et al 2003 Lamin A truncation in Hutchinson-Gilford progeria. Science 300:2055

De Sandre-Giovannoli A, Chaouch M, Kozlov S et al 2002 Homozygous defects in *LMNA*, encoding lamin A/C nuclear-envelope proteins, cause autosomal recessive axonal neuropathy in human (Charcot-Marie-Tooth Disorder Type 2) and mouse. Am J Hum Genet 70:726–736

di Barletta MR, Ricci E, Galluzzi G et al 2000 Different mutations in the LMNA gene cause autosomal dominant and autosomal recessive Emery-Dreifuss muscular dystrophy. Am J Hum Genet 66:1407–1412

Emery AEH 2000 Emery-Dreifuss muscular dystrophy — a 40 year retrospective. Neuromusc Disord 10:228-232

Eriksson M, Brown WT, Gordon LB et al 2003 Recurrent de novo point mutations in lamin A cause Hutchinson-Gilford progeria syndrome. Nature 25:25

Fatkin D, MacRae C, Sasaki T et al 1999 Missense mutations in the rod domain of the lamin A/C gene as causes of dilated cardiomyopathy and conduction-system disease. N Engl J Med 341:1715-1724

Forissier JF, Bonne G, Bouchier C et al 2003 Apical left ventricular aneurysm without atrio-ventricular block due to a lamin A/C gene mutation. Eur J Heart Fail 5:821–825

Garg A, Speckman RA, Bowcock AM 2002 Multisystem dystrophy syndrome due to novel missense mutations in the amino-terminal head and alpha-helical rod domains of the lamin A/C gene. Am J Med 112:549–555

Goizet C, Ben Yaou R, Demay L et al 2004 A new mutation of lamin A/C gene leading to autosomal dominant axonal neuropathy, muscular dystrophy, cardiac disease and leukonychia. J Med Genet 41:e39

Hoffmann K, Dreger CK, Olins AL et al 2002 Mutations in the gene encoding the lamin B receptor produce an altered nuclear morphology in granulocytes (Pelger-Huet anomaly). Nat Genet 31:410–414

Manilal S, Nguyen TM, Sewry CA, Morris GE 1996 The Emery-Dreifuss muscular dystrophy protein, emerin, is a nuclear membrane protein. Hum Mol Genet 5:801–808

Muchir A, Bonne G, van der Kooi A J et al 2000 Identification of mutations in the gene encoding lamins A/C in autosomal dominant limb girdle muscular dystrophy with atrioventricular conduction disturbances (LGMD1B). Hum Mol Genet 9:1453–1459

Muchir A, van Engelen BG, Lammens M et al 2003 Nuclear envelope alterations in fibroblasts from LGMD1B patients carrying nonsense Y259X heterozygous or homozygous mutation in lamin A/C gene. Exp Cell Res 291:352–362

Muchir A, Medioni J, Laluc M et al 2004 Nuclear envelope alterations in fibroblasts from patients with muscular dystrophy, cardiomyopathy, and partial lipodystrophy carrying lamin A/C gene mutations. Muscle Nerve 30:444–450

Novelli G, Muchir A, Sangiuolo F et al 2002 Mandibuloacral dysplasia is caused by a mutation in LMNA encoding lamins A/C. Am J Hum Genet 71:426–431

Ostlund C, Worman H J 2003 Nuclear envelope proteins and neuromuscular diseases. Muscle Nerve 27:393–406

Sebillon P, Bouchier C, Bidot LD et al 2003 Expanding the phenotype of LMNA mutations in dilated cardiomyopathy and functional consequences of these mutations. J Med Genet 40:560–567

Sewry CA, Brown SC, Mercuri E et al 2001 Skeletal muscle pathology in autosomal dominant Emery-Dreifuss muscular dystrophy with lamin A/C mutations. Neuropathol Appl Neurobiol 27:281–290

Shackleton S, Lloyd DJ, Jackson SN et al 2000 LMNA, encoding lamin A/C, is mutated in partial lipodystrophy. Nat Genet 24:153–156

Taylor MR, Fain PR, Sinagra G et al 2003 Natural history of dilated cardiomyopathy due to lamin A/C gene mutations. J Am Coll Cardiol 41:771–780 van der Kooi AJ, Bonne G, Eymard B et al 2002 Lamin A/C mutations with lipodystrophy, cardiac abnormalities, and muscular dystrophy. Neurology 59:620–623

Vigouroux C, Caux F, Capeau J, Christin-Maitre S, Cohen A 2003 LMNA mutations in atypical Werner's syndrome. Lancet 362:1585–1586

Vigouroux C, Auclair M, Dubosclard E et al 2001 Nuclear envelope disorganization in fibroblasts from lipodystrophic patients with heterozygous R482Q/W mutations in lamin A/C gene. J Cell Sci 114:4459–4468

Vytopil M, Benedetti S, Ricci E et al 2003 Mutation analysis of the lamin A/C gene (LMNA) among patients with different cardiomuscular phenotypes. J Med Genet 40:e132

Waterham HR, Koster J, Mooyer P et al 2003 Autosomal recessive HEM/Greenberg skeletal dysplasia is caused by 3 beta-hydroxysterol delta 14-reductase deficiency due to mutations in the lamin B receptor gene. Am J Hum Genet 72:1013–1017

DISCUSSION

Fatkin: I think the gender difference in the mouse model is very interesting. In our experience looking at families with lamin A/C mutations, we haven't noticed any gender differences in the severity of the clinical phenotype. You have a much larger database of families: have you noticed any gender differences?

Bonne: It is hard to have a clear idea of the gender difference. We examine the mice every day for any symptoms. Patients come to the clinic once they are sick and we lack a clear retrospective analysis of all cases. We have just started to combine all our data to see whether there is a difference, but it will be difficult to get a clear answer.

Davies: Related to this question, does the phenotype always breed true in families? I get the impression that there is a strong genotype–phenotype correlation in this particular disease as opposed to some of the other forms of muscular dystrophy.

Bonne: It is not clear.

Capeau: For familial partial lipodystrophy (FPLD) it seems that the situation is the reverse of what was found in the mouse model. There is an increased occurrence and also severity of the disease in human female patients as compared with males. When we screen for *LMNA* mutations in a whole family, we know who is mutated and who is not. When we compare siblings, sisters are more severely affected than brothers. This is true for lipodystrophy. We are sometimes unable to see any lipodystrophy in male patients even where it is clear in the female patients. The same is true for the metabolic alterations.

Bonne: I should clarify that my previous answer was just for muscular dystrophy and cardiomyopathy.

Young: Regarding the gender difference, if my memory serves correctly, there is a similar finding in mice with males being more severely affected with defects in the mitochondrial oxidation of fatty acids.

Wilson: I'd like to ask Gisèle Bonne and Colin Stewart to compare and contrast the phenotypes of lamin A deletion. I thought the lamin A knockout mouse had no overt cardiac dilation. You then created a missense homozygous mutant mouse that showed cardiac problems. How do you explain this difference?

Bonne: What I know is that the knockout gets sick much earlier than our mice. The cardiac phenotype is clearly a dilation. The systolic function is altered first. The heart is not able to contract correctly, and it then becomes dilated and the diastolic function is altered afterwards. This is an EDMD mutation.

Stewart: Wayne Giles can give a clearer answer than I can about the cardiac properties of the lamin nulls.

Giles: At the relatively early stage that we have studied; in the heterozygotes many animals showed quite minimal left ventricular dilations, but did have histological signs of cardiomyopathy. They have a small heart. I'll mention some of the electrical differences in my paper (Grattan et al 2005, this volume).

Fatkin: We have done *in vivo* studies of cardiac function in the homozygous lamin A/C knockout mice. Using echocardiography and micromanometry we have found that these mice develop dilated cardiomyopathy by 4–6 weeks of age. We also found intrinsic defects in contractility in isolated myocytes. I think these mice will be a useful model for studying the mechanisms of cardiac dysfunction associated with LMNA mutations.

Goldman: What happens to the myofibrils? You are isolating cardiac muscle cells and saying that they are not functioning normally. This is interesting. What is the organization of the myofibrillar network? Do they have normal morphology? Are they oriented properly within cells?

Fatkin: Electron microscopy studies show that the myofibril and sarcomere organization in the heart are relatively normal. The most striking finding is altered nuclear morphology.

Goldman: Can you speculate on how this would interfere with the normal contractile process?

Fatkin: When we looked in detail at cardiomyocyte morphology, we found that the desmin connections to the nuclear surface were disrupted and there was disorganization of the cytoskeletal desmin network. It is possible that these desmin changes may alter cytoskeletal tension and reduce the mechanical efficiency of force transduction in the cytoskeleton. Defective force transmission is one of the mechanisms that has been proposed for contractile defects in a number of other forms of familial dilated cardiomyopathy.

Bonne: It would be interesting to compare the defects of these different mice because the mutations are different. If we can dissect in more detail the differences in phenotype, we may learn something.

Shackleton: What effect does the H222P mutation have on lamin protein localization or function?

Bonne: Unfortunately, we don't have any cells from the patient carrying this mutation. Some families are not open to research. From the mouse mutant we haven't yet got any results from cell culture experiments. The *in situ* localization in heart and skeletal muscle looks similar to controls.

Gruenbaum: Did you test whether the H222P mutation affects lamin assembly *in vitro*?

Bonne: We haven't yet looked at this.

Julien: It is surprising that the heterozygotes don't develop any symptoms.

Bonne: So far we have done ECGs on 18-month old heterozygote mice. They look the same as controls.

Davies: What levels of lamin A/C protein do the heterozygotes have?

Bonne: The levels of proteins are similar to controls. There is no obvious reduction.

Giles: I'd like to return to the issue of the overall contractile function of the heart. When assessing cardiac function in a mouse using single cells, it is important to pace these cells as fast as possible. We aren't able to pace them at anything like the normal heart rate in our experiments done at room temperature. Nonetheless, there is an interesting difference. The chamber of the ventricle that has to do the most work, the left ventricle, is compromised, whereas the right ventricle is not. This is an interesting difference. We think that there must be some link to the overall energy supply of the myocytes. We study the ventricle of young animals because we need to; the homozygotes die early. We are probably not analysing these deficiencies at quite as advanced a stage of disease as others are.

Julien: We are assuming that the disease here is cell autonomous. But there is a lesson to be learned from transgenic mice expressing superoxide dismutase (SOD) mutants linked to amyotrophic lateral sclerosis. Everyone assumed that motor neuron disease would be due to defects within motor neurons, and it turns out that expressing the mutant forms of SOD in the motor neuron did not drive the disease. The chamber difference here may tell us that other cell types might be having an effect on this disease.

Worman: We have discussed genotype–phenotype correlations with the lamin gene. How about the same clinical diseases other than EDMD caused by mutations in *LMNA* as well as other genes? Does the limb girdle muscular dystrophy that people with lamin mutations have look like any of the other numerous limb girdle muscular dystrophies caused by mutations in other proteins? Does the CMT II caused by lamin A/C mutation look like the CMTs caused by mutations in other proteins? Can you say, for example, that the one with the lamin mutation looks a lot like the one with the neurofilament mutation?

Levy: What we can tell about neurofilaments is that they are involved in other types of CMT. This is one of the reasons why we checked the lamin gene for mutations in another CMT disorder. It is now well known that neurofilament light chain interacts directly with another protein called myotubularin-related protein 2 (MTMR2), which is also involved in other CMT disorders.

Goldman: What does this MTMR2 do?

Levy: It belongs to a superfamily of proteins involved in phosphoinositol metabolism. Also, I have a point concerning the genotype–phenotype correlations. There are no clear phenotype–genotype correlations for each mutation. If you take all the muscle disorders, it seems that mutations are spread out over all the gene, but essentially in the rod domain. If you take the other phenotypes involving adipose, blood and skin tissues, they are more-or-less located in globular domains, either in the N-terminal or the C-terminal domains.

Goldman: Gisèle Bonne, you said that the rod mutations in EDMD gave no nuclear phenotype.

Bonne: I don't agree with Nicolas Lévy regarding the clinical phentotype and the genotype localization. Mutations leading to muscular dystrophy/cardiomyopathy are spread all along the gene in the rod and in the C-terminal domain. As for the cellular phenotype we observed, in the few cells that we analysed from patients with mutations affecting the rod domain, we didn't find any nuclear abnormalities.

Goldman: On the basis of our own findings with progeria, we have observed that the longer cells are passaged in culture, the more defects are apparent. You can't see anything early on, but if you wait, the defects accumulate. I should mention a rod mutation that gives an extraordinary phenotype. This is the progeria E145A mutation in segment 2B of the rod domain, which results in the most extensively lobulated nucleus that we have seen to date. This is interesting because most point mutations in the rod in all of the other intermediate filament diseases, especially skin blistering diseases produce nuclear structural defects.

Bonne: With our study, what we found may be due to fixation techniques. We use a very mild fixation procedure, and using this we find a difference between the domains.

Goldman: You don't need to fix a cell to look at the nucleus as it is readily visible in most live cells using phase contrast microscopy.

Bonne: When other groups use a more rigorous fixation procedure, they find abnormalities in cells of patients with rod domain mutations. We still need to confirm this. Nevertheless, using mild fixation conditions we see a difference. This may say something about structural differences. It is hard to draw clear conclusions from just a few cases.

Levy: I have a comment about the nuclei in progeria cells. In our experience, we have observed nuclear shape deformities at very early stages in cell culture.

Goldman: We can see changes early, but the percentage of cells with changes increases enormously with time in culture. When you derive cells from a biopsy, the number of generations these cells have already gone through is variable. We were concerned when we looked at progeria cells at early passage number in culture as their nuclei looked quite normal. Later, they became abnormal especially with respect to their shapes. Someone should do this with EDMD.

Bonne: To continue on this subject of a proportion of cells showing nuclear abnormalities, we have had the opposite experience, at least in cells with the homozygous nonsense mutation. During the first passages in culture more than 80% of the cells were abnormal. With increased time in culture the percentage of abnormally shaped nuclei decreased.

Goldman: But it is well established that fibroblasts tend to overgrow the culture. With progeria, there is premature senescence of cells in culture. Normal human cells from a biopsy can be grown for many passages with no problem. With progeria we have problems keeping things going.

Stewart: Have these cells been transformed, or are they transformable, with telomerase?

Oshima: We did this. The nuclear shape becomes much more normal. This is because those cells which can take up h-Tert and then become immortalized are the ones that are normal-looking to begin with.

Goldman: So are you saying that you can rescue the cells from progeria patients or the atypical progeria patients?

Oshima: Yes, both. But when you want to do experiments you need to shut down h-Tert. In laminopathies even patients with the same mutation show considerable differences in their clinical course. Polymorphisms within the lamin gene might affect the severity of the symptoms.

Bonne: In our screening process we now note every polymorphism known in the lamin A/C genes. We have just started this, but we haven't done any retrospective analysis on phenotype–genotype correlations, to see whether a patient with more polymorphisms may have a different course of disease. It is something we'd like to do, but it will be hard work.

Oshima: Somehow the CNS and epithelium are spared, despite the fact that they have high lamin expression. Why is this?

Goldman: In progeria, as we discussed earlier, premature baldness could be an epithelial cell problem.

Oshima: I was thinking more of the intestinal wall or the CNS. Progeria patients don't get dementia.

Goldman: Several people also mentioned skin problems. Are these dermal or epidermal?

Oshima: It is keratinocytes.

Goldman: These are epithelial cells, so there is epithelial cell involvement.

Burke: I have a question about cells from the functional human knockout. You don't see any lamin A fragments. Is this because the protein itself may be unstable? Have you actually looked for mRNA?

Bonne: There is no mRNA. We have done RT-PCR on the various cells, and there is no laminA/C mRNA in the homozygous cells.

Burke: I have come across an EDMD patient in Florida who has essentially a silent mutation in exon 4. It appears to cause spurious splicing. In his case, mRNA is undetectable.

Bonne: We did not screen patient mRNA each time we found a stop codon or splice site mutation. First we did a Western blot experiment looking for the truncated protein. After this we looked for mRNA and screened various sites: N-terminal, C-terminal and rod domain. So far we have found nothing.

Wilson: You previously estimated that about half of the EDMD patients have no mutations in lamin A or emerin; what is this number currently?

Bonne: People with clear-cut symptoms of EDMD in our panel were screened for emerin or lamin. We didn't find anything. There is still a group of about a third of the patients without a known mutation.

Wilson: Have you screened for other mutations?

Bonne: We have started screening for candidate genes. There is a long list of genes in front of us that we need to screen for. We want to look for a large deletion of lamin, at least. The processes we use involve DHPLC and sequencing, and by amplification of each individual exon we might miss large deletions involving several exons or part of them.

Fatkin: If you have a group of families with dilated cardiomyopathy the ones that are most likely to have *LMNA* mutations are the ones with progressive AV conduction block before the development of dilated cardiomyopathy. It is interesting that most of the familial dilated cardiomyopathies without conduction disease are caused by sarcomeric or cytoskeletal protein mutations. One thing we haven't really got to grips with is why emerin and lamin mutations cause conduction disease. This is a subject that needs to be explored.

Bonne: A mouse model might help here, as our mice seem to develop conduction defects quite early in the process of cardiac disease. This may be the first point to understand: why this specific model has this conduction defect. You are right: all the dilated cardiomyopathies with other mutated genes so far have no conduction defects.

Gerace: I have a general question about dilated cardiomyopathy. Is there an increased level of cell proliferation, or is there an increased number of cells there? Or is it just the tissue organization that is distorted.

Bonne: I am not sure whether the 19 different genes identified as affecting dilated cardiomyopathy have been checked individually for increased numbers of cells. I would probably argue for dilation of the tissue itself. I am not sure that there is an increased number of cells.

Gerace: How can this happen? How would you explain this change in tissue structure?

Goldman: It has profound implications.

Wilson: When the heart gets larger (dilates), is this because the existing tissue enlarges, or because the chambers inside the heart enlarge?

Bonne: It seems to be a general process for all the dilated cardiomyopathies. There are two different things. There is hypertrophic cardiomyopathy in which there is wall thickening while the chamber remains the same, and there is dilated cardiomyopathy in which the chamber dilates but the wall thickness either remains the same or gets thinner.

Fatkin: There are two important points here. The textbook answer about the cell proliferation is that once you have a mature heart there is a finite number of cells. However, there are some recent data suggesting that there may be pools of stem cells in the heart that may have regenerative capacity. For a diagnosis of dilated cardiomyopathy, there needs to be both ventricular dilation and reduced contractile function. The cause of ventricular dilation depends on the aetiology of the disease in question. For example, there can be a primary systolic defect that causes secondary dilation; or the presence of ventricular dilation can itself impair contractile function.

Lee: It's a basic question, but no one actually knows what the cells are doing relative to each other when the heart dilates. There is a broad distribution in the size of myocytes. We have just worked out a method by which we can do histology on the whole heart at once, at sub-micron resolution. We are currently using this to study Colin Stewart's knockout mice and it may help us understand how dilation occurs.

Wilson: When you are doing this can you also assess the distribution of connective tissue cells?

Lee: Yes. We can scan in three dimensions and see the connective tissue, myocytes and vasculature. To do the whole mouse heart takes a long time, but one can study small volumes and see the connective tissue.

Julien: Someone mentioned earlier that the desmin organization was affected in mice bearing the lamin A mutation. It would be interesting to compare the histopathology with the desmin knockout mice, because these also exhibit calcified deposits.

Bonne: I have seen a nice picture of muscle fibres from desmin knockout mice. There was a batch of aggregated nuclei. I have never seen so many nuclei at the same place in a muscle fibre.

Fatkin: The desmin knockout mice look quite different to lamin A/C knockout mice. From a very early age the desmin-null hearts show large patches of cytoskeletal disorganization, necrosis and calcification. These mice develop dilated cardiomyopathy with a compensatory hypertrophic response fairly early on. In the lamin knockout mice, however, there is very little cytoskeletal disorganization at light microscopy level, and a hypertrophic response is not observed.

Goldman: Where were the nuclei? Were they mislocalized?

Fatkin: We didn't look at that specifically.

Goldman: How did you know that desmin was not interacting with the nuclear surface?

Fatkin: We looked at the nuclear connections of desmin with immunogold-labelled desmin antibodies and electron microscopy. In the wild-type mice, there were tight desmin filament attachments near the nuclear pores. In the homozygous mice, there was marked dispersion of the immunogold particles with a gap between the nuclear surface and adjacent myofibrils.

Goldman: Remember, though, that desmin also anchors to intercalated discs and connects to the myofibrils. It is not an accepted fact that IFs such as desmin bind to the nuclear surface. It would be nice to know a little more specifically about what the changes are in the perinuclear region.

Reference

Grattan MJ, Kondo C, Thurston J et al 2005 Skeletal and cardiac muscle defects in a murine model of Emery-Dreifuss muscular dystrophy. In: Nuclear organization in development and disease. Wiley, Chichester (Novartis Found Symp 264) p 118–139

Muscular dystrophies related to the cytoskeleton/nuclear envelope

Kristen Nowak, Karl McCullagh, Ellen Poon and Kay E Davies[1]

Department of Human Anatomy and Genetics, University of Oxford, South Parks Road, Oxford OX1 3QX, UK

Abstract. Mutations in genes encoding proteins expressed in skeletal muscle cause a significant number of human diseases. Neuromuscular diseases are often severely debilitating for affected individuals, frequently leading to a shortened life span. Identifying the cause of these muscle diseases has provided insight not only into disease pathogenesis and muscle dysfunction, but also into the normal function of muscle. In 1987, dystrophin became the first disease-related human gene to be identified by positional cloning. Dystrophin is an integral component of the membrane-attached cytoskeleton of muscle fibres, with mutations in this gene causing Duchenne and Becker muscular dystrophy. One group of proteins known as the dystrophin-associated protein complex (DAPC), is believed to provide a molecular link between the actin cytoskeleton and the extracellular matrix in muscle cells, thereby sustaining sarcolemmal integrity during muscle contraction. Mutations in many members of the DAPC cause a variety of diseases, emphasising the importance of these genes. Another group of important proteins in skeletal muscle is the intermediate filament family, which provides mechanical strength and a supporting framework within the muscle cell. They anchor actin thin filaments through their expression at the Z-disk in sarcomeres, which in turn interact with myosin thick filaments to cause muscle contraction. This chapter will explore the protein components of the DAPC and the intermediate filament complex, highlighting a novel protein, which links the two, syncoilin. Human diseases and studies of existing animal models caused by mutations in these genes will also be described.

2005 Nuclear organization in development and disease. Wiley, Chichester (Novartis Foundation Symposium 264) p 98–117

The dystrophin associated protein complex (DAPC)

At approximately 2.4 million bp long, and encompassing about 0.5% of the human genome, dystrophin is one of the largest genes currently known. The 427 kDa dystrophin protein is expressed in skeletal and cardiac muscle and in brain

---

[1]This paper was presented at the symposium by Kay E Davies to whom all correspondence should be addressed.

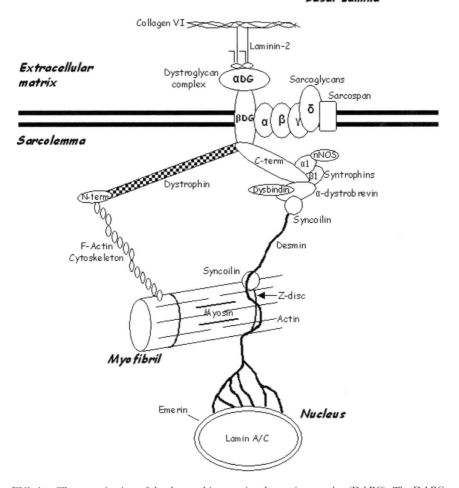

FIG. 1.   The organization of the dystrophin-associated protein complex (DAPC). The DAPC
in skeletal muscle is composed of a number of proteins located within and around the
sarcolemmal membrane. The intermediate filament syncoilin links the DAPC through the
binding partners α-dystrobrevin and desmin with the muscle cytoskeletal architecture possibly
providing support to maintain muscle integrity during contractions.

(Hoffman et al 1987, Koenig et al 1987). In skeletal muscle, dystrophin is located at
the cytoplasmic face of the sarcolemma, the depths of the postsynaptic folds of the
neuromuscular junction (NMJ) and at the myotendinous junction (MTJ) (Bonilla
et al 1988, Watkins et al 1988, Byers et al 1991). Dystrophin interacts with other
proteins such as the dystroglycans, dystrobrevins, syntrophins, sarcoglycans and

sarcospan to constitute the DAPC (Yoshida & Ozawa 1990, Campbell & Kahl 1989, Ervasti & Campbell 1993). The organization of the DAPC is shown in Fig. 1. Dystrophin binds the DAPC via an α-helical coiled-coil domain at the C-terminus (Blake et al 1995), whilst the N-terminal region of the protein is involved with binding to actin filaments (Norwood et al 2000). These data suggest that dystrophin serves as a molecular link between the actin cytoskeleton and the extracellular matrix (as dystroglycan interacts with laminin), thereby sustaining sarcolemmal integrity by providing the plasma membrane with support during muscle contraction (Hack et al 2000).

Mutations in dystrophin and dystrophin-associated proteins are now well established as major causes of inherited muscular dystrophies, a group of clinically heterogeneous diseases characterized by muscle wasting. The fact that these diseases exist and are often severe, demonstrates the importance of the DAPC in muscle function. One of the most common neuromuscular disorders affecting man is Duchenne Muscular Dystrophy (DMD; OMIM 300377, Online Mendelian Inheritance in Man, *http://www.ncbi.nlm.nih.gov/entrez/query.fcgi?db=OMIM*), which along with Becker Muscular Dystrophy (OMIM 300376) is caused by mutations within the dystrophin gene. DMD affects 1 in 3500 boys (Emery 1993), and although patients appear clinically normal at birth, the first clinical symptoms being a waddling gait and difficulty in climbing stairs appear when the child reaches 2–5 years of age (Dubowitz 1978, Jennekens et al 1991). Over time pseudohypertrophy and proximal limb muscle weakness develops, culminating in wheelchair dependence by the age of 12 and death in early 20s, most often due to diaphragm and cardiac involvement (Emery 1993).

The naturally occurring dystrophic mouse, *mdx*, is the best characterized mouse model of a muscular dystrophy caused by a mutation in dystrophin (Sicinski et al 1989). Clinically, the *mdx* phenotype is milder than that observed in human DMD patients, as while human patients undergo rapid and progressive muscle weakness and die prematurely, *mdx* mice do not suffer from obvious weakness and live a normal life span (De la Porte et al 1999, Gillis 1999). Like DMD patients, *mdx* mice undergo repeated cycles of degeneration and regeneration, however this regeneration is sustained and thus compensates for the dystrophic process (De la Porte et al 1999, Gillis 1999). The mouse model that most closely resembles the phenotype of DMD patients is both dystrophin and utrophin negative (Deconinck et al 1997, Grady et al 1997), utrophin being an autosomal homologue of dystrophin. The most appropriate animal model available due to its size is that of the Golden Retriever muscular dystrophy (GRMD) dog (Cooper et al 1988, Valentine et al 1992).

There is concomitant decrease in the levels of other members of the DAPC in DMD and *mdx* muscles lacking dystrophin, suggesting that dystrophin is crucial for the assembly of this protein complex and that the DAPC may contribute to the pathogenesis of DMD (Ervasti et al 1990, Ohlendieck & Campbell 1991, Matsumura et al 1993). Furthermore, in *mdx* muscle, mechanical properties are compromised and the incidence of sarcolemmal rupturing during contraction increases, indicating that the DAPC is necessary to protect against muscle damage (Pasternak et al 1995). To further dissect the mechanism of DAPC assembly in muscle, transgenic experiments were performed using different domains of dystrophin. In *mdx* mice expressing an N-terminal truncated form of dystrophin, the DAPC is restored to the sarcolemma (Cox et al 1994, Greenberg et al 1994). However, the dystrophic phenotype still persists. In contrast, a dystrophin minigene lacking a large section of the rod domain successfully rescued *mdx* muscle from dystrophy (Deconinck et al 1996). This demonstrates the importance of the link between the DAPC and the extracellular matrix.

As already mentioned, dystrophin interacts with laminin-2 by binding to the dystroglycans. Laminin-2 (also known as merosin) is a heterotrimer and is the main laminin found in muscle, being expressed in the basement membrane. Mutations in the gene for laminin-2, *LAMA2*, can lead to congenital merosin-deficient muscular dystrophy type 1A (OMIM 607855). Complete loss-of-function mutations appear to cause the more severe form of the disease with onset in the neonate, whilst a milder form of the disease is indicative of missense mutations (Tezak et al 2003). A severe, progessive muscular dystrophy is seen in the homozygous *dy* mouse (Michelson et al 1955), which has a mutation leading to abnormal splicing of the laminin α2 chain (Xu et al 1994). Moll et al (2001) were able to restore muscle function in the *dy* mouse by expressing a minigene of agrin. Agrin binds to both the basement membrane and to α-dystroglycan, and thus expression of the mini-gene stabilizes the link between the DAPC and the integrins (Moll et al 2001).

α- and β-dystroglycan are widely expressed proteins, with both isoforms being heavily glycosylated and arising from a single post-translationally modified polypeptide (Winder 2001). Whilst β-dystroglycan is a vital membrane protein, α-dystroglycan is only associated with the membrane through an interaction with the extracellular portion of β-dystroglycan. By binding to cytoplasmic proteins, the dystroglycans provide an essential link from the actin cytoskeleton to the basement membrane. No human diseases have been associated with mutations within the dystroglycan gene, but this is perhaps not surprising as knockout mice for this gene are embryonic lethal, indicating the importance of dystroglycan (Williamson et al 1997).

The dystroglycans interact with the sarcoglycan proteins, with five different sarcoglycan genes currently identified, each of which codes for a transmembrane protein. α- (50 kDa) and γ-sarcoglycan (35 kDa) are expressed predominantly and almost exclusively in striated muscle, whilst β- (43 kDa) and δ-sarcoglycan (35 kDa) are expressed more widely in a variety of tissues (Lim et al 1995,

Noguchi et al 1995, Bonnemann et al 1995, Nigro et al 1996a,b, Jung et al 1996). ε-Sarcoglycan (50 kDa) is widely distributed in muscle and non-muscle cells in both embryonic and adult tissue (Ettinger et al 1997, McNally et al 1998), including the brain, smooth muscle (Straub et al 1999), and peripheral nerve (Imamura et al 2000).

The limb girdle muscular dystrophies (LGMDs) are a group of diseases that mainly involve the muscles of the shoulder and pelvic girdles (Fanin et al 1997). Within the LGMDs exists the sarcoglycanopathies, a heterogenous group of autosomal recessive diseases caused by mutations in four of the five sarcoglycan genes: α (LGMD2D, OMIM 608099), β (LGMD2E, OMIM 604286), γ (LGMD2C, OMIM 253700) and δ (LGMD2F, OMIM 601287) (Roberds et al 1994, Lim et al 1995, Noguchi et al 1995, Piccolo et al 1996, Bonnemann et al 1995, Nigro et al 1996a,b). Mutations within ε-sarcoglycan have been identified in patients with myoclonus-dystonia syndrome, OMIM 159900 (Zimprich et al 2001). Many hamster and mouse models of differing progressive muscular dystrophy severities exist for the sarcoglycan LGMDs (e.g. Nigro et al 1997, Straub et al 1998, Duclos et al 1998, Hack et al 1998, Araishi et al 1999).

What is known as the cytoplasmic sub-complex of the DAPC consists of the syntrophins and dystrobrevins. To date, two dystrobrevin isoforms have been identified, α- and β-, each encoded by a separate gene (Ambrose et al 1997, Peters et al 1997, Sadoulet-Puccio et al 1997a, Loh et al 1998). Alternative splicing of the C-terminus produces five α-dystrobrevin isoforms. Three of these are found in muscle: α-dystrobrevin 1 (94 kDa), α-dystrobrevin 2 (62 kDa) and α-dystrobrevin 3 (42 kDa) (Nawrotzki et al 1998). The α-dystrobrevin isoforms differ significantly in their localization in muscle. α-dystrobrevin 1 is concentrated at the NMJ with no or weak sarcolemmal distribution while α-dystrobrevin 2 is highly abundant at the sarcolemma and is also enriched at the NMJ (Nawrotzki et al 1998, Peters et al 1998). α-dystrobrevins form complex interactions with other members of the DAPC. *In vitro* studies have shown an association between the coiled-coil domains of α-dystrobrevin 1 and 2 and dystrophin (Blake et al 1995, Sadoulet-Puccio et al 1997b). Grady et al (1999) reported that mice deficient in α-dystrobrevin exhibit both skeletal and cardiac myopathies. Interestingly, whilst the expression of the signalling molecule neuronal nitric oxide synthase (nNOS) is disrupted and signalling impaired in the absence of α-dystrobrevin, other structural components of the DAPC are reserved (Grady et al 1999).

Dystrobrevin has been shown to associate with the sarcoglycan–sarcospan complex, possibly providing the link between syntrophin and nNOS (Yoshida et al 2000). No human diseases have been associated with mutations in the sarcospan gene. The 25 kDa sarcospan protein has four transmembrane domains with both intracellular N- and C-termini, and is expressed predominantly in skeletal and cardiac muscle (Crosbie et al 1997). Mice that are null for sarcospan do not develop muscular dystrophy, most likely due to retained expression of all sarcoglycans at the sarcolemma (Lebakken et al 2000). Sarcospan has been found to be absent from the muscle of some patients with γ-sarcoglycanopathy (Crosbie et al 2000). This finding suggests that correct localisation of γ-sarcoglycan into the sarcoglycan complex is essential for sarcospan binding.

Two isoforms of syntrophin (58 kDa) are expressed at the sarcolemma in skeletal muscle (α1- and β1-), whilst three isoforms are present at the NMJ (Kramarcy & Sealock 2000). The syntrophins are signalling proteins due to the presence of two domains of approximately 100 amino acids (pleckstrin homology domains), and have been shown to bind to the stress-activated protein kinase 3, nNOS, MAST205, serine/threonine kinases and proteins contained within sodium channels (Rando 2001). The predominant isoform of syntrophin expressed in muscle is α1-syntrophin, but mice that are knocked-out for this isoform do not show any obvious phenotype (Hosaka et al 2002). However it appears that α1-syntrophin is important for regeneration as these knockout mice show aberrant NMJ formation and hypertrophy during this process (Hosaka et al 2002).

Dysbindin is a 40 kDa ubiquitously expressed protein that has been shown to be up-regulated in the skeletal muscle of *mdx* mice, presumably as a result of it binding to α-dystrobrevin and β-dystrobrevin (Benson et al 2001). Dysbindin has also been identified as being involved in the biogenesis of lysosome-related organelles complex 1, BLOC1 (Li et al 2003). A mutation in dysbindin causes Hermansky-Pudlak syndrome (OMIM 203300), which is characterized by prolonged bleeding, oculocutaneous albinism and lysosomal ceroid storage. Other protein members of BLOC1 have been associated with Hermansky-Pudlak syndrome in mice e.g. pallidin (Huang et al 1999) and cappuccino (Gwynn et al 2000).

The intermediate filament protein family

Intermediate filament (IF) proteins are characterized by their ability to form 10 nm diameter filaments. Together with microfilaments (actin) and microtubules (tubulin), these three networks form the basis of the higher eukaryotic cytoskeleton. IF proteins share a common structural organization which consists of an N-terminal head domain, a central rod domain and a C-terminal domain. The N- and C-terminal domains show a great degree of variability and are postulated to be responsible for cell-type specific roles by interacting with extra- or intracellular proteins. In contrast, the central rod domain is highly conserved. IF proteins are classified into six groups on the basis of sequence homology within this central region (Fuchs & Hanukoglu 1983, Fuchs & Weber 1994, Fuchs 1997a,b). Several IF proteins have been identified in muscle at various stages of development. In mature fibres, desmin, synemin, paranemin, nestin and cytokeratin constitute the main IF proteins identified, whilst vimentin is mostly present in immature muscle.

Desmin is the main IF protein found in skeletal, cardiac and smooth muscle (Tokuyasu et al 1983). In normal muscle it is concentrated at the sarcolemma, Z-lines, NMJ and MTJ, where it is responsible for holding adjacent myofibrils together and attaching myofibrils to the sarcolemma (Askanas et al 1990, Small et al 1992, Tidball 1992). Desmin is also one of the earliest known myogenic markers. In mice it is first detected at 8.25 days post coitum in the neorectoderm, where vimentin, nestin and keratin are also transiently expressed (Furst et al 1989, Schaart et al 1989). It is also found at low levels in satellite cells (Allen et al 1991) and replicating blasts (Kaufman & Foster 1988). Since desmin expression precedes most muscle-specific structural genes and transcription factors such as MyoD and myogenin, it is proposed to play a role in myogenic commitment and differentiation.

Desmin-related myopathies (OMIM 601419) are a group of often hereditary myopathies marked by accumulation of desmin within the cytoplasm of skeletal and cardiac muscle cells. This myopathy can lead to skeletal muscle weakness, heart arrhythmias and restrictive heart failure (for review see Goebel 1997). Despite mutations being identified in the desmin gene for some patients, a mutation has not been detected for the majority of cases (Vicart et al 1996), hence mutations or abnormalities of desmin interactors are thought to contribute to these diseases. An example is the chaperone-like protein αB-crystallin, which has been found to be mutated in some patients with desmin-related myopathy (Vicart et al 1998). Desmin-null animals exhibit myopathy and cardiomyopathy (Li et al 1996, Milner et al 1996, Li et al 1997).

Desmin can self-assemble into homofilaments, but can also form heterofilaments with vimentin, synemin (Bellin et al 1999) and paranemin (Schweitzer et al 2001). Vimentin is known as the precursor of desmin since it is present in myotubes and in primary and secondary generation myofibres but is replaced by desmin when terminal differentiation occurs (Furst et al 1989). Up-regulation of vimentin has also been reported in several muscular dystrophies and is attributed to regeneration (Gallanti et al 1992). However mice homozygous for a null vimentin mutation show no obvious phenotype (Colucci-Guyon et al 1994). As a type III IF protein, vimentin shares significant sequence similarity with desmin and can functionally interact with binding partners of desmin, such as synemin and paranemin (Hemken et al 1997, Bellin et al 1999).

Synemin (also known as desmuslin; Mizuno et al 2001) and paranemin were first identified via their interactions with desmin and vimentin filaments (Granger & Lazarides 1980). In adult skeletal muscle, both proteins co-localize with desmin at the Z-lines, sarcolemma, NMJ and MTJ (Granger & Lazarides 1980, Breckler & Lazarides 1982, Price & Lazarides 1983, Carlsson et al 2000). Synemin and paranemin both contain the rod domain typical of IF proteins and were subsequently classified to be type VI IF proteins. An important feature of synemin and paranemin is their extended C-terminal domains, which are believed to mediate cell-type specific interactions. The tail domain of synemin has already been shown to bind α-actinin, a myofibrillar protein and vinculin, a costameric protein (Bellin et al 1999, 2001). This interaction is believed to play a role in maintaining myofibrillar Z-line alignment.

In similarity to synemin and paranemin, nestin is a type VI IF protein and has a predicted mass of approximately 200 kDa (Lendahl et al 1990, Dahlstrand et al 1992). Nestin has been shown to co-localize with desmin and vimentin filaments in cultured G6 human fetal skeletal muscle cells (Sejersen & Lendahl 1993, Sjoberg et al 1994). Expression is highest during muscle development but its level decreases after birth (Sejersen & Lendahl 1993). In mature muscle, it is restricted to the NMJ and MTJ (Carlsson et al 1999). No human diseases have been associated with mutations in nestin.

Cytokeratins are type I/II IF proteins and are expressed in developing muscle and muscle tumours (Langbein et al 1989, Kosmehl et al 1990). O'Neill et al (2002) recently reported the presence of cytokeratins at the costameres of mature muscle. Unlike synemin, paranemin and nestin, there has been no report of heterofilament formation between keratins and desmin/vimentin. In cells expressing vimentin and keratin, distinct IF structures are formed (Franke et al 1979).

Syncoilin — the missing link

Syncoilin is highly expressed in skeletal muscle and heart and was first identified via its interaction with α-dystrobrevin in muscle (Newey et al 2001). In adult skeletal muscle, syncoilin is present as a single 64 kDa protein and is localised to the sarcolemma, Z-lines and NMJ. In the heart, two protein products of 55 and 64 kDa are observed and may either represent products of alternative splicing or proteolytic cleavage. Expression is seen predominantly at the Z-lines and intercalated discs, but the contribution of each protein product is unclear (Newey et al 2001).

Syncoilin was proposed to be a member of the IF protein superfamily based on sequence analysis, with it being most similar to type III and IV IFs such as α-internexin (Newey et al 2001). Although syncoilin has the central coiled-coil domains typical of IF proteins, it failed to form filamentous structures in transfected cells, suggesting that other factors may be required to initiate heteropolymeric filament formation (Newey et al 2001).

By using the C-terminal region of syncoilin as the bait for a yeast two-hybrid screen, Poon et al (2002) identified desmin as a binding partner of syncoilin

(Poon et al 2002). This puts forward the interesting scenario that syncoilin links the extracellular matrix by tethering the DAPC via α-dystrobrevin to the IF cytoskeleton. This indicates that syncoilin is likely to be important for muscle function due to it's association with both important protein families, and thus it is likely to contribute to the maintenance of muscle integrity.

As yet no human diseases caused by mutations within syncoilin have been identified. A syncoilin knock-out mouse has been made by this laboratory and is currently being analysed. Studies of syncoilin expression in various muscle diseases have shown that syncoilin is upregulated in the skeletal muscle of animals in which the organisation of the DAPC is perturbed (Newey et al 2001), or where desmin has aggregated (Howman et al 2003). For muscular dystrophy models, the increase in syncoilin expression is thought to relate to the amount of muscle pathology, since the highest increase was recorded in the very severely affected *dko* mice, followed by the mildly dystrophic *mdx* mice. The least affected utrophin knock-out (*uko*) mice had syncoilin levels that are similar to control mice. Newey et al (2001) therefore proposed that the up-regulation of syncoilin is a mechanism to protect against muscle damage. Exactly how this is achieved is unclear at present.

Summary

Since mutations within the dystrophin gene were found to cause Duchenne and Becker muscular dystrophy 16 years ago, the genetic cause of a considerable number of other muscle diseases has been unravelled. In many cases, knowledge of the disease-causing mutations and genes has led to a greater understanding of interactions within the muscle cell. Two important structural protein complexes for muscle are the DAPC and the IF family. While the DAPC components dystrophin, laminin 2 and the sarcoglycans cause muscular dystrophies in humans, mutations in the IF protein desmin cause myopathies. As a member of both the DAPC and the IF family, syncoilin is believed to form an essential and unique link between the extracellular matrix and the cytoskeleton of muscle fibres. The localization of syncoilin at the sarcolemma, Z-lines and the NMJ in muscle cells, and studies showing that it is up-regulated in pathological muscle, suggests that syncoilin serves diverse and important roles.

References

Allen RE, Rankin LL, Greene EA et al 1991 Desmin is present in proliferating rat muscle satellite cells but not in bovine muscle satellite cells. J Cell Physiol 149:525–535
Ambrose HJ, Blake DJ, Nawrotzki RA, Davies KE 1997 Genomic organization of the mouse dystrobrevin gene: comparative analysis with the dystrophin gene. Genomics 39:359–369
Araishi K, Sasaoka T, Imamura M et al 1999 Loss of the sarcoglycan complex and sarcospan leads to muscular dystrophy in β-sarcoglycan-deficient mice. Hum Mol Genet 8:1589–1598
Askanas V, Bornemann A, Engel WK 1990 Immunocytochemical localization of desmin at human neuromuscular junctions. Neurology 40:949–953

Bellin RM, Sernett SW, Becker B et al 1999 Molecular characteristics and interactions of the intermediate filament protein synemin. Interactions with alpha-actinin may anchor synemin-containing heterofilaments. J Biol Chem 274:29493–29499

Bellin RM, Huiatt TW, Critchley DR, Robson RM 2001 Synemin may function to directly link muscle cell intermediate filaments to both myofibrillar Z-lines and costameres. J Biol Chem 276:32330–32337

Benson MA, Newey SE, Martin-Rendon E et al 2001 Dysbindin, a novel coiled-coil-containing protein that interacts with the dystrobrevins in muscle and brain. J Biol Chem. 276: 24232–24241

Blake DJ, Tinsley JM, Davies KE et al 1995 Coiled-coil regions in the carboxy-terminal domains of dystrophin and related proteins: potentials for protein-protein interactions. Trends Biochem Sci 20:133–135

Bonilla E, Samitt CE, Miranda AF et al 1988 Duchenne muscular dystrophy: deficiency of dystrophin at the muscle cell surface. Cell 54:447–452

Bonnemann CG, Modi R, Noguchi S et al 1995 Beta-sarcoglycan (A3b) mutations cause autosomal recessive muscular dystrophy with loss of the sarcoglycan complex. Nat Genet 11:266–273

Breckler J, Lazarides E 1982 Isolation of a new high molecular weight protein associated with desmin and vimentin filaments from avian embryonic skeletal muscle. J Cell Biol 92: 795–806

Byers TJ, Kunkel LM, Watkins SC 1991 The subcellular distribution of dystrophin in mouse skeletal, cardiac, and smooth muscle. J Cell Biol 115:411–421

Campbell KP, Kahl SD 1989 Association of dystrophin and an integral membrane glycoprotein. Nature 338:259–262

Carlsson L, Li Z, Paulin D, Thornell LE 1999 Nestin is expressed during development and in myotendinous and neuromuscular junctions in wild type and desmin knock-out mice. Exp Cell Res 251:213–223

Carlsson L, Li ZL, Paulin D, Price MG et al 2000 Differences in the distribution of synemin, paranemin, and plectin in skeletal muscles of wild-type and desmin knock-out mice. Histochem Cell Biol 114:39–47

Colucci-Guyon E, Portier MM, Dunia I et al 1994 Mice lacking vimentin develop and reproduce without an obvious phenotype. Cell 18:679–694

Cooper BJ, Winand NJ, Stedman H et al 1988 The homologue of the Duchenne locus is defective in X-linked muscular dystrophy of dogs. Nature 334: 154–156

Cox GA, Sunada Y, Campbell KP, Chamberlain JS 1994 Dp71 can restore the dystrophin-associated glycoprotein complex in muscle but fails to prevent dystrophy. Nat Genet 8:333–339

Crosbie RH, Heighway J, Venzke DP et al 1997 Sarcospan, the 25-kDa transmembrane component of the dystrophin-glycoprotein complex. J Biol Chem 272:31221–31224

Crosbie RH, Lim LE, Moore SA et al 2000 Molecular and genetic characterization of sarcospan: insights into sarcoglycan-sarcospan interactions. Hum Mol Genet 9:2019–2027

Dahlstrand J, Zimmerman LB, McKay RD, Lendahl U 1992 Characterization of the human nestin gene reveals a close evolutionary relationship to neurofilaments. J Cell Sci 103(Pt 2):589–597

Deconinck N, Rafael JA, Beckers-Bleukx G et al 1996 Functional protection of dystrophic mouse (mdx) muscles after adenovirus-mediated transfer of a dystrophin minigene. Proc Natl Acad Sci USA 93:3570–3574

Deconinck AE, Rafael JA, Skinner JA et al 1997 Utrophin-dystrophin-deficient mice as a model for Duchenne muscular dystrophy. Cell 90:717–727

De la Porte S, Morin S, Koenig J 1999 Characteristics of skeletal muscle in mdx mutant mice. Int Rev Cytol 191:99–148

Dubowitz V 1978 Muscle disorders in childhood. Major Probl Clin Pediatr 16:iii–xiii, 1–282

Duclos F, Straub V, Moore SA et al 1998 Progressive muscular dystrophy in α-sarcoglycan-deficient mice. J Cell Biol 142:1461–1471

Emery AEH 1993 Duchenne muscular dystrophy. Oxford Monographs of Medical Genetics, No. 24. Oxford University Press, Oxford

Ervasti JM, Ohlendieck K, Kahl SD et al 1990 Deficiency of a glycoprotein component of the dystrophin complex in dystrophic muscle. Nature 345:315–319

Ervasti JM, Campbell KP 1993 A role for the dystrophin-glycoprotein complex as a transmembrane linker between laminin and actin. J Cell Biol 122:809–823

Ettinger AJ, Feng GP, Sanes JR 1997 Epsilon-sarcoglycan, a broadly expressed homologue of the gene mutated in limb-girdle muscular dystrophy 2d. J Biol Chem 272:32534–32538

Fanin M, Duggan DJ, Mostacciuolo ML et al 1997 Genetic epidemiology of muscular dystrophies resulting from sarcoglycan gene mutations. J Med Genet 34:973–977

Franke WW, Schmid E, Breitkreutz D et al 1979 Simultaneous expression of two different types of intermediate sized filaments in mouse keratinocytes proliferating in vitro. Different 14:35–50

Fuchs E 1997a Keith R Porter lecture 1996. Of mice and men: genetic disorders of the cytoskeleton. Mol Biol Cell 8:189–203

Fuchs E 1997b The cytoskeleton and disease: genetic disorders of intermediate filaments. Annu Rev Genet 30:197–231

Fuchs E, Hanukoglu I 1983 Unraveling the structure of the intermediate filaments. Cell 34:332–334

Fuchs E, Weber K 1994 Intermediate filaments: structure, dynamics, function, and disease. Annu Rev Biochem 63:345–382

Furst DO, Osborn M, Weber K 1989 Myogenesis in the mouse embryo: differential onset of expression of myogenic proteins and the involvement of titin in myofibril assembly. J Cell Biol 109:517–527

Gallanti A, Prelle A, Moggio M et al 1992 Desmin and vimentin as markers of regeneration in muscle diseases. Acta Neuropathol (Berl) 85:88–92

Gillis JM 1999 Understanding dystrophinopathies: an inventory of the structural and functional consequences of the absence of dystrophin in muscles of the mdx mouse. J Muscle Res Cell Motil 20:605–625

Goebel HH 1997 Desmin-related myopathies. Curr Opin Neurol 10:426–429

Grady RM, Teng H, Nichol MC et al 1997 Skeletal and cardiac myopathies in mice lacking utrophin and dystrophin: a model for Duchenne muscular dystrophy. Cell 90:729–738

Grady RM, Grange RW, Lau KS et al 1999 Role for alpha-dystrobrevin in the pathogenesis of dystrophin-dependent muscular dystrophies. Nat Cell Biol 1:215–220

Granger BL, Lazarides E 1980 Synemin: a new high molecular weight protein associated with desmin and vimentin filaments in muscle. Cell 22:727–738

Greenberg DS, Sunada Y, Campbell KP et al 1994 Exogenous Dp71 restores the levels of dystrophin associated proteins but does not alleviate muscle damage in mdx mice. Nat Genet 8:340–344

Gwynn B, Ciciotte SL, Hunter SJ et al 2000 Defects in the cappuccino (cno) gene on mouse chromosome 5 and human 4p cause Hermansky-Pudlak syndrome by an AP-3-independent mechanism. Blood 96:4227–4235

Hack AA, Ly CT, Jiang F et al 1998 Gamma-sarcoglycan deficiency leads to muscle membrane defects and apoptosis independent of dystrophin. J Cell Biol 142:1279–1287

Hack AA, Groh ME, McNally EM 2000 Sarcoglycans in muscular dystrophy. Microsc Res Tech 48:167–180

Hemken PM, Bellin RM, Sernett SW et al 1997 Molecular characteristics of the novel intermediate filament protein paranemin. Sequence reveals EAP-300 and IFAPa-400 are highly homologous to paranemin. J Biol Chem 272:32489–32499

Hoffman EP, Monaco AP, Freener CC, Kunkel LM 1987 Conservation of the Duchenne muscular dystrophy gene in mice and humans. Science 238:347–350

Howman EV, Sullivan N, Poon EP, Britton JE, Hilton-Jones D, Davies KE 2003 Syncoilin accumulation in two patients with desmin-related myopathy. Neuromuscul Disord 13:42–48

Hosaka Y, Yokota T, Miyagoe-Suzuki Y et al 2002 Alpha1-syntrophin-deficient skeletal muscle exhibits hypertrophy and aberrant formation of neuromuscular junctions during regeneration. J Cell Biol 158:1097–1107

Huang L, Kuo YM, Gitschier J 1999 The pallid gene encodes a novel, syntaxin 13-interacting protein involved in platelet storage pool deficiency. Nat Genet 23:329–332

Imamura M, Araishi K, Noguchi S, Ozawa E 2000 A sarcoglycan-dystroglycan complex anchors Dp116 and utrophin in the peripheral nervous system. Hum Mol Genet 9:3091–3100

Jennekens FG, ten Kate LP, de Visser M, Wintzen AR 1991 Diagnostic criteria for Duchenne and Becker muscular dystrophy and myotonic dystrophy. Neuromuscul Disord 1:389–391

Jung D, Duclos F, Apostol B et al 1996 Characterization of delta-sarcoglycan, a novel component of the oligomeric sarcoglycan complex involved in limb-girdle muscular dystrophy. J Biol Chem 271:32321–32329

Kaufman SJ, Foster RF 1988 Replicating myoblasts express a muscle-specific phenotype. Proc Natl Acad Sci USA 85:9606–9610

Koenig M, Hoffman EP, Bertelson CJ et al 1987 Complete cloning of the Duchenne muscular dystrophy (DMD) cDNA and preliminary genomic organization of the DMD gene in normal and affected individuals. Cell 50:509–517

Kosmehl H, Langbein L, Katenkamp D 1990 Transient cytokeratin expression in skeletal muscle during murine embryogenesis. Anat Anz 171:39–44

Kramarcy NR, Sealock R 2000 Syntrophin isoforms at the neuromuscular junction: developmental time course and differential localization Mol Cell Neurosci 15:262–274

Langbein L, Kosmehl H, Kiss F et al 1989 Cytokeratin expression in experimental murine rhabdomyosarcomas. Intermediate filament pattern in original tumors, allotransplants, cell culture and re-established tumors from cell culture. Exp Pathol 36:23–36

Lebakken CS, Venzke DP, Hrstka RF et al 2000 Sarcospan-deficient mice maintain normal muscle function. Mol Cell Biol 20:1669–1677

Lendahl U, Simmerman LB, McKay RD 1990 CNS stem cells express a new class of intermediate filament protein. Cell 60:585–595

Li Z, Colucci-Guyon E, Pincon-Raymond M et al 1996 Cardiovascular lesions and skeletal myopathy in mice lacking desmin. Dev Biol 175:362–366

Li Z, Mericskay M, Agbulut O et al 1997 Desmin is essential for the tensile strength and integrity of myofibrils but not for myogenic commitment, differentiation, and fusion of skeletal muscle. J Cell Biol 139:129–144

Li W, Zhang Q, Oiso N et al 2003 Hermansky-Pudlak syndrome type 7 (HPS-7) results from mutant dysbindin, a member of the biogenesis of lysosome-related organelles complex 1 (BLOC-1). Nat Genet 35:84–89

Lim LE, Duclos F, Broux O et al 1995 Beta-sarcoglycan — characterization and role in limb-girdle muscular dystrophy linked to 4q12. Nat Genet 11:257–265

Loh NY, Ambrose HJ, Guay-Woodford LM et al 1998 Genomic organization and refined mapping of the mouse beta-dystrobrevin gene. Mamm Genome 9:857–862

Matsumura K, Tome FM, Ionasescu V et al 1993 Deficiency of dystrophin-associated proteins in Duchenne muscular dystrophy patients lacking COOH-terminal domains of dystrophin. J Clin Invest 92:866–871

McNally EM, Ly CT, Kunkel LM 1998 Human epsilon-sarcoglycan is highly related to alpha-sarcoglycan (adhalin), the limb girdle muscular dystrophy 2D gene. FEBS Lett 422:27–32

Michelson AM, Russell ES, Harman PJ 1995 Dystrophia muscularis: a hereditary primary myopathy in the house mouse. Proc Natl Acad Sci USA 41:1079–1084

Milner DJ, Weitzer G, Tran D et al 1996 Disruption of muscle architecture and myocardial degeneration in mice lacking desmin. J Cell Biol 134:1255–1270

Mizuno Y, Puca AA, O'Brien KF et al 2001 Genomic organization and single-nucleotide polymorphism map of desmuslin, a novel intermediate filament protein on chromosome 15q26.3. BMC Genet 2:8 (free full text article on *www.biomedcentral.com/1471-2156/2/8*)

Moll J, Barzaghi P, Lin S et al 2001 An agrin minigene rescues dystrophic symptoms in a mouse model for congenital muscular dystrophy. Nature 413:302–307

Nawrotzki R Loh NY, Ruegg et al 1998 Characterisation of alpha-dystrobrevin in muscle. J Cell Sci 111:2595–2605

Newey SE, Howman EV, Ponting CP et al 2001 Syncoilin, a novel member of the intermediate filament superfamily that interacts with alpha-dystrobrevin in skeletal muscle. J Biol Chem 276:6645–6655

Nigro V, Moreira ED, Piluso G et al 1996b Autosomal recessive limb-girdle muscular dystrophy, LGMD2F, is caused by a mutation in the delta-sarcoglycan gene. Nat Genet 14:195–198

Nigro V, Piluso G, Belsito A et al 1996a Identification of a novel sarcoglycan gene at 5q33 encoding a sarcolemmal 35 Kda glycoprotein. Hum Mol Genet 5:1179–1186

Nigro V, Okazaki Y, Belsito A et al 1997 Identification of the syrian hamster cardiomyopathy gene. Hum Mol Genet 6:601–607

Noguchi S, McNally EM, Ben Othmane K et al 1995 Mutations in the dystrophin-associated protein gamma-sarcoglycan in chromosome 13 muscular dystrophy. Science 270:819–822

Norwood FL, Sutherland-Smith AJ, Keep NH, Kendrick-Jones J 2000 The structure of the N-terminal actin-binding domain of human dystrophin and how mutations in this domain may cause Duchenne or Becker muscular dystrophy. Structure Fold Des 8:481–491

Ohlendieck K, Campbell KP 1991 Dystrophin-associated proteins are greatly reduced in skeletal muscle from *mdx* mice. J Cell Biol 115:1685–1694

O'Neill A, Williams MW, Resneck WG et al 2002 Sarcolemmal organization in skeletal muscle lacking desmin: evidence for cytokeratins associated with the membrane skeleton at costameres. Mol Biol Cell 13:2347–2359

Pasternak C, Wong S, Elson EL 1995 Mechanical function of dystrophin in muscle cells. J Cell Biol 128:355–361

Peters MF, O'Brien KF, Sadoulet-Puccio HM et al 1997 beta-dystrobrevin, a new member of the dystrophin family. Identification, cloning, and protein associations. J Biol Chem 272:31561–31569

Peters MF, Sadoulet-Puccio HM, Grady MR et al 1998 Differential membrane localization and intermolecular associations of alpha-dystrobrevin isoforms in skeletal muscle. J Cell Biol 142:1269–1278

Piccolo F, Jean Pierre M, Leturcq F et al 1996 A founder mutation in the gamma-sarcoglycan gene of gypsies possibly predating their migration out of India. Hum Mol Genet 5:2019–2022

Poon E, Howman EV, Newey SE, Davies KE 2002 Association of syncoilin and desmin: linking intermediate filament proteins to the dystrophin-associated protein complex. J Biol Chem 277:3433–3439

Price MG, Lazarides E 1983 Expression of intermediate filament-associated proteins paranemin and synemin in chicken development. J Cell Biol 97:1860–1874

Rando TA 2001 The dystrophin-glycoprotein complex, cellular signalling and the regulation of cell survival in the muscular dystrophies. Muscle Nerve 12:1575–1594

Roberds SL, Leturcq F, Allamand V et al 1994 Missense mutations in the adhalin gene linked to autosomal recessive muscular dystrophy. Cell 78:625–633

Sadoulet-Puccio HM, Feener CA, Schaid DJ, Thibodeau SN, Michels VV, Kunkel LM 1997a The genomic organization of human dystrobrevin. Neurogenetics 1:37–42

Sadoulet-Puccio HM, Rajala M, Kunkel LM 1997b Dystrobrevin and dystrophin: an interaction through coiled-coil motifs. Proc Natl Acad Sci USA 94:12413–12418

Schaart G, Viebahn C, Langmann W, Ramaekers F 1989 Desmin and titin expression in early postimplantation mouse embryos. Development 107:585–596

Schweitzer SC, Klymkowsky MW, Bellin RM et al 2001 Paranemin and the organization of desmin filament networks. J Cell Sci 114:1079–1089

Sejersen T, Lendahl U 1993 Transient expression of the intermediate filament nestin during skeletal muscle development. J Cell Sci 106:1291–1300

Sicinski P, Geng Y, Ryder-Cook AS et al 1989 The molecular basis of muscular dystrophy in the *mdx* mouse: a point mutation. Science 244:1578–1580

Sjoberg G, Jiang WQ, Ringertz NR et al 1994 Colocalization of nestin and vimentin/desmin in skeletal muscle cells demonstrated by three-dimensional fluorescence digital imaging microscopy. Exp Cell Res 214:447–458

Small JV, Furst DO, Thornell LE 1992 The cytoskeletal lattice of muscle cells. Eur J Biochem 208:559–572

Straub V, Duclos F, Venzke DP et al 1998 Molecular pathogenesis of muscle degeneration in the δ-sarcoglycan-deficient hamster. Am J Pathol 153:1623–1630

Straub V, Ettinger AJ, Durbeej M et al 1999 epsilon-sarcoglycan replaces alpha-sarcoglycan in smooth muscle to form a unique dystrophin-glycoprotein complex. J Biol Chem 274:27989–27996

Tezak Z, Prandini P, Boscaro M et al 2003 Clinical and molecular study in congenital muscular dystrophy with partial laminin alpha 2 (LAMA2) deficiency. Hum Mutat 21:103–111

Tidball JG 1992 Desmin at myotendinous junctions. Exp Cell Res 199:206–212

Tokuyasu KT, Dutton AH, Singer SJ 1983 Immunoelectron microscopic studies of desmin (skeletin) localization and intermediate filament organization in chicken skeletal muscle. J Cell Biol 96:1727–1735

Valentine BA, Winand NJ, Pradhan D et al 1992 Canine X-linked muscular dystrophy as an animal model of Duchenne muscular dystrophy: a review. Am J Med Genet 42:352–356

Vicart P, Dupret JM, Hazan J et al 1996 Human desmin gene: cDNA sequence, regional localization and exclusion of the locus in a familial desmin-related myopathy. Hum Genet 98:422–429

Vicart P, Caron A, Guicheney P et al 1998 A missense mutation in the alphaB-crystallin chaperone gene causes a desmin-related myopathy. Nat Genet 20:92–95

Watkins SC, Hofman EP, Slayter HS, Kunkel LM 1988 Immunoelectron microscopic localization of dystrophin in myofibres. Nature 333:863–866

Williamson RA, Henry MD, Daniels KJ et al 1997 Dystroglycan is essential for early embryonic development: disruption of Reichert's membrane in Dag1-null mice. Hum Mol Genet 6:831–841

Winder SJ 2001 The complexities of dystroglycan. Trends Biochem Sci 26:118–124

Xu H, Wu X-R, Wewer UM, Engvall E 1994 Murine muscular dystrophy caused by a mutation in the laminin alpha-2 (Lama2) gene. Nat Genet 8:297–302

Yoshida M, Ozawa E 1990 Glycoprotein complex anchoring dystrophin to sarcolemma. J Biochem (Tokyo) 108:748–752

Yoshida M, Hama H, Ishikawa-Sakurai M et al 2000 Biochemical evidence for association of dystrobrevin with the sarcoglycan-sarcospan complex as a basis for understanding sarcoglycanopathy. Hum Mol Genet 9:1033–1040

Zimprich A, Grabowski M, Asmus F et al 2001 Mutations in the gene encoding epsilon-sarcoglycan cause myoclonus-dystonia syndrome. Nat Genet 29:66–69

DISCUSSION

Goldman: Syncoilin is an interesting protein. You didn't give us a lot of detail about its structure. You said it was IF-like, yet it can't form filaments. Can it copolymerize with or bind to desmin filaments?

Davies: We have not done those sorts of experiments in detail. All we have done is co-transfections to see whether they would form heterofilaments. We have no evidence that this occurs.

Goldman: So the only evidence for binding *in vitro* is from the yeast two-hybrid assay.

Davies: Yes, and from co-immunoprecipitations (co-IPs). This was why it was important to do the knockout mice. We needed to see that taking desmin away would influence the localization of syncoilin. This could be an indirect link, though.

Goldman: So it is present at the cell surface, where filaments presumably interact with the actin–desmin complex. Is there any evidence that it shares similarity with any of the plaque proteins associated with IFs such as desmoplakin?

Davies: No. By this I mean that they won't be as similar as they are to the type 3 IF proteins. We searched through the whole genome for similar proteins and the highest score came out for the type 3 IFs.

Goldman: Is that because of the heptad repeats?

Davies: That's correct.

Goldman: I don't know if you have analysed this in detail yet, but in that region there are certain subcompartments that you could consider to be IF-like. It is important to know whether there is a 1A/1B region of the rod, or if there is a linker group. Do you know whether there are interruptions in the heptad repeat?

Davies: No.

Goldman: That should be easy to analyse, and it would be extremely useful in predicting whether syncoilin could form a coiled coil with type 3 proteins, or whether it is an IFAP (IF-associated protein). It looks to me as if you have an IFAP, which would be very interesting.

Davies: That is what we believe it is from the evidence that we have. However, we haven't rigorously done all of those experiments to look at the association. We had terrible problems just with the solubility of this protein. These are very difficult experiments to do.

Goldman: You also stated that in some cases you have aggregates of desmin but you don't have a mutation in desmin.

Davies: There are two classes of that type of muscle disease. They all accumulate desmin, which is why they are called desminopathies. But some of them have mutations in desmin and some don't. Our hypothesis was that in this latter group there might be a mutation in a desmin-associated protein. Because syncoilin is found in those inclusions that are so characteristic of these disorders, it is a prime candidate.

Goldman: It could also be a problem in the way that desmin is modified post-translationally. For example, intermediate filaments are excellent substrates for many kinases.

Wilson: Does the non-coiled-coil part of syncoilin have any predicted functional or structural domains?

Davies: No.

Wilson: Was there any problem with nuclear positioning in syncoilin mutants? Might syncoilin couple nuclei to structures at the cell surface?

Davies: There is no evidence for this.

Julien: The stoichiometry of the IF subunit is very important for filament assembly. For example, if you overexpress a neurofilament of high molecular weight NF-H subunit you can disrupt the assembly of peripherin. In a situation where syncoilin is overexpressed with desmin, would this provoke a change in the aggregation of desmin?

Davies: We have tried this. Under no experimental condition could we find a situation where syncoilin disrupted desmin filaments.

Julien: But these proteins are present in aggregates in disease.

Davies: Yes, but we shouldn't get too hooked up on this: there are a lot of other molecules also expressed in those aggregates. It is not just desmin and syncoilin.

Stewart: Do you find those aggregates on every muscle fibre?

Davies: No, just on some of them.

Stewart: Is there any other factor correlating with the occurrence of the aggregates, or any explanation as to why the aggregates are found is just a few cells?

Davies: It is a characteristic that has been found in this disease. Nothing is known about the cause of it. I should also say that in development, syncoilin appears to be expressed before desmin. Originally desmin was knocked out because it was thought to have something to do with the development of muscle. We expected that mouse to be dead, and it wasn't. Perhaps it is the syncoilin that comes in, and this may be the thing that forms the structures.

Wilson: You said that there are mutations in desmin that lead to overexpression of other proteins. What is the nature of those mutations? Do they make sense?

Davies: No.

Bonne: Other mutations in desmin lead to no accumulations at all. There is no correlation with accumulation and the point mutations. Point mutations are also found in dilated cardiomyopathy. There is no proof so far of any big accumulation of desmin.

Wilson: Mutating any component of this complex might disrupt the whole thing.

Davies: That is not strictly true. There is no accumulation of other binding partners of dystrophin. They stay the same.

Bonne: In myopathy with accumulation of myofibril components, you can have accumulation of a lot of different components. So far no one knows exactly why they accumulate, and we don't have all the mutations identified so far for all these disorders.

Davies: The difference with syncoilin is that in certain myopathies there is increased expression of syncoilin, whereas you don't get accumulation of desmin or the others. Syncoilin is different in this respect.

Gerace: In relation to that question, is the half-life of desmin in muscle cells known?

Goldman: Very few people have looked at this. In the case of keratin and vimentin it appears that the half-life is 14–16 h in cultured cells. The half-life is probably equivalent in the other types of intermediate filaments. What is the half-life of lamins?

Worman: Probably around 14 h.

Julien: For neurofilaments it has been suggested that they may spend 3 years in very long axons.

Goldman: I don't believe that.

Wilson: What is special about the neuromuscular junction, in terms of syncoilin localization? You said syncoilin was mostly at the neuromuscular junction.

Davies: It depends on which antibody you use. It is clearly expressed fairly highly at the neuromuscular junction, but other antibodies show a much greater staining of the sarcolemma than the one I showed.

Wilson: Does syncoilin aggregate similarly at other locations?

Davies: We haven't looked.

Starr: Does syncoilin completely co-localize with the bungaratoxin staining?

Davies: That is an interesting question. My view would be that it is around the bungarotoxin. In neuromuscular junction preps there is a slight displacement.

Starr: There have to be some specialized membranes at the neuromuscular junction. Nuclei cluster underneath the neuromuscular junction.

Morris: Presumably it also associates with utrophin.

Davies: Everything we say about dystrophin we have to say is true of utrophin. We can't dissect the differences unless we look at the knockout mice. We have looked at *Mdx* but we haven't looked in utrophin knockout mice at the localization of syncoilin.

Bonne: How is syncoilin localized at the Z discs?

Davies: It is a perfect overlay with desmin and actinin.

Starr: I have a more general question about neuromuscular junctions. Have people looked at neuromuscular junctions specifically in dystrophic tissues? Do the nuclei cluster normally there in dystrophic tissues?

Bonne: It is difficult to look at this in patient tissues, because there are several teams specializing on the neuromuscular junction. A lot of features that can be seen by muscle biopsy are non-specific and they can be found in any kind of muscular dystrophy. Neuromuscular junctions have been extensively analysed in myasthenic syndromes, but for muscular dystrophy, I am not sure whether any team has looked systematically at the neuromuscular junction. It is a common dystrophic feature that we can observe.

Starr: It is interesting that all these multiple components localize there.

Davies: There is a reduction of the clusters in *Mdx* mice. This is visible by EM. If you take away utrophin and dystrophin, the neuromuscular junctions still function but you can actually see the reduction in the number of folds in the NMJs.

Wilson: Has the NMJ been studied systematically in mouse? Muscle cells can be very long and are only innervated in one place. Thus the chance of finding the NMJ is quite low.

Bonne: In the Myology Institute there is analysis of NMJ in biopsies. But it is hard to draw a conclusion from very few data.

Davies: There is also a problem in that the *Mdx* mouse is not a good model for the DMD because it is so mildly affected. I don't think you can say that what you see at the NMJ in the *Mdx* is going to be related to what occurs in human. The difference, of course, is that the mouse gets compensation from utrophin.

Goldman: In cases with aggregates of desmin have you looked at microtubule organization?

Davies; We haven't.

Goldman: MTs and IFs, especially type III IFs, are closely associated with each other in numerous types of cells. It has been shown using reagents that disrupt either MTs or IFs that their organization and distribution is mutually dependent. It might be interesting to look at MTs. When you see the aggregates, are they perinuclear?

Davies: It depends which patient you look at. Where the aggregates form is dependent on the particular patient.

Goldman: Overall, morphologically these have the general appearance of the aggregates seen in motor neuron disease and also Parkinson's disease. In Parkinson's disease they are not filaments. Are they filaments in the aggregates you see, or just an accumulation of protein?

Davies: I think they are just an accumulation of protein, but I suppose they could be filaments. We don't have access to those patients.

Goldman: Has anyone looked at this by electron microscopy?

Davies: Not as far as I know.

Goldman: It would be important to know if they are aggregates of filaments, or aggregates of protein.

Davies: I think they are aggregates of protein, because there are many different myofibrillar proteins in there.

Goldman: Are they ubiquitinated? They could be similar to an aggresome, and it would be interesting to know whether protein degradation is going on. Aggresomes form when there is lots of substrate to be degraded because it is ubiquitinated and there aren't enough proteosomes to break down all the proteins.

Davies: I am not sure anyone has looked at this. What is seen in an aggresome is a very specific aggregation of a small number of proteins, but what is seen here is an aggregation of a lot of different proteins.

Goldman: But they are all associated with each other. So it could be a large Mallory-like body or aggresome. These were first seen in alcoholic cirrhosis of the liver, and they are large accumulations of proteins such as keratin. One way to begin to determine whether these are the structures that you are seeing — which is important to know — is just to stain with a proteasome antibody.

Davies: We could certainly do that. The basic question we were asking had nothing to do with the morphology of those aggregates. We are not looking at desminopathies. We were just interested to see whether syncoilin paralleled desmin, and would therefore be a candidate for mutations in those particular diseases.

Julien: How can IF filament proteins affect the microtubule motors?

Goldman: Because they are major cargoes for motors. If you look at dynein and kinesin, the major cargo in most cells that have IFs is the IF protein.

Julien: So when aggregates form they may sequester motor molecules

Goldman: Yes. There are now mutations in motor proteins that lead to aggregation of IFs.

Julien: What about the opposite situation: do mutations in IFs affect motor function?

Goldman: I don't know if there are any of these, but people haven't looked very carefully. IFs are a major cargo moved by motors along MT tracks.

Hutchison: When you looked at changes in the association of syncoilin with the cytoskeleton using a fractionation technique, which cells did you use?

Davies: C2C12 cells.

Hutchison: Would you see additional changes if you compared myoblasts to myotubes?

Davies: I haven't done that experiment.

Goldman: Do you see syncoilin in prefusion myoblasts and in myotubes?

Davies: Yes.

Goldman: In prefusion myoblasts vimentin is expressed. If syncoilin binds to any type 3 IF protein that would be very interesting. It would be interesting to find out where in the desmin or vimentin structure syncoilin binds. It looks as if it has a long α helical stretch — it seemed that 80% of the protein is competent to form a coiled coil. You don't know yet whether it forms a dimer or an oligomeric complex.

Davies: I am showing very recent data, so we haven't done those experiments.

Wilkins: Are there subtle defects in any of these mutants in smooth muscle? MDs are picked up on the basis of what they do in skeletal muscles. Have people looked at effects in other kinds of muscle?

Davies: Not in human patients. The vast majority of the pathology is in the skeletal muscle, which is where the attention has been concentrated.

Wilkins: Are there any subtle defects?

Davies: There may be. You would have expected people to look extensively at the heart, but there is still a dispute about whether even the absence of dystrophin has an effect on the heart. This is remarkable. Most of the patients do have abnormal ECGs by the age of 18. This is curious when we are thinking of therapy: if we cure the skeletal myopathy we might then have to think about the cardiomyopathy. We still haven't sorted out what it is doing in skeletal muscle, so that is our focus for the time being.

Goldman: It will be interesting to see whether syncoilin is another IF linker protein. It could potentially also have an actin binding domain.

Davies: We would have seen that and there is no evidence for it.

Skeletal and cardiac muscle defects in a murine model of Emery-Dreifuss muscular dystrophy

M. J. Grattan, C. Kondo, J. Thurston, P. Alakija, B. J. Burke*, C. Stewart†, D. Syme‡ and W. R. Giles§[1]

*Faculty of Medicine, University of Calgary, Calgary, Canada, *Department of Anatomy and Cell Biology, University of Florida, Gainesville, FL, USA, †Laboratory of Cancer and Developmental Biology, NCI at Frederick, Frederick, MD, USA, ‡Department of Zoology, University of Calgary, Calgary, Canada, and §Departments of Bioengineering and Medicine, University of California, San Diego, La Jolla, CA, 92093-0412, USA*

Abstract. Previous histological findings, physiological data, and behavioral observations on the A-type lamin knockout mouse ($Lmna^{-/-}$) suggest that important aspects of this model resemble the human Emery-Dreifuss muscular dystrophy (EDMD) phenotype. The main goal of our experiments was to study skeletal and cardiac muscle function in this murine model to obtain the semiquantitative data needed for more detailed comparisons with human EDMD defects. Measurements of the mechanical properties of preparations from two different skeletal muscle groups, the soleus and the diaphragm, were made *in vitro*. In addition, records of the electrocardiogram, and measurements of heart rate variability were obtained; and phasic contractions (unloaded shortening) of enzymatically isolated ventricular myocytes were monitored. Soleus muscles from $Lmna^{-/-}$ mice produced less force and work than control preparations. In contrast, force and work production in strips of diaphragm were not changed significantly. Lead II electrocardiograms from conscious, restrained $Lmna^{-/-}$ mice revealed slightly decreased heart rates, with significant prolongations of PQ, QRS, and 'QT' intervals compared with those from control recordings. These ECG changes resemble some aspects of the ECG records from humans with EDMD; however, the cardiac phenotype in this $Lmna^{-/-}$ mouse model appears to be less well-defined/developed. Ventricular myocytes isolated from $Lmna^{-/-}$ mice exhibited impaired contractile responses, particularly when superfused with the β-adrenergic agonist, isoproterenol (1 μM). This deficit was more pronounced in myocytes isolated from the left ventricle(s) than in myocytes from the right ventricle(s). In summary, tissues from the $Lmna^{-/-}$ mouse exhibit a number of skeletal and cardiac muscle deficiencies, some of which are similar to those which have been reported in studies of human EDMD.

2005 Nuclear organization in development and disease. Wiley, Chichester (Novartis Foundation Symposium 264) p 118–139

---

[1]This paper was presented at the symposium by W. R. Giles to whom correspondence should be addressed.

118

Our main goal was to study the phenotype of the A-type lamin knockout mouse ($Lmna^{-/-}$), in order to determine the extent to which the skeletal and cardiac muscle defects in this murine model resemble human Emery-Dreifuss muscular dystrophy (EDMD). This genetically modified mouse ($Lmna^{-/-}$) has been proposed as an animal model for this devastating, and often terminal human neuromuscular disease (Manilal et al 1999, Sullivan et al 1999, Mounkes et al 2001, Taylor et al 2003), and recent studies on murine hearts strongly support this view (Fatkin et al 1999, Nikolova et al 2004).

EDMD, like the most common form of muscular dystrophy, Duchenne muscular dystrophy, is characterized by onset in childhood or juvenile years (cf. Emery 2002, Engvall & Wewer 2003). In contrast, myotonic dystrophy exhibits adult onset. At present, none of these neuromuscular disorders have a cure. Each can result in premature death, which often is due to either cardiac or respiratory muscle compromise or failure.

EDMD was first described and documented in 1966 (cf. Emery 2002). It can be transmitted as an X-linked recessive or an autosomal disorder, and it arises from mutations in two proteins which are associated with the inner nuclear envelope, emerin and A-type lamins (Bione et al 1994, Ostlund & Worman 2003). EDMD usually develops in individuals between five and seven years of age; and manifests first as muscle weakness in the legs and arms and joint stiffness, as well as significant cardiac defects. However, the cardiac pathology is usually not evident until early adulthood. At that time it includes dilated cardiomyopathy, often with conduction defects, sometimes ending with sudden death (Fatkin et al 1999, Arbustini et al 2002, Becane et al 2000). Muscle specimens from individuals diagnosed with EDMD have a characteristic histological profile. Skeletal muscle fibres are markedly non-uniform in size and usually have increased numbers of centrally located nuclei, together with interstitial fibrosis. In affected progeny the levels of serum creatine kinase can be elevated (3–10 times higher than normal). However, many of these phenotypic descriptors are quite variable, not only between kindred with different mutations, but also in different families with identical mutations (cf. Brodsky et al 2000). It is now well-known that mutations in the A-type lamin gene in different locations can result in a number of different phenotypes, including EDMD with dilated cardiomyopathy and conduction defects, as well as a significant familial partial lipodystrophy in some progeny (Arbustini et al 2002, Bonne et al 2000, Fatkin et al 1999).

At present, it is not known how these well-documented mutations in the A-type lamins or emerin (cf. Clements et al 2000) lead to the EDMD phenotype. The A-type lamins are positioned in close apposition to the inner nuclear membrane, and emerin is also an inner nuclear membrane protein (Manilal et al 1999, Morris & Manilal 1999, Bonne et al 2000, Gruenbaum et al 2000, Hutchison et al 2001, Stuurman et al 1998). The nuclear lamina is adjacent to the inner nuclear membrane and chromatin; it has been proposed that this structure can mechanically stabilize the nucleus in strongly contractile cells (Lippincott-Schwartz 2002). It also plays an important role in organizing chromatin and in modulating DNA replication. Experiments in *Drosophila melanogaster* have shown that when nuclear lamins are absent there tends to be a clustering of these nuclear pore complexes (Lenz-Bohme et al 1997).

The first murine model of EDMD was reported in 1999 (Fatkin et al 1999, Sullivan et al 1999, Wehnert & Muntoni 1999, Emery 2002). Affected animals all exhibited marked muscular defects early in life. All null ($Lmna^{-/-}$) homozygotes died by eight weeks and showed marked dystrophy in the muscles surrounding the femur and perivertebral areas. However, creatine kinase levels were unchanged in these animals. Skeletal muscles from affected progeny also showed variations in fibre diameter and uniformity, as well as increased numbers of nuclei. No detailed characterization of the mechanical function of the skeletal muscles in these animals has been reported. Histological studies have revealed structural defects in the heart of affected progeny including significant transmural thinning in the left ventricles. A second model of EDMD has been described: this genetically altered mouse lacks the ability to produce a protein (ZMPSTE-24), which is involved in processing of prelamin A. These progeny exhibited stunted growth at seven weeks and had a lifespan of only approximately 20 weeks (Pendas et al 2002).

As indicated, the murine model which is the focus of this study, the $Lmna^{-/-}$ mouse, was first described in 1999 (Wehnart & Muntoni 1999, Fatkin et al 1999, Sullivan et al 1999). No detailed studies of limb or respiratory muscle mechanics or of cardiac electrophysiology have yet been published, although the recent paper from the Fatkin laboratory, (Nikolova et al 2004) provides comprehensive information of the cardiomyopathy which develops and some data on ventricular excitation–contraction coupling.

Methods

Breeding pairs of adult mice, heterozygous for the A-type lamin gene, were obtained from Dr C. L. Stewart of the National Cancer Institute (Frederick, MD, USA), and a colony of 10 pairs of heterozygous animals was established. Genotyping was performed using DNA obtained from ear clippings. In the chosen experimental design $Lmna^{-/-}$ mice were compared to age-matched $Lmna^{+/+}$ progeny. All animals were between four and six weeks of age. At three weeks all $Lmna^{-/-}$ progeny showed definite signs of musculoskeletal pathology including splayed hind limbs, a hunched posture, and inability to hang from their forelimbs. Most of these null animals did not survive beyond eight weeks of age.

Strips from the soleus and the diaphragm muscles were prepared to study the mechanical properties of the skeletal muscles. These two muscles were chosen since the soleus consists of both slow and fast muscle fibres in approximately equal proportions, while the diaphragm (which has been studied in detail in other murine models of muscular dystrophy) is comprised of about 10% slow fibres and 90% fast fibres. After these muscles were removed from anaesthetized animals, strips were dissected free and placed in a temperature-controlled recording/ superfusion chamber. Each strip was attached by one tendon to a fixed post and by the other to the force transducer which was connected to a servocontrolled motor arm. In this way, muscle length could be controlled while programmed stimuli were applied from silver–silver chloride electrodes. Data were recorded on a personal computer and analyzed using custom software written in Labview. In most experiments, muscle strips were stimulated with 0.5 ms stimuli at 150 Hz, and at 1.5 times threshold. Stimuli were applied either individually or in trains (e.g. 10–50 Hz lasting 200–400 ms) to elicit twitch and tetanus force, respectively. In each experiment the measurements consisted of twitch kinetics and maximal force at tetanus as a function of muscle length. Active work was obtained using the work loop method (Josephson 1985). Force–velocity measurements were made and muscle stiffness was characterized by first tetanizing the muscle to maximum force and then quickly stretching it by approximately 0.5% of its resting length.

A custom-designed apparatus was used to obtain lead I or lead II ECG data from restrained, but unanaesthetized animals. This approach in principle resembles that of Chu et al (2001). It provided high-resolution measurements of ECG parameters, and hence, heart rate and heart rate variability. Proprietary recognition software yielded on-line detailed characterization of standard ECG parameters, including P-R, PQ and QRS duration and 'T-wave' morphology. In the case of a rodent heart, the 'QT' immediately follows the QRS complex (i.e. there is no isoelectric segment) and may reflect early repolarization (Gussak et al 2000, Knollmann et al 2001, Danik et al 2002, Farkas & Curtis 2003). Measurements of ventricular mechanics using the Langendorff approach (Larsen et al 1999) proved difficult in $(-/-)$ progeny. Therefore, isolated single myocytes were obtained from either the right or the left ventricle using enzymatic dispersion, as previously described (Bouchard et al 1995, Clark et al 1996) and contractility was assessed qualitatively as cell-length shortening in response to field stimulation. In these experiments, myocytes were challenged with isoproterenol to obtain a qualitative descriptor of the positive c-AMP-dependent inotropic reserve of these myocytes (Poon et al 1998, 2001).

Results

The initial experiments in this study compared the mechanical properties of two different skeletal muscles from control and $Lmna^{-/-}$ animals. The results are illustrated in Figs 1 and 2. Figure 1A illustrates twitch force development in the

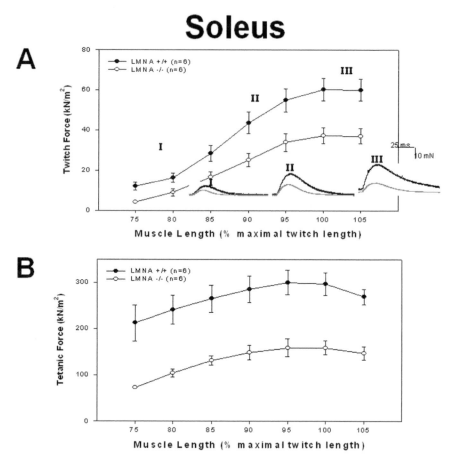

FIG. 1. *In vitro* measurements of twitch and tetanic force from strips of adult mouse soleus muscle. Panel A shows twitch force at muscle lengths ranging from 75–105% of the maximal length for twitch force development. In both panels of this figure closed circles (●) correspond to data from control animals ($Lmna^{+/+}$), and open circles (○) denote results obtained from $Lmna^{-/-}$ animals. Twitch force was recorded in response to stimuli applied at 0.1 Hz. Tetanic force (Panel B) was measured in response to a stimulus train lasting 300–400 ms, applied at 150 Hz. Note the marked depression of twitch force and reduced ability of strips of soleus muscle from the $Lmna^{-/-}$ animals to generate tetanic force.

strips of soleus muscle as a function of muscle length, where I_{max} or 100 denotes the muscle length at which the largest twitch was recorded in responses to stimuli in a train applied at 0.1 Hz (temp. 30 °C). Panel B of Fig. 1 summarizes the corresponding tetanic force data. Tetani were developed in response to a 300 ms train of stimuli applied at 50 Hz. Note that the results from both of these sets of

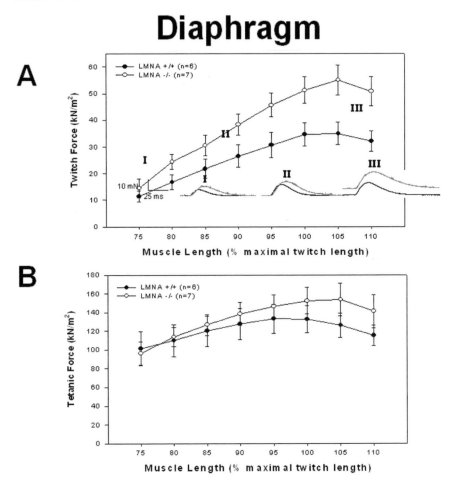

FIG. 2. Measurements of twitch and tetanus force in strips from the diaphragm of control ($Lmna^{+/+}$) or lamin A knockout ($Lmna^{-/-}$) adult mice. Twitch recordings are shown in Panel A and results from measurements of tetanic force as a function of muscle length are shown in Panel B. Note that diaphragm muscle strips showed no significant deficit in these mechanical parameters in the $Lmna^{-/-}$ animals.

experiments demonstrate substantially reduced force-generating capacity in the soleus muscles from the $Lmna^{-/-}$ progeny.

Results from corresponding twitch and tetanus experiments on strips of muscle from the diaphragm are shown in Figs 2A and 2B, respectively. In contrast to the soleus muscle findings, there were no significant changes in the tetanic force in strips of diaphragm from the control vs. $Lmna^{-/-}$ animals. Although twitch

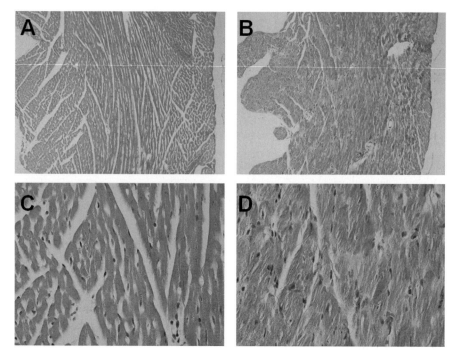

FIG. 3.   Histological assessment of left ventricular muscle from the hearts of control animals (Panels A and C), and lamin knockout progeny (Panels B and D). In each panel thin sections of the lateral left ventricular wall stained with Gamori's trichrome are shown. Panels A and B are at 100×magnification and Panels C and D are at 400×magnification. These findings document a significant development of fibrosis in the $Lmna^{-/-}$ hearts, and provide an indication of the transmural thinning in the left ventricle, as has been described previously (see text).

force in the diaphragm muscle strips from the $Lmna^{-/-}$ animals showed a tendency to exceed that from controls, none of these differences were statistically significant.

After documenting some aspects of the skeletal muscle deficits in $Lmna^{-/-}$ animals, a detailed assessment of the ventricular phenotype was begun. Histological assessments of $Lmna^{-/-}$ (Fig. 3) confirmed quite extensive fibrosis in the left ventricles of $Lmna^{-/-}$ animals. Transmural thinning of the left ventricular wall was also a consistent observation. Both findings are in agreement with previous more detailed analysis which has defined the significant ventricular cardiomyopathy in these affected progeny (Fatkin et al 1999, Sullivan et al 1999, Nikolova et al 2004).

The excitation–contraction coupling of the ventricular myocardium was assessed by studying field-stimulated myocytes obtained by enzymatic dispersion of tissue from right or left ventricles. These results are summarized in Fig. 4. Panel A shows cell shortening (as a percentage of resting length); Panel B denotes the maximum rate of shortening; and Panel C shows maximum rate of relaxation. These measurements were made in the steady-state in response to a train of 30 field stimuli applied at 1, 2, and 3 Hz. Note that myocytes from the left ventricles of $Lmna^{-/-}$ progeny exhibited markedly depressed rates of shortening and relaxation, but that these effects were not observed in myocytes obtained from right ventricular tissue from the same hearts.

The data in Fig. 5 provide further evidence for a ventricular chamber-specific deficit in the $Lmna^{-/-}$ animals. Cell shortening data from myocytes from right and left ventricular (RV, LV) tissue are compared (i) under control conditions and (ii) in the presence of an effective concentration of a β-adrenergic receptor agonist, isoproterenol (10^{-7} M). Note that control myocytes and those from RV and tissue respond in the expected pattern following both application of isoproterenol: (i) an enhanced rate and amount of myocyte shortening, (ii) an increased rate of relaxation. In contrast, myocytes from LV exhibited significantly decreased contractile responses as judged by these criteria.

The final set of experiments provided an initial characterization of the electrophysiological parameters of the heart. Lead II ECG records were obtained from unanaesthetized animals in a plexiglass tube, an approach somewhat similar to one described by Chu et al (2001). Panel A of Fig. 6 shows an ECG complex with the fiducial markers which our pattern recognition program employs. This analysis revealed that $Lmna^{-/-}$ animals exhibit (i) significantly increased RR intervals (i.e. reduced heart rate) (approximately 82 vs. 119 ms), prolonged PR intervals (approximately 28 vs. 35 ms); and broadened QRS complexes (approximately 10 vs. 12.2 ms). In addition, as shown in Panel B (from a different set of earlier experiments,) $Lmna^{-/-}$ animals exhibited reduced heart rate variability, this was particularly apparent in the so-called high-frequency (approximately 1 Hz) range, which is thought to be modulated by vagal innervation/responsiveness (Wickman et al 1998, Gehrmann et al 2000). These preliminary HRV measurements were obtained from animals anaesthetized with isofluorane (see figure legends).

Discussion

Results from these experiments reveal substantial deficits in the ability of soleus muscle strips to generate tetanic tension. Mechanical function was characterized *in vitro* in terms of phasic (twitch) and tonic (tetani) tension. No such deficits were identified when the same mechanical tests were applied to strips excised from the diaphragm. This marked difference may be due to the different histological properties of these tissues, and/or it may be more closely related to changes in muscle energetics (see below). More detailed measurements of

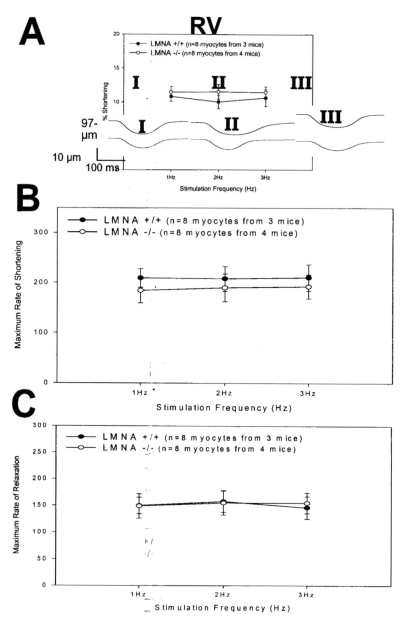

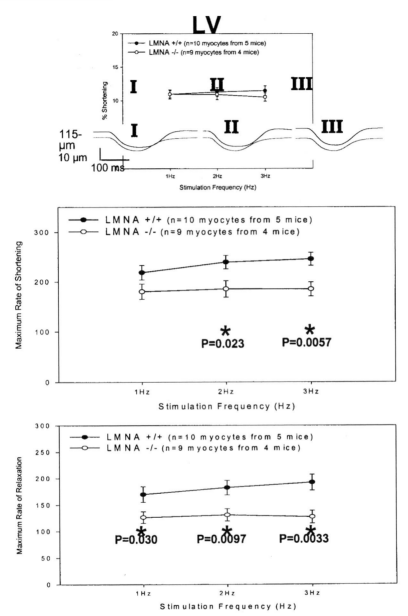

FIG. 4.   Measurements of ventricular contractility as judged by unloaded shortening records from myocytes isolated from the right and left ventricles of control and *Lmna*$^{-/-}$ animals. Contractions of single, isolated myocytes were elicited by field stimulation at 1, 2, or 3 Hz. (A) myocyte shortening as a percent of resting length. (B) Maximum rate of shortening. (C) Maximum rat of relaxation. Note that myocytes from the LV of *Lmna*$^{-/-}$ progeny show significant reduction in maximal rates of shortening and relaxation.

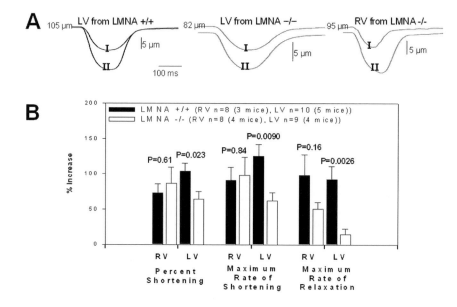

FIG. 5. Comparison of responses of control and $Lmna^{-/-}$ field-stimulated ventricular myocytes to bath-applied β-adrenergic agonist (isoproterenol, 10^{-7} M). Panel A shows data from control animals (left) and $Lmna^{-/-}$ progeny (right). The two superimposed traces compare control recordings (I) with data obtained after bath application of isoproterenol (10^{-7} M) (II). The histograms in Panel B summarize these findings. Note that the positive inotropic response of the left ventricular myocytes to isoproterenol (10^{-7} M) is compromised in $Lmna^{-/-}$ progeny; while responses from myocytes obtained from the *right* ventricle of the same hearts did *not* show any significant deficit in the positive inotropic responses to isoproterenol (10^{-7} M).

mechanical parameters, together with histological analyses and heat measurements (cf. Gibbs 2003) from these muscle preparations are needed before any further mechanistic insights can be obtained.

Our measurements of the physiological responses of the ventricular myocytes confirm and extend previous reports (see Nikolova et al 2004). Sullivan et al (1999) and Fatkin et al (1999) have previously described a significant cardiomyopathy resulting from deficiencies of lamin A and C in humans and in mice. Previous papers have drawn attention to the cardiac electrophysiological (ECG) changes in other murine models of muscular dystrophy (Saba et al 1999), although some of these phenotypes are quite variable in onset and heritability. More detailed studies like those done previously on other genetically altered mice will be needed (Berul et al 1996, Baker et al 2000). The recent paper by Nikolova et al (2004) describes in detail the mechanical deficits of the LV function in $Lmna^{-/-}$

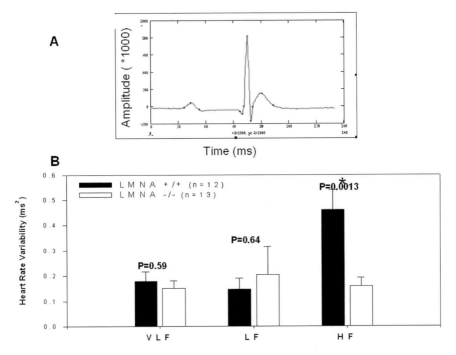

FIG. 6.   Characterization of lead II electrocardiogram parameter from control animals and
lamin A knockout progeny ($Lmna^{-/-}$). Panel A shows a typical lead II electrocardiogram
obtained using our proprietary device for recording the ECG (and respiratory activity) from
restrained, but unanaesthetized animals. A typical P wave, QRS complex, and 'T' wave of the
ECG are illustrated, along with the fiducial waveform markers, which this pattern recognition
program provides. $Lmna^{-/-}$ progeny showed significant lengthening of the RR intervals and
consistent increases in the PR and QRS interval durations data (not shown). Panel B of this figure
shows data obtained from a different set of experiments in which heart rate variability was studied
in animals anaesthetized with isofluorane (0.75%, 1 litre O_2 per min). These are presented in
terms of high frequency (HF) low frequency (LF) and very low frequency (VLF) power of
heart rate variance in control and $Lmna^{-/-}$ animals. These results showed significant
reduction in HF power, consistent with compromised vagal modulation of cardiac parameter
activity under control conditions.

animals on the basis of changes in unloaded shortening of enzymatically isolated
myocytes.

Our findings on myocyte shortening confirm the results of Fatkin et al (1999),
and Nikolova et al (2004) while also adding significant new insights. Thus, as
shown in Figs 4 and 5 it is not the ventricular myocardium *per se* which is
compromised in the $Lmna^{-/-}$ animals. Instead the deficit is mainly, and perhaps
only, in myocytes from the left ventricle. Most right ventricular myocytes show
normal contractile function in response to field stimulation and following

stimulation with the β-adrenergic agonist isoproterenol. A plausible explanation, or interpretation, of this finding is that an important element of the $Lmna^{-/-}$ phenotype could include compromise of cellular (perhaps glycolytic) metabolism and/or alterations in cellular microanatomy. For example, intracellular creatinase kinase levels are altered (cf. Ozawa et al 1999, Harrison et al 2003, Saha & Ruderman 2003, Janssen et al 2003); and/or essential elements for β-receptor signalling (Petrofski & Koch 2003) cAMP production, compartmentalization, or downstream phosphorylation effects on ion channels, contractile proteins, or elements of the sarcoplasmic ventriculum could be altered.

The results obtained from our lead II ECG recordings would be expected to be based on clinical data reporting in terms of reduced heart rate and significant alterations in PR interval and QRS direction in patients. Our $Lmna^{-/-}$ animals failed to live beyond approximately 8 weeks, but the cause of death was not determined. Application of telemetry in the animals is challenging, since the smallest available transceiver is comparable in volume to the abdominal volume of these very small, stunted animals. It was for this reason that our ECG analysis was done on unanaesthetized animals. Heart rate variability was obtained from these data and another set of results obtained from animals under isofluorane anaesthesia. Heart rate variability in the high frequency ranges was significantly reduced in $Lmna^{-/-}$ animals (Fig. 6). This pattern of results which has been described in the setting of murine myotonic dystrophy (Inuoe et al 1995) suggests alterations in autonomic control of cardiac pacemaker activity. Additional experiments are needed in which more extensive analysis of the atropine and propanolol- sensitivity of the RR interval is undertaken in an attempt to make the distinction between alterations in autonomic modulation of cardiac pacing (cf. Shusterman et al 2002) as opposed to a primary defect in myogenic parameter activity. In addition, voltage-sensitive dye mapping could also be used to study activation patterns and some aspects of tissue-specific action-potential waveforms (Nygren et al 2000, 2003).

Acknowledgements

Grants-in-aid from the Canadian Institutes of Health Research and the Heart and Stroke Foundation of Canada are gratefully acknowledged. M. Grattan and W. Giles received stipends from the Alberta Heritage Foundation for Medical Research, and Dr Giles held a Research Chair endowed by the Heart Foundation of Alberta and Northwest Territories.

References

Arbustini E, Pilotto A, Repetto A et al 2002 Autosomal dominant dilated cardiomyopathy with atrioventricular block: a Lamin A/C defect-related disease. J Am Coll Cardiol 39:981–990
Becane HM, Bonne G, Varnous S et al 2000 High incidence of sudden death with conduction system and myocardial disease due to lamins A and C gene mutation. Pacing Clin Electrophysiol 23:1661–1666

Baker LC, London B, Choi B-R, Koren G, Salama G 2000 Enhanced dispersion of repolarization and refractoriness in transgenic mouse hearts promotes reentrant ventricular tachycardia. Circ Res 86: 396–407

Berul CI, Aronovitz MJ, Wang PJ, Mendelsohn ME 1996 In vivo cardiac electrophysiology studies in the mouse. Circulation 94:2641–2648

Bione S, Maestrini E, Rivella S et al 1994 Identification of a novel X-linked gene responsible for Emery-Dreifuss muscular dystrophy. Nat Genet 8:323–327

Bonne G, Mercuri E, Muchir A et al 2000 Clinical and molecular genetic spectrum of autosomal dominant Emery- Dreifuss muscular dystrophy due to mutations of the lamin A/C gene. Ann Neurol 48:170–180

Brodsky GL, Muntoni F, Miocic S, Sinagra G, Sewry C, Mestroni L 2000 Lamin A/C gene mutation associated with dilated cardiomyopathy with variable skeletal muscle involvement. Circulation 101:473–476

Bouchard RA, Clark RB, Giles WR 1995 Effects of action potential duration on excitation-contraction coupling in rat ventricular myocytes. Action potential voltage-clamp measurements. Circ Res 76:790–801

Clark RB, Bouchard RA, Giles WR 1996 Action potential duration modulates calcium influx, Na^+–Ca^{2+} exchange, and intracellular calcium release in rat ventricular myocytes. Ann N Y Acad Sci 779:417–429

Clements L, Manilal S, Love DR, Morris GE 2002 Direct interaction between emerin and lamin A. Biochem Biophys Res Commun 267:709–714

Chu V, Otero JM, Lopez O, Morgan JP, Amende I, Hampton TG 2001 Method for non-invasively recording electrocardiograms in conscious mice. BMC Physiol 1:6–10

Danik S, Cabo C, Chiello C, Kang S, Wit AL, Coromilas J 2002 The correlation of repolarization of ventricular monophasic action potential with ECG in the murine heart. Am J Physiol 283:H372–H381

Drury E 1967 Carleton's histological techniques, 4th edn. Oxford University Press, Oxford, p 133–137

Emery AE 2002 The muscular dystrophies. Lancet 359:687–695

Engvall E, Wewer U 2003 The new frontier in muscular dystrophy research: booster genes. FASEB J 17:1579–1584

Fahrenkrog B, Aebi U 2003 The nuclear pore complex: nucleocytoplasmic transport and beyond. Nat Rev 4:757–766

Farkas A, Curtis MJ 2003 Does QT widening in the Langendorff-perfused rat heart represent the effect of repolarization delay or conduction slowing? J Cardiovasc Pharmacol 42:612–621

Fatkin D, MacRae C, Sasaki T et al 1999 Missense mutations in the rod domain of the lamin A/C gene as causes of dilated cardiomyopathy and conduction-system disease. N Engl J Med 341:1715–1724

Gehrmann J, Hammer PE, Maguire CT, Wakimoto H, Triedman JK, Berul CI 2000 Phenotypic screening for heart rate variability in the mouse. Am J Physiol Heart Circ Physiol 279:H733–H740

Gibbs CL 2003 Cardiac energetics: sense and nonsense. Clin Exp Pharmacol Physiol 30:598–603

Gussak I, Chaitman BR, Kopecky SL, Nerbonne J 2000 Rapid ventricular repolarization in rodents: electrocardiographic manifestations, molecular mechanisms, and clinical insights. J Electrocardiol 33:159–170

Gruenbaum Y, Wilson KL, Harel A, Goldberg M, Cohen M 2000 Nuclear lamins — structural proteins with fundamental functions. J Struct Biol 129:313–323

Harrison GJ, van Wijhe MH, de Groot B, Dijk FJ, Gustafson LA, van Beek JHGM 2003 Glycolytic buffering affects cardiac bioenergetic signaling and contractile reserve similar to creatine kinase. Am J Physiol Heart Circ Physiol 285:H883–H890

Hutchison CJ, Alvarez-Reyes M, Vaughan OA 2001 Lamins in disease: why do ubiquitously expressed nuclear envelope proteins give rise to tissue-specific disease phenotypes? J Cell Sci 114:9–19

Inoue K, Ogata H, Matsui M et al 1995 Assessment of autonomic function in myotonic dystrophy by spectral analysis of heart-rate variability. J Auton Nerv Syst 55:131–134

Janssen E, Terzic A, Wieringa B, Dzeja PP 2003 Impaired intracellular energetic communication in muscles from creatine kinase and adenylate kinase (M-CK/AK1) double knock-out mice. J Biol Chem 278:30441–30449

Josephson RK 1985 Mechanical power output from striated muscle during cycle contraction. J Exp Biol 114:493–512

Knollmann BC, Katchman AN, Franz MR 2001 Monophasic action potential recordings from intact mouse heart: validation, regional heterogeneity, and relation to refractoriness. J Cardiovasc Electrophysiol 12:1286–1294

Larsen TS, Dani R, Severson DL et al 1999 Response of the isolated working mouse heart to changes in preload and heart rate. Pflügers Arch 437:979–985

Lippincott-Schwartz J 2002 Ripping up the nuclear envelope. Nature 416:31–32

Lenz-Bohme B, Wismar J, Fuchs S et al 1997 Insertional mutation of the Drosophila nuclear lamin Dm0 gene results in defective nuclear envelopes, clustering of nuclear pore complexes, and accumulation of annulate lamellae. J Cell Biol 137:1001–1006

Manilal S, Sewry CA, Pereboev A et al 1999 Distribution of emerin and lamins in the heart and implications for Emery-Dreifuss muscular dystrophy. Hum Mol Genet 8:353–359

Morris GE, Manilal S 1999 Heart to heart: from nuclear proteins to Emery-Dreifuss muscular dystrophy. Hum Mol Genet 8:1847–1851

Mounkes LC, Burke B, Stewart CL 2001 The A-type lamins: nuclear structural proteins as a focus for muscular dystrophy and cardiovascular diseases. Trends Cardiovasc Med 11:280–285

Nikolova V, Leimena C, McMahon AC et al 2004 Defects in nuclear structure and function promote dilated cardiomyopathy in lamin A/C-deficient mice. J Clin Invest 113:357–369

Nygren A, Clark RB, Belke D, Kondo C, Giles WR, Witkowski FX 2000 Voltage-sensitive dye mapping of activation and conduction in adult mouse hearts. Ann Biomed Eng 28:958–967

Nygren A, Kondo C, Clark RB, Giles WR 2003 Voltage-sensitive dye mapping in Langendorff-perfused rat hearts. Am J Physiol Heart Circ Physiol 284:H892–H902

Ostlund C, Worman HJ 2003 Nuclear envelope proteins and neuromuscular diseases. Muscle Nerve 27:393–406

Ozawa E, Hagiwara Y, Yoshida M 1999 Creatine kinase, cell membrane and Duchenne muscular dystrophy. Mol Cell Biochem 190:143–151

Pendas AM, Zhou Z, Cadinanos J et al 2002 Defective prelamin A processing and muscular and adipocyte alterations in Zmpste24 metalloproteinase-deficient mice. Nat Genet 31:94–99

Petrofski JA, Koch WJ 2003 The β-adrenergic receptor kinase in heart failure. J Mol Cell Cardiol 35:1167–1174

Poon BY, Ward CA, Giles WR, Kubes P 1998 Emigrated neutrophils regulate ventricular contractility via α4 integrin. Circ Res 84:1245–1251

Poon BY, Ward CA, Cooper CB, Giles WR, Burns AR, Kubes P 2001 α_4-Integrin mediates neutrophil-induced free radical injury to cardiac myocytes. J Cell Biol 152:857–866

Saba S, Vanderbrink BA, Luciano B et al 1999 Localization of the sites of conduction abnormalities in a mouse model of myotonic dystrophy. J Cardiovasc Electrophysiol 10:1214–1220

Saha AK, Ruderman NB 2003 Malonyl-CoA and AMP-activated protein kinase: an expanding partnership. Mol Cell Biochem 253:65–70

Shusterman V, Usiene I, Harrifal C et al 2002 Strain-specific patterns of autonomic nervous system and heart failure susceptibility in mice. Am J Physiol Heart Circ Physiol 282:H2076–H2083

Stuurman N, Heins S, Aebi U 1998 Nuclear lamins: their structure, assembly, and interactions. J Struct Biol 122:42–66

Sullivan T, Escalante-Alcalde D, Bhatt H et al 1999 Loss of A-type lamin expression compromises nuclear envelope integrity leading to muscular dystrophy. J Cell Biol 147:913–920

Taylor MRG, Fain PR, Sinagra G et al 2003 Natural history of dilated cardiomyopathy due to lamin A/C gene mutations. J Am Coll Cardiol 41:771–780

Wehnert M, Muntoni F 1999 60th ENMC International Workshop: non X-linked Emery-Dreifuss Muscular Dystrophy 5-7 June 1998, Naarden, The Netherlands. Neuromuscul Disord 9:115–121

Wickman K, Nemec J, Gedler S J, Clapham D E 1998 Abnormal heart rate regulation in GIRK4 knockout mice. Neuron 20:103–114

DISCUSSION

Oshima: Can abnormalities in parasympathetic and sympathetic neurons alone cause cardiomyopathy in the long-run?

Giles: That's an interesting question: can alterations in the autonomic nervous system cause cardiomyopathy? I am not an expert in cardiomyopathy, but I am not aware that they can.

Lee: A large pulse of sympathetic tone can injure the heart. For example, many people who have strokes also develop depressed cardiac function just before they die. That is because the sympathetic mediators injure the myocytes themselves. However, I don't know of a chronic neuropathy that causes it.

Worman: Listening to your paper made me think of ageing and the heart. People with EDMD have early conduction system delay problems. You showed an excess of fibrosis in the heart. We know that as patients age they often get second and third degree heart block. Is there any connection between what is seen early in EDMD and what is seen in a normal ageing heart?

Giles: I can't answer that question. However, I am particularly interested in the changes of ion channel physiology that occur as a function of ageing. There are classes of K$^+$ channels that are essential for repolarizing the heart, which change dramatically as a function of development and ageing.

Worman: So there are certain types of K$^+$ channels which are expressed less in ageing.

Giles: Yes, usually a functional down-regulation. As far as we can tell this is not particularly important in repolarizing the action potential of a single cell, but it gives rise to a change of around 40% in the safety factor for repolarization.

Hutchison: Does the difference that you measured between cardiomyocytes from the left and right ventricles reflect an intrinsic difference in the organization of myocytes in those ventricles? Or does it arise because you are taking the myocytes from an animal that already has cardiomyopathy?

Giles: None of the differences in autoimmune regulation or 'work' capacity that I have just described have been described in a normal mouse. Certainly, we think that what I have just described to you is a result of the pathophysiology of this knockout animal.

Bonne: At what stage of the disease do you do your analyses? How old are the mice?

Giles: These animals only live for 6–8 weeks. In this study we wanted to play it a little safe, so our results were all recorded between 5.5 and 6 weeks.

Gerace: How much cell death has occurred in these muscles at the time you are looking? How much does cell death contribute to the altered parameters that you measure?

Giles: We have no data on this.

Morris: You showed some histological changes in the contractile muscle of the mouse heart, but did you look at the SA node system for any changes that might produce the cardiac conduction problems associated with EDMD?

Giles: No, we have done histological analysis of the left and right ventricle, but not of supraventricular structures.

Fatkin: One of the things that no one seems to have looked at yet is the cause of the conduction abnormalities. We don't know yet whether there are structural and/ or functional defects in the conduction system. This comes back to the ageing/cell death question. As normal people get older there can be degenerative changes in the nodal tissue that give rise to conduction defects. It is possible that the conduction abnormalities associated with *LMNA* mutations will turn out to be related to localized fibrosis. The heart rate variability results are very interesting and may give some important information as to the mechanisms of sudden death.

Morris: The only papers I have seen published on hearts with conduction defects are in myotonic dystrophy (Nguyen et al 1988). People have studied quite a lot of postmortem hearts and have seen a lot of atrial degeneration that might have caused earlier conduction defects, but I haven't seen anything like this for EDMD.

Fatkin: In human laminopathies, one of the early papers from the Seidman lab showed AV node fibrosis (Kass et al 1994). We don't know whether this is the case in these mice. In terms of apoptosis we have found that apoptosis is present in the left ventricle late in the disease. This is not a big surprise because we know that severe heart failure *per se* can promote apoptosis. We did not look specifically for apoptosis in the conduction system.

Goldman: What exactly does fibrosis entail?

Fatkin: It can be an active process with activation of fibrotic pathways, or it can be replacement fibrosis following cell death. Basically, it is an accumulation of extracellular matrix (ECM) material.

Goldman: Is there a proliferation of fibroblasts or is it just accumulation of ECM components?

Fatkin: In dilated cardiomyopathy, both of those mechanisms have been implicated.

Lee: Yes, it involves both. The normal heart has almost the same number of fibroblasts and myocytes, but myocytes don't secrete a lot of matrix. No one knows why the extracellular matrix synthesis pathway is turned on in a degenerating heart. It is a critical problem.

Goldman: Is there also an increase in fibroblast number?

Lee: Yes. The fibrosis of cardiomyopathy is often accompanied by more fibroblasts.

Goldman: Has anyone suggested that this might be a wound-healing response? When you repress or down-regulate lamin A it is possible that you drive the cell to begin cycling. Most cells that are expressing A-type lamins, on the basis of what we know about development, might be in a state where they are out of cycle. Could it be that the loss of lamin A and the presence of only B-type lamins has something to do with hypertrophy and overgrowth of some subpopulation of cells?

Lee: This is an important point. In humans, like many higher-order mammals, the heart heals with fibrosis. In a relatively few amphibian species, such as the newt, the heart heals without fibrosis. If we could understand why this happens we would understand a lot about prevention of sudden cardiac death, since fibrosis is an important cause of cardiac rhythm disturbances.

Goldman: In culture we always see A-type lamins in dividing cells, but are they expressed in progenitor cells of real tissues?

Gerace: Liver regeneration, or skin.

Goldman: Do basal cells in skin have A- and B-type lamins?

Hutchison: Yes, they do. In skin, you start off with cells that don't have A-type lamins, which are presumably the progenitors. However, as soon as they enter the cell cycle they express lamins A and C.

Goldman: Do the known progenitor cells of the skin express these?

Hutchison: Yes. The majority have A-type lamins.

Capeau: We have seen a patient with premature ageing and a R133L lamin mutation. He had some skin lesions. There was increased collagen with shortened collagen fibres. He presented with what was called atypical Werner syndrome (Chen et al 2003). Perhaps this skin lesion has something to do with premature ageing.

Wilson: Connective tissue cells and the stem cells that give rise to them might be very relevant to laminopathy and tissue repair. Skeletal muscle injuries can lead to scarring (dense connective tissue formation), which interferes with muscle fibre regeneration. The Huard lab showed that the mechanism of scarring involves TGFβ signalling, which determines whether regenerating stem cells fuse to the muscle cell to repair it versus becoming a fibroblast. This fate decision is influenced by the surrounding inflammatory response, which can ruin the repair process (Li et al 2004). TGFβ and other signalling pathways might be relevant for laminopathies. Maybe laminopathies involve 'bad' fate decisions by stem cells that can either choose to repair muscle, or build connective tissue. Signalling or gene regulation that depends on lamins, might go wrong in laminopathy patients. In other words, lamins might be involved in cell fate decisions or knowledge of proper fate.

Shanahan: I think you have to be careful with the heart defect. Fibrosis in the heart is caused by any sort of injury to the heart. Cardiac myocytes are terminally differentiated and can't repair the heart. Fibroblasts are there to repair the heart. If you have an infarct, you will get a response — wound healing — which is what you would have in any tissue repair process. If you are looking at the lamin hearts, you have to be careful to look at whether there was an injury there first, of whatever kind, whether mechanical or oxidative stress. These cells could be more susceptible to stress, and this is why you are getting an accelerated scar formation, rather than there being anything wrong with the scar formation itself, which is the normal response to damage.

Goldman: We will hear something about mechanical stress and lamin A knockout cells later.

Giles: There is no electrophysiological sign of any inflammation after infarct in the hearts we have been looking at. For completely different reasons we study acute inflammation with these technologies. In myocytes, acute inflammation gives rise to a very interesting and quick 'signature' in terms of overall conduction defects in the ventricle. These aren't seen in the lamin $(-/-; +/-)$ progeny. My knowledge of the literature about this phenotype and others exhibiting muscular dystrophy is that one cardiac deficit is quite localized in or near the AV node. Our novel observation is that when the ventricular myocardial muscle needs to work hard, a prominent defect is seen. To explain this I would invoke muscle metabolism e.g. creatine kinase deficiencies.

Shanahan: The scarring is something that would happen in any sort of tissue that has a defect. It is probably completely secondary to the original phenotype.

Lee: The cardiac conduction phenotype is relatively specific for lamin and emerin-related diseases. We see an awful lot of degenerative heart disease of many aetiologies and usually you don't see AV block until the very end. Eventually I think it will be shown that the conduction defect is something very special at the molecular level: it is probably transcriptional and not simply degenerative. Diane Fatkin pointed out that this association in humans is very clear.

Morris: Are there any proteins or genes that are specifically associated with the conduction system in the heart?

Giles: Both in terms of the origin of the cells from neural crest and also some aspects of functions of the cells, neurofilaments seem to be more importantly involved in the conduction system than the myocytes.

Julien: Didn't you analyse the NF-L knockout mice?

Giles: Yes. For other reasons we have studied a line of neurofilament knockout mice provided by Jean-Pierre Julien. When we stimulated the left versus the right vagus nerve we saw a defect in the ability of the right vagus nerve to influence the heart electrophysiology, but no defect in the left vagus.

Goldman: Would this be an electrophysiological defect?

Giles: We were studying the PR interval of the electrocardium. That is, we are studying conduction across the AV node. We were stimulating the nervous 'supply' to that region of the heart to ask two questions. First, is there a lengthening of PR? The answer was 'yes', in normals but 'no' in mutants. Second, can we produce something that is very commonly observed in humans in a number of pathophysiological circumstances, an actual AV block? This can be recapitulated in the mouse, but apparently not in the neurofilament knockout.

Goldman: Which knockout was this?

Julien: The NF-L knockout mouse which has no neurofilaments in the axons.

Morris: Are you suggesting that the conduction defects in X-linked EDMD might originate in the nerve supply to the heart?

Giles: I mentioned these results not so much to try to make that conclusion, but to remind the audience to consider the autonomic nervous system as well as the other more conventional targets, sites and mechanisms.

Ellis: In answer to Glen Morris' question, there was a paper a few years ago in which the transcription factor HF-1b in mouse heart was knocked out (Nguyen-Tran et al 2000). This is involved in separating the conducting cells from the mechanical cells in the heart. These mice died from sudden cardiac death. This is evidence for the involvement of a transcription factor.

Giles: My colleague at UCSD was the primary author of that paper. My laboratory did the electrophysiology experiments.

Goldman: Is it mechanically activated?

Ellis: No, it's a developmental transcription factor that turns on very early. It's a member of the SP-1 transcription factor gene family. Very early on in heart development the cells failed to go one way for the conductive tissue and another way for mechanical cells: there was just a mixture so the AV node system didn't form properly.

Wilson: You said that the K^+ channels which change during development and ageing are connected to the cytoskeleton. Can you expand on that?

Giles: The K^+ channels involved are the K_v4 family, and are important for early repolarization. The connection appears to be to F-actin. One observation is that when actin is disrupted, then this class of K^+ channels is no longer under autonomic hormonal control.

Stewart: From what I recollect the sarcoplasmic reticulum (SR) is a modified version of the ER. We know the ER is connected to the nuclear envelope. Some things I have seen in the literature indicate that both the SR and nuclear envelope have very important functions in regulating Ca^{2+} entry and exit. Is it possible that disruptions to the ER would affect Ca^{2+} efflux and influx?

Giles: That is a possibility. If one looked at this in detail, in principal one could assess gain and release of Ca^{2+} from the SR. There is no doubt that in other types of MD these Ca^{2+} channels are compromised.

Ellis: In EDMD we still don't know whether the heart defect is a primary or secondary defect. In knockout lamin A mice has anyone looked at components of the intercalated discs? Are all the proteins which are supposed to be there still present in the right amounts?

Giles: We haven't studied this. In experiments in the literature reporting electrophysiological effects of the connexins, changes in QRS duration are observed, although this is only a 30–40% change.

Fatkin: We have some Ca^{2+} transient data on the cardiac myocytes. We found that the baseline and peak amplitude of the transients were similar in the homozygotes and wild-type mice. There was prolonged relaxation, which is a fairly non-specific feature: it can be secondary to the presence of dilated cardiomyopathy. These findings suggest that Ca^{2+} release is probably normal although it is still possible that there might be altered Ca^{2+} affinity.

Bonne: I have a comment about heart disease in EDMD. It is said that the cardiac disease is the same in X-linked EDMD and laminopathies. In a workshop we attended last September, various cardiologists were discussing whether the cardiac diseases were the same in the two forms of the disease. Although there aren't any solid numbers, it seems that in laminopathies there is a higher frequency of sudden deaths, whereas in EDMD this is rarer. There are subtle differences in the cardiac disease in these two disorders.

Morris: Isn't it also the case that you rarely, if ever, have to perform heart transplants in X-linked EDMD, whereas this commonly has to be done in laminopathy patients (Becane et al 2000). The cardiomyopathy is obviously much more severe in patients with lamin A/C mutations.

Bonne: There is still a lot of work to be done on this aspect to see whether it is clearly different or not. There are some hints that perhaps it is slightly different.

References

Becane HM, Bonne G, Varnous S et al 2000 High incidence of sudden death with conduction system and myocardial disease due to lamins A and C gene mutation. Pacing Clin Electrophysiol 23:1661–1666

Chen L, Lee L, Kudlow BA et al 2003 LMNA mutations in atypical Werner's syndrome. Lancet 362:440–445

Kass S, MacRae C, Graber HL et al 1994 A gene defect that causes conduction system disease and dilated cardiomyopathy maps to chromosome 1p1-1q1. Nat Genet 7:546–551

Li Y, Foster W, Deasy BM et al 2004 Transforming growth factor-$\beta 1$ induces the differentiation of myogenic cells into fibrotic cells in injured skeletal muscle: a key event in muscle fibrogenesis. Am J Pathol 164:1007–1019

Nguyen HH, Wolfe JT, Holmes DR, Edwards WD 1988 Pathology of the cardiac conduction system in myotonic dystrophy: a study of 12 cases. J Am Coll Cardiol. 11:662–671

Nguyen-Tran VT, Kubalak SW, Minamisawa S et al 2000 A novel genetic pathway for sudden cardiac death via defects in the transition between ventricular and conduction system cell lineages. Cell 102:671–682

Multiple pathways tether telomeres and silent chromatin at the nuclear periphery: functional implications for Sir-mediated repression

Angela Taddei, Marc R. Gartenberg, Frank R. Neumann, Florence Hediger and Susan M. Gasser[1]

University of Geneva, Department of Molecular Biology and NCCR Frontiers in Genetics, Quai, Ernest-Ansermet 30, CH-1211 Geneva 4, Switzerland

Abstract. The positioning of chromosomal domains within interphase nuclei is thought to contribute to establishment and maintenance of epigenetic control. Using GFP-tagged chromosomal domains, LexA-fusion targeting and live microscopy, we investigated mechanisms through which chromatin can be anchored to the nuclear envelope (NE). We find that a subdomain of the silencing information regulator Sir4 (Sir4^{PAD}) and yKu80 are sufficient to tether a chromatin region to the nuclear periphery, independently of their silencing function. Sir4^{PAD} interacts with Esc1 (Establishes Silent Chromatin 1), a large acidic protein, localized at the nuclear periphery in the absence of Sir4 and yKu. Sir4 also binds to the periphery through yKu80, whose perinuclear ligand is unidentified. Both pathways are involved in the localization of natural telomeres. To show that silent chromatin can also mediate anchorage, we uncoupled the *HMR* silent mating-type locus from the chromosome using inducible site-specific recombination. Real-time cytological techniques reveal the position and dynamics of the excised locus. We show that the silent *HMR* ring associates with the NE in a SIR-dependent manner, while derepressed excised rings move without detectable constraint throughout the nucleoplasm. Dual anchoring pathways thus cooperate to generate high concentrations of SIR proteins in perinuclear foci, which in turn promote repression.

2005 Nuclear organization in development and disease. Wiley, Chichester (Novartis Foundation Symposium 264) p 140–165

It is now well established that chromatin domains can have specified positions within a eukaryotic nucleus. It is also clear, however, that even the most precisely

---

[1]This paper was presented at the symposium by Susan M. Gasser to whom all correspondence should be addressed. Current address: Friedrich Miescher Institute for Biomedical Research, Marlbeerstrasse 66, CH-4058 Basel, Switzerland.

localized chromatin participates in constant, albeit spatially constrained motion (Chubb & Bickmore 2003, Gasser 2002, Marshall 2002). Haploid budding yeast has provided an excellent model system in which to examine the impact of nuclear organization and chromatin dynamics on the regulation of transcription. In interphase nuclei, the 32 budding yeast telomeres are usually clustered at the nuclear envelope (NE) in three to six groups. These foci sequester histone-binding silencing factors (Sir2, Sir3 and Sir4; Gotta & Gasser 1996), which confer a variegated and heritable repression of telomere-proximal genes (called telomere position effect or TPE; reviewed in Huang 2002). The Sir complex is recruited to telomeres mainly through interaction between Sir4 and DNA binding factors such as the yKu heterodimer, which binds the very end of the chromosome, and Rap1, which binds the TG repeats (Martin et al 1999, Mishra & Shore 1999).

The clustering of telomeres is thought to facilitate TPE by maintaining a high local concentration of Sir proteins (Maillet et al 1996). On the other hand, the grouping of repressed domains may have no function *per se*, and simply result from interactions of integral chromatin proteins with the nuclear envelope. Here we review recent advances made in the characterization of the mechanisms that mediate anchoring at the yeast nuclear envelope. We find that silencing and anchoring at the nuclear periphery are separable: domains can be anchored in the absence of repression and silent chromatin itself can anchor to the periphery through a novel nuclear envelope-associated factor called Esc1. We discuss evidence arguing that spatial clustering within the nucleus can promote heterochromatin formation or propagation.

How to monitor locus position and chromatin mobility

Because much of our analysis of the function of chromosome positioning relies on the use of live fluorescence microscopy, we first summarize the methods used to study chromosome dynamics. Rapid time-lapse microscopy has shown that chromatin in the yeast interphase nucleus is mobile, while specific chromosomal structural elements, such as telomeres and centromeres, counteract this by providing sites of anchorage (Chubb & Bickmore 2003, Gasser 2002, Marshall 2002). The ability to map the position and to follow the movement of specific chromosomal loci is possible due to the development of a GFP-tagged lac repressor-operator system for site recognition (Belmont 2001). This system exploits the high affinity and highly specific interaction of the bacterial lac-repressor (lac^i) for a DNA sequence called lac operator (lac^{op}). Through the directed insertion of an array of lac operators (usually 256 copies or $\sim 10\,kb$) and expression of a fusion construct between lac^i and a fluorescent protein such as GFP, defined DNA loci can be visualized in living cells (Fig. 1).

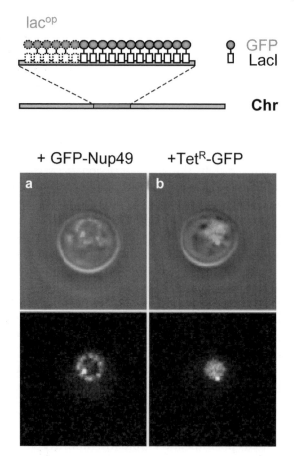

FIG. 1.   Visualization of a chromosomal locus and the analysis of chromosomal locus position.
Budding yeast cells are modified to express the lac^i-GFP fusion which binds the inserted lac^{op}
array (see scheme) along with either a Nup49-GFP (a) or a Tet^R-GFP fusion (b). Shown are
representative images of G1 phase cells tagged at Tel 14-L (a; Hediger et al 2002b) or near
*MAT*a the mating type locus at Chr 3 (b).

The tracking of a single DNA locus can be monitored relative to global nuclear
movement, to correct for background motion. This is achieved through the
integration of a GFP-Nup49- fusion which generates a characteristic ring of
nuclear pores, with which the background movement of the nucleus can be
monitored (Belgareh & Doye 1997). Alternatively a TetR-GFP fusion in the
absence of an integrated Tet^{op} array can be used to give a low intensity
background signal (Fig. 1).

Interphase chromatin is highly dynamic

By tracking the movement of various chromosomal loci at 1.5 s intervals, we find that yeast chromatin is extremely dynamic, moving with the same amplitude as chromatin in *Drosophila* and mammalian cells. In the systems explored to date, fluorescence and transmission imaging were combined to characterize the movements of the tagged locus with respect to division cycles or differentiation pathways. Both were found to influence chromatin mobility (reviewed in Gasser 2002, Chubb & Bickmore 2002, Marshall 2002). In general two types of chromatin motion can be distinguished: smaller, saltatory movements $<0.2\,\mu m$ that occur every 1–2 s, and larger, more rapid movements (i.e. $>0.5\,\mu m$ in a 10.5 s interval; Heun et al 2001a). These larger movements are characteristic of euchromatic domains, or of sites in metabolically active yeast that are distant from telomeres and centromeres. Such movements ($>0.5\,\mu m$) represent half the radius of a yeast nucleus, a distance equivalent to $\sim 100\,kb$ of chromatin in its physiological compaction state (Bystricky et al 2004). Tracking of these movements shows that they are generally limited to a space less than the full nuclear volume.

The correlation of increased interphase dynamics of chromatin with changes in cellular metabolism (i.e. the depletion of glucose), makes it unlikely that chromatin movement results from simple diffusion (Heun et al 2001a). We suggest that the movement reflects the action of ATP-dependent enzymes as they activate transcription or remodel nucleosomes. This would be consistent with the lack of mobility detected in stationary phase cells where transcriptional activity drops significantly, or in cells in which ATP is depleted (Heun et al 2001a).

The extent of chromatin movement

When thousands of measurements for a GFP-tagged locus are compared over time, either as a relative movement between two spots in a diploid, or the movement of one locus relative to the nuclear center, the spatial constraints imposed on the movement can be estimated by plotting the mean squared displacement (MSD) over fixed intervals of time (Marshall et al 1997; Fig. 2). These MSD plots reach a plateau at larger time intervals which allows one to calculate the radius of constraint on a given tagged locus, and to compare this reliably among different chromosomal loci in different species. The movement of euchromatic sites in flies, man and yeast is confined to volumes of ~ 0.5–$0.7\,\mu m$ in radius, significantly less than even the radius of a yeast nucleus ($1\,\mu m$). Interestingly, in budding yeast, an excised ring of chromatin was shown to move without detectable constraint throughout the nucleus, suggesting that there is no inherent barrier to chromatin movement in yeast nucleoplasm (Gartenberg et al 2004 and Fig. 2).

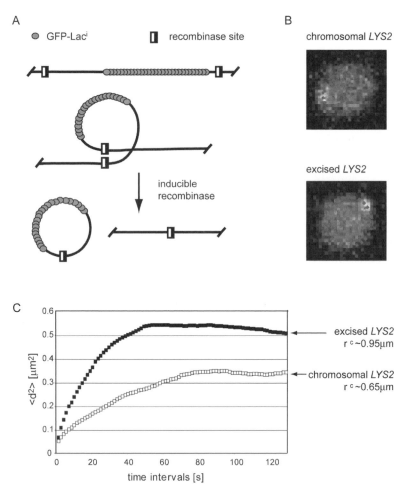

FIG. 2.   Chromatin tracking reveals spatial constraints on chromatin dynamics. The genomic *LYS2* locus was tagged with a lac^{op} insert and flanked by repeats to mediate excision by an inducible recombinase (A). Using a LSM510 (Zeiss) confocal microscope 300 sequential image stacks of GFP fluorescence were captured at 1.5 s intervals in G1 phase yeast cells carrying either the chromosomal or the excised *LYS2* locus. The images were first aligned based on their nuclear centers and a track was determined in two dimensions after projection of the 3D stack of images (see tracks in B). To quantify spatial constraints on these movements 8 movies (each with 300 sequential image stacks) were subjected to the Mean Squared Displacement (MSD) analysis (see C). After alignment of nuclear centres the absolute position of the focus was determined for each frame and then the square of the displacement of the locus for all intervals from 0–150 s (Δt sec) was computed using the formula: $\Delta d^2 = \{d(t) - d(t + \Delta t)\}^2$. The average of all Δd^2 values for each Δt value are plotted against Δt. The slope is the derivative of the diffusion coefficient and a plateau indicates a radius of spatial constraint (r^c) as determined from modelling randomly generated values within specific sized surfaces (F. R. Neumann and M. Blaszczyk, personal communication).

In view of the dynamic character of chromatin, it seems likely that anchorage sites may function as important regulators of transcription. For example, the sites at which yeast telomere cluster could contribute to the proper replication of chromosomal ends or to the regulation of silencing. Long-range interactions between blocks of heterochromatin in other species also probably facilitate a mass action assembly of repressors with their binding sites (Henikoff 1996). In various cells, recruitment and stable association of damaged DNA sites into internal foci has been correlated with relatively fixed sites of repair (Lisby et al 2001, Liu et al 1999). Finally, it has been proposed that the insulators that create boundaries between active and inactive chromatin domains may also function by tethering these sites near nuclear pores (Gerasimova & Corces 1998, Ishii et al 2002).

Redundant mechanisms of chromatin attachment at the nuclear periphery

The anchorage of yeast chromatin appears to occur by two pathways: either through microtubules that attach the centromere with the spindle pole body or through the tethering of DNA to the NE. We focus here on the latter mechanism.

Yeast telomere clusters appear to define relatively stable subnuclear compartments, based on images taken at fixed time-points (Gotta et al 1996). Individually tagged telomeres, as well as GFP-labelled foci, nonetheless move along the nuclear envelope, significantly more than an integral NE component like the spindle pole body (Heun et al 2001a,b, Hediger et al 2002a). Telomere movement is primarily lateral, a pattern that could arise from a reversible association with multiple dispersed binding sites along the inner NE. A multiplicity of anchorage sites may allow telomeres to be trapped within a perinuclear zone. In both wild-type and *yku*-deficient cells, entire clusters of telomeres appear to be displaced from the periphery (Gotta et al 1996, Laroche et al 1998), suggesting that telomere anchorage can be distinguished from the telomere–telomere interactions that lead to clustering.

The analysis of GFP-tagged native telomeres in living yeast cells has revealed two partially redundant pathways that mediate anchorage: one requires the heterodimeric yKu factor, and the second is mediated by one or more of the Sir proteins (Hediger et al 2002b). Both yKu and Sir proteins are integral components of the telosome and of subtelomeric chromatin, and could thus play a direct role in tethering. The Sir-dependent pathway correlates with chromatin-mediated transcriptional repression, may require higher order chromatin structures, and is able to tether a native telomere in S phase in the absence of yKu. The yKu pathway, on the other hand, anchors a fraction of telomeres throughout interphase, even in the absence of TPE (i.e. in Sir-deficient strains,

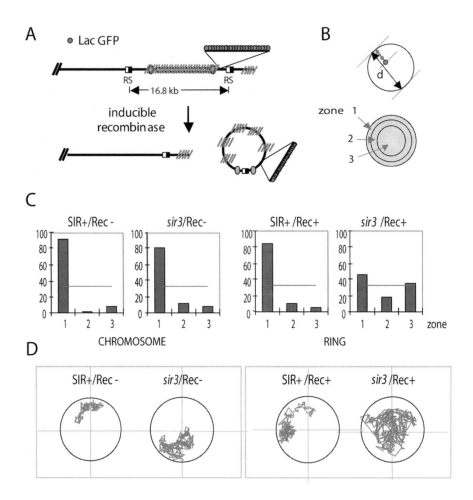

FIG. 3.   (A) Experimental design: site-specific recombination at engineered target sites (*RS* sites) uncouples the lac^{op}-tagged *HMR* locus from the chromosome (see Gartenberg et al 2004, for details of strain construction and imaging). (B) In an equatorial section of a GFP-tagged locus and GFP-visualized nuclear pores distances between *HMR* spot and Nup49-GFP ring (x) were normalized to the nuclear diameter (y) and binned according to three zones of equal surface area. If a spot is located randomly within the nucleus it will show a distribution of 33% in every zone. (C) Distribution (plotted as percent of total) of *HMR* in the three zones for each strain. n, number of cells. Bar at 33% represents an idealized random distribution. (D) Tracking of the movement of the GFP-tagged HMR locus over 5 min timelapse imaging at 1.5 s intervals. The track is shown in pale grey and is calculated after alignment of nuclear centers among all frames. Only the derepressed and excised locus samples a large fraction of the nucleoplasm.

see Fig. 3 and Hediger et al 2002b). Thus, anchoring does not necessarily correlate with repression (Hediger et al 2002b, Tham et al 2001). For telomeres that are less dependent on the yKu anchorage pathway, the Sir-dependent pathway helps tether telomeres in S phase, in a manner that correlates with repression. Such analysis, extended to three independently tagged telomeres, confirms the generality of these redundant anchorage mechanisms, which in addition are subject to cell cycle regulation.

Subtelomeric repeats can provide 'anti-silencing' and 'anti-anchor' functions

Given this redundancy, what regulates the use of one anchor or the other? Preliminary evidence suggests that subtelomeric sequence organization itself influences telomere positioning in subtle ways. This is supported by two sets of data: first there is significant variation in the anchoring efficiencies among native telomeres, and second, truncated telomeres do not behave the same as ends that contain subtelomeric repeats (Hediger et al 2002b, Tham et al 2001). With respect to the first point, we note that the native Tel 6R end, which has the X but not the Y' and STR elements, is displaced throughout the cell cycle in a *yku* deficient strain; Tel 14L which has Y', X and STR sequences, on the other hand, is sensitive to yKu deletion only in G1 phase. This residual S phase anchoring reflects the contributions of the Sir-dependent pathway. We predict that not only a telomere's insensitivity to *YKU* deletion, but also its anchoring efficiency in wild-type S phase cells, is likely to correlate with the variability of native subtelomeric repression (Fourel et al 1999, Pryde & Louis 1999).

Intriguingly, it was found that a truncated Tel 6R (which has TG repeats but no X or Y' elements) remains anchored at the periphery in the absence of the *YKU* gene products, while the native Tel 6R is released by their deletion. This residual anchoring is Sir4-dependent, even though the efficiency of TPE is low. Thus, the Sir4–Rap1 interaction is sufficient to tether the truncated, but not the native telomere. To explain this we suggest that the X subtelomeric element interferes with the Sir-dependent pathway, acting as an anti-anchor and rendering native telomere positioning sensitive to the loss of yKu (Hediger et al 2002b). This is consistent with recent observations showing that subtelomeric sequences either disfavour repression of reporter genes or cause an abrupt drop in their propagation (Fourel et al 1999, Pryde & Louis 1999).

Silent chromatin directs its own anchorage independently of a chromosomal context

Not only telomeres but the two silent yeast mating-type loci, which are positioned within 25 kb of telomeres, are found at or near the nuclear periphery. *HML*, for instance, frequently colocalizes with telomeric clusters (Heun et al 2001b, Laroche et al 2000). In order to study whether the native *HMR* locus is intrinsically tethered to the nuclear periphery, or whether it requires an adjacent telomere for anchorage, we used an inducible site-specific recombination to uncouple the locus from its normal chromosomal context (Cheng et al 1998). The region containing *HMR*, including silencers and a 256 lac operator array (*lac^op*), was flanked by a pair of target sites for the R site-specific recombinase (Fig. 3). Galactose induced expression of the recombinase yielded 95% excision within 2 hours in a silencing-competent strain. The locus could be monitored relative to GFP-tagged nuclear pores, and by comparing strains that had or had not undergone excision, the impact of chromosomal context on location and dynamics of *HMR* could be evaluated.

The unexcised *HMR* in a strain that lacks recombinase (SIR+/Rec−) remained closely linked to the nuclear envelope. In a silencing-defective *sir3* deletion strain (*sir3*/Rec−), the perinuclear fraction of *HMR* dropped only slightly from 91 to 81% (Fig. 3). This indicates that a silencing-independent mechanism anchors Tel 3R. When *HMR* was uncoupled from the chromosome, the excised ring remained at the nuclear envelope in 86% of silencing-competent cells (SIR+/Rec+, Fig. 3), suggesting that targeting of *HMR* to the nuclear periphery is an intrinsic property of the locus and does not depend on linkage to a telomere. Importantly, when the ring is excised in a *sir3* deletion strain, it distributes throughout the nucleoplasm (*sir3*/Rec+). Its unconstrained mobility can be demonstrated by time-lapse tracking (Fig. 3), and allows us to conclude that immobilization of the excised *HMR* locus at the NE depends on its silent state, unlike telomere anchoring.

Candidates for perinuclear anchorage sites

Because yeast has no intermediate filament proteins like the nuclear lamina, and because neither yKu nor Sir proteins have membrane spanning domains, other NE components must be responsible for the perinuclear tethering telomeres and silent chromatin. By exploiting an *in vivo* chromatin relocalization assay (see Fig. 4) we have been able to assay proteins for domains that determine subnuclear position independently of their interaction with silencing factors. Minimal protein domains that are sufficient to relocate a tagged locus to the NE include YKu80, Sir4^PAD and the C-terminal domain of Esc1 (Andrulis et al 2002). Esc1 is a protein that interacts with Sir4 and is required for the stable mitotic partitioning of a Sir4-bound plasmid (Andrulis et al 2002). Importantly, this protein is localized on the nucleoplasmic surface of the inner bilayer of the NE, positioned primarily between pores and excluded from the nucleolus (Taddei et al 2004). Esc1 localization is independent of Ku and SIR genes, arguing that it is a structural element, and not bound to the periphery through chromatin. Esc1 nonetheless contributes both to efficient TPE and to the yKu-independent anchoring of Tel 14-L and truncated Tel 6R. Most importantly, the relocalization of a chromatin domain through the targeting of the PAD subdomain of Sir4 requires the presence of either Esc1 or Ku, arguing that these ligands constitute redundant pathways for the anchoring of both telomeres and silent chromatin through Sir4 (Taddei et al 2004).

In another study, Mlp1 and Mlp2 were proposed to form a bridge between yKu and nuclear pores (Galy et al 2000). However, careful analysis using both live GFP imaging and *in situ* hybridization showed no influence of the double *mlp1 mlp2* deletion on telomere positioning nor on dynamics when intact nuclei were analysed (Hediger et al 2002a, Hediger & Gasser 2002). Thus, the pore-associated Mlp proteins are also unlikely to participate as the anchor through which yKu tethers telomeres to the NE. Consistently, TPE has been shown by several groups to be intact in the double *mlp1 mlp2* mutant, which is not the case in *yku* or *esc1* mutants (Hediger et al 2002a, Andrulis et al 2002). Given their perinuclear localization and coiled-coil structure, Mlp proteins may be involved in other aspects of nuclear envelope function, such as aspects of mRNA biogenesis or selective export (Green et al 2003, Vinciguerra & Stutz 2004), and a general role in the distribution of NE components is also possible. To date, however, there is no support for the model that telomeres are anchored to nuclear pores, as nuclear pore complexes can be clearly distinguished from most telomeric foci in stained wild-type cells (Gotta et al 1996), and telomeres remain evenly distributed in a *nup133Δ* mutant despite a dramatic clustering of nuclear pores (Hediger et al 2002a).

Testing the role of perinuclear anchoring in repression

It has been proposed that the anchorage of telomeres at the nuclear periphery concentrates silencing factors thus favouring silencing and causing its accumulation in this nuclear subcompartment (Maillet et al 1996). This model is based on several observations: the first is that normal cellular Sir protein concentrations are limiting for repression (Renauld et al 1993, Buck & Shore 1995, Maillet et al 1996). Secondly, we know that silencer-flanked reporter genes are more efficiently repressed when inserted near the telomere (Thompson et al 1994, Shei & Broach 1995, Maillet et al 1996, Marcand et al 1996). Third, it was shown that the delocalization of Sir factors from telomeres, as well as their overexpression, is able to restore repression at loci distant from telomeres (Maillet et al 1996, 2001). This confirms that telomeric foci sequester repressors from other sites of action.

In support of a functional role for telomere clustering, the Sternglanz laboratory showed that by artificially 'tethering' a reporter gene to the yeast NE through a membrane-spanning polypeptide, transcriptional repression could be favoured (Andrulis et al 1998). This was shown to require the presence of at least one

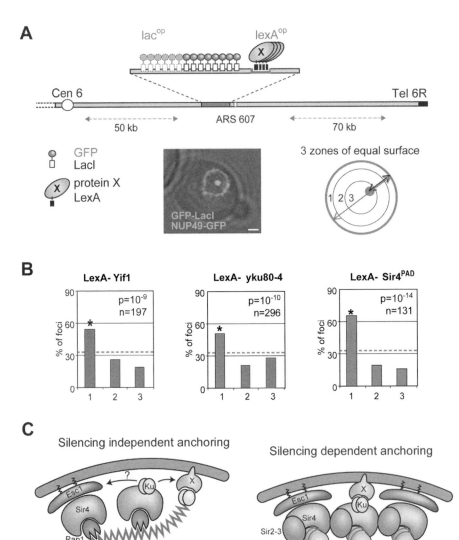

silencer element, a specific *cis*-acting sequence that nucleates Sir-dependent repression. It was proposed that the NE 'tether' was able to compensate for the crippled silencer and to promote Sir-dependent repression because it placed the weak silencer near a zone of high Sir-protein concentration. Indeed, consistent with this model, NE tethering of this reporter was found to promote silencing only if telomeres were able to sequester Sir proteins in perinuclear pools (data not shown).

We propose that silencing depends on anchoring because Sir protein concentrations are limiting within the nucleoplasm. Ironically, this restriction on Sir protein availability stems at least in part from the recruitment of Sir proteins into pools at the nuclear periphery. From this it follows that silencing can be achieved without (or prior to) NE anchoring if Sir proteins can be attracted to the reporter gene in sufficient amounts to achieve a threshold concentration at a given nucleation site. If Sir proteins are not recruited with sufficient efficiency to nucleate repression, then relocalization to a zone of high Sir concentration should help the silencing process. All these predictions have been tested and found to be true (data not shown). Such results imply that relocalization to the NE *per se* should not be sufficient to silence a locus, and that the accumulation of Sir proteins is the critical limiting factor. This could be shown to be true for truncated telomeres in *yku* mutants (Hediger et al 2002b, Tham et al 2001) and for reporter genes associated with a weak silencer (Taddei et al 2004).

Anchoring can cause induce and result from transcriptional repression

Do changes in subnuclear position cause or result from transcriptional silencing? As summarized above, previous studies clearly implicated yKu and the Sir proteins

---

FIG. 4.   A yKu and Sir-dependent relocalization identifies chromatin anchors in yeast. (A) A yeast strain bearing lac^{op} repeats and LexA binding sites at an internal locus on the right arm of Chr 6 (Chr6^{int}) and a GFP-Nup49 fusion is transformed with plasmids expressing different LexA-protein fusions. Typically the position of the locus is scored as being among one of three concentric zones of equal surface by using the ratio of the spot to pore measurement (black) over the nuclear diameter (grey) (Hediger et al 2002b). (B) LexA.Yif1 LexA-Sir4PAD and LexA-yku80 are able to relocate the locus to the NE. Bar graphs present the percentage of spots (y axis; n = number of cells analysed) per zone (x axis). (C) We show schematically dual pathways of anchoring mediated by Ku and an unknown protein (x) and Sir4-Esc1 at yeast telomeres both in the absence and presence of SIR-mediated repression. Rap1-bound Sir4 can bring a telomere to the NE through interaction with either Ku or Esc1. Sir4^{PAD} promotes the perinuclear anchoring of silent chromatin through either pathway. During G1 phase in wild-type yeast cells the yKu-mediated pathway is necessary for native telomere anchoring suggesting that the Sir-Esc1 pathway may be less efficient in this phase of the cell cycle. In S phase cells on the other hand the Sir-mediated anchoring pathway functions efficiently (Hediger et al 2002b).

in anchoring, but it was unclear whether the loss of telomere position was direct, due to the elimination of the chromatin anchoring polypeptide, or indirect, due to derepression. Using subdomains or mutant proteins in a novel anchoring assay, we have now been able to identify yKu and the Sir4^{PAD} domain as minimal anchors that can tether chromatin to NE in the absence of transcriptional repression (Taddei et al 2004). When targeted to a tagged locus, the Sir4^{PAD} domain can relocate this DNA to the NE without repressing transcription, through two partially redundant pathways. The first requires Esc1, and the other yKu70/yKu80. Although this shows that anchoring can be mediated through a subdomain of a silent regulatory protein, it also provides a means for silent chromatin to localize itself to the nuclear periphery. This has been demonstrated by showing that an excised *HMR* mating-type locus that is anchored at the nuclear periphery in a silencing-dependent manner, is also released to move freely in cells lacking Esc1 and Ku (Gartenberg et al 2004).

These results lead to a model that explains how a Sir-rich compartment can form spontaneously in the nucleus, to favour and maintain Sir-mediated repression (see Fig. 5). Central to this model is the demonstration of silencing-independent anchorage *via* yKu or Sir4. This would provide a means for telomeres to accumulate at the nuclear envelope prior to the formation of a repressed state. The anchoring of telomeric repeats creates a large number of potential Sir4 protein binding sites within a restricted volume, due to the presence of 20–25 Rap1 consenses on each telomere end. The ORC sites within subtelomeric repeats can also contribute to the efficiency of Sir4 accumulation. The resulting high density of Sir4-binding chromatin would in turn attract Sir2 and Sir3 through protein–protein interactions. It is not clear whether only the presence of Sir proteins, or the presence of another molecular catalyst is needed to promote formation of silent chromatin. Nonetheless, a local concentration of silencing factors is clearly a necessary prerequisite for repression. As silent chromatin spreads, the number of Sir4 molecules bound naturally increases and reinforces the interaction with the NE. This creates a feedback loop that is initiated by a limited number of yKu or Sir4 proteins bound at given site (Fig. 5). Consistent with this model, chromatin immunoprecipitation experiments showed that Sir4 is the first component bound in the assembly process of silent chromatin at *HM* loci and at telomeres (Bourns et al 1998, Luo et al 2002, Rusche et al 2002, Hoppe et al 2002).

This clustering-feedback loop provides a mechanism for self-organization of a subnuclear compartment in which silencing factors and repressed domains accumulate. We envision such mechanisms as also being relevant for centromeric satellite and heterochromatin sequences in higher eukaryotes, whose clustering correlates with high concentrations of heterochromatin factors like HP1.

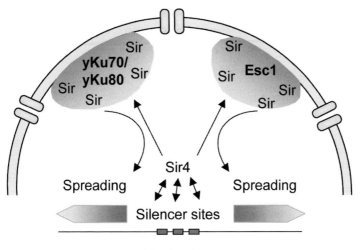

Nucleation

FIG. 5.   Model for the role of chromatin anchoring in the promotion of silencing. We propose that Sir4 is first recruited at the nucleation center by DNA binding proteins that can bind Sir proteins. These include Rap1 ORC Abf1 and yKu. The presence of Sir4 at the locus will then bring it to the nuclear periphery through one of the two Sir4 anchoring pathways (yKu or Esc1) where the high local concentrations of Sir proteins will help silencing complexes assemble and spread. The anchoring of silent loci at the periphery will increase the concentration of Sir proteins and reinforce the silencing of other loci within this region. In addition yKu bound to chromosome ends can independently recruit telomeres to the NE.

Generalizing a model for heterochromatin formation and clustering

The tethering of chromatin, and quite often heterochromatin, at the nuclear envelope is a universal phenomenon that has long been observed in higher eukaryotes (Ferreira et al 1997, Marshall et al 1996, Spann et al 2002). In most eukaryotes, the rigid, perinuclear lamin meshwork underlies the NE, held in place by membrane-spanning lamin receptors (Holaska et al 2002). In flies, points of contact between the lamina meshwork and chromosomes were monitored at irregular intervals along the chromosomal arm (Marshall et al 1996). Transcriptionally inactive chromatin appears to be enriched at the lamin–nucleoplasm interface, except at nuclear pores, which are subtended by less condensed chromatin. Yeast, plants and many protozoa lack nuclear lamins. The absence of this perinuclear intermediate filament network undoubtedly accounts for the sensitivity of yeast nuclei to nonionic detergents, and may render yeast nuclear integrity more dependent on pore components. Despite these differences, interaction in *trans* — or the clustering of heterochromatin — is likely to contribute to gene regulation and nuclear organization in mammalian cells, as in yeast. Indeed, the relocalization of tissue-specific genes to centromeric heterochromatin has been demonstrated for Ikaros-controlled genes in mouse lymphocytes (Fisher & Merkenschlager 2002). To decipher the role of nuclear compartmentation in regulated gene expression, the genetic approaches offered by budding and fission yeast systems are likely to provide valuable paradigms.

Acknowledgements

We thank Thierry Laroche, and Drs Patrick Heun, Kerstin Bystricky, Francoise Stutz and Marek Blaszczyk for having shared results and ideas, and all members of the Gasser laboratory for helpful discussions. Our research is supported by the Swiss National Science Foundation and the NCCR programme 'Frontiers in Genetics' and a Novartis Sabbatical Fellowship to M.R.G. A.T. is supported by an EMBO fellowship.

References

Andrulis ED, Neiman AM, Zappulla DC, Sternglanz R 1998 Perinuclear localization of chromatin facilitates transcriptional silencing. Nature 394:592–595
Andrulis ED, Zappulla DC, Ansari A et al 2002 Esc1, a nuclear periphery protein required for Sir4-based plasmid anchoring and partitioning. Mol Cell Biol 22:8292–8301
Belgareh N, Doye V 1997 Dynamics of nuclear pore distribution in nucleoporin mutant yeast cells. J Cell Biol 136:747–759
Belmont AS 2001 Visualizing chromosome dynamics with GFP. Trends Cell Biol 11:250–257
Bourns BD, Alexander MK, Smith AM, Zakian VA 1998 Sir proteins, Rif proteins, and Cdc13p bind Saccharomyces telomeres in vivo. Mol Cell Biol 18:5600–5608
Buck SW, Shore D 1995 Action of a RAP1 carboxy-terminal silencing domain reveals an underlying competition between HMR and telomeres in yeast. Genes Dev 9:370–384
Bystricky K, Heun P, Gehlen L, Langowski J, Gasser SM 2004 Long range compaction and flexibility of interphase chromatin in budding yeast analysed by high resolution imaging techniques. Proc Natl Acad Sci USA, in press
Cheng TH, Li YC, Gartenberg MR 1998 Persistence of an alternate chromatin structure at silenced loci in the absence of silencers. Proc Natl Acad Sci USA 95:5521–5526
Chubb JR, Bickmore WA 2003 Considering nuclear compartmentalization in the light of nuclear dynamics. Cell 112:403–406
Ferreira J, Paolella G, Ramos C, Lamond AI 1997 Spatial organization of large-scale chromatin domains in the nucleus:a magnified view of single chromosome territories. J Cell Biol 139:1597–1610
Fisher AG, Merkenschlager M 2002 Gene silencing, cell fate and nuclear organisation. Curr Opin Genet Dev 12:193–197
Fourel G, Revardel E, Koering CE, Gilson E 1999 Cohabitation of insulators and silencing elements in yeast subtelomeric regions. EMBO J 18:2522–2537
Galy V, Olivo-Marin JC, Scherthan H, Doye V, Rascalou N, Nehrbass U 2000 Nuclear pore complexes in the organization of silent telomeric chromatin. Nature 403:108–112
Gartenberg MR, Neumann FR, Laroche T, Blaszczyk M, Gasser SM 2004 Sir-mediated repression can occur independently of chromosomal and subnuclear contexts. Cell 119: in press
Gasser SM 2002 Visualizing chromatin dynamics in interphase nuclei. Science 296:1412–1416
Gerasimova TI, Corces VG 1998 Polycomb and trithorax group proteins mediate the function of a chromatin insulator. Cell 92:511–521

Gotta M, Gasser SM 1996 Nuclear organization and transcriptional silencing in yeast. Experientia 52:1136–1147

Gotta M, Laroche T, Formenton A, Maillet L, Scherthan H, Gasser SM 1996 The clustering of telomeres and colocalization with Rap1, Sir3, and Sir4 proteins in wild-type Saccharomyces cerevisiae. J Cell Biol 134:1349–1363

Green DM, Johnson CP, Hagan H, Corbett AH 2003 The C-terminal domain of myosin-like protein 1 Mlp1p is a docking site for heterogeneous nuclear ribonucleoproteins that are required for mRNA export. Proc Natl Acad Sci USA 100:1010–1015

Hediger F, Gasser SM 2002 Nuclear organization and silencing:putting things in their place. Nat Cell Biol 4:E53–55

Hediger F, Dubrana K, Gasser SM 2002a Myosin-like proteins 1 and 2 are not required for silencing or telomere anchoring, but act in the Tel1 pathway of telomere length control. J Struct Biol 140:79–91

Hediger F, Neumann FR, Van Houwe G, Dubrana K, Gasser SM 2002b Live imaging of telomeres. yKu and Sir proteins define redundant telomere-anchoring pathways in yeast. Curr Biol 12:2076–2089

Henikoff S 1996 Dosage-dependent modification of position-effect variegation in Drosophila. Bioessays 18:401–409

Heun P, Laroche T, Shimada K, Furrer P, Gasser SM 2001a Chromosome dynamics in the yeast interphase nucleus. Science 294:2181–2186

Heun P, Laroche T, Raghuraman MK, Gasser SM 2001b The positioning and dynamics of origins of replication in the budding yeast nucleus. J Cell Biol 152:385–400

Holaska JM, Wilson KL, Mansharamani M 2002 The nuclear envelope, lamins and nuclear assembly. Curr Opin Cell Biol 14:357–364

Hoppe GJ, Tanny JC, Rudner AD et al 2002 Steps in assembly of silent chromatin in yeast:Sir3-independent binding of a Sir2/Sir4 complex to silencers and role for Sir2-dependent deacetylation. Mol Cell Biol 12:4167–4180

Huang Y 2002 Transcriptional silencing in Saccharomyces cerevisiae and Schizosaccharomyces pombe. Nucleic Acids Res 30:1465–1482

Ishii K, Arib G, Lin C, Van Houwe G, Laemmli UK 2002 Chromatin boundaries in budding yeast. The nuclear pore connection. Cell 109:551–562

Laroche T, Martin SG, Gotta M et al 1998 Mutation of yeast Ku genes disrupts the subnuclear organization of telomeres. Curr Biol 8:653–656

Laroche T, Martin SG, Tsai-Pflugfelder M, Gasser SM 2000 The dynamics of yeast telomeres and silencing proteins through the cell cycle. J Struct Biol 129:159–174

Lisby M, Rothstein R, Mortensen UH 2001 Rad52 forms DNA repair and recombination centers during S phase. Proc Natl Acad Sci USA 98:8276–8282

Liu Y, Li M, Lee EY, Maizels N 1999 Localization and dynamic relocalization of mammalian Rad52 during the cell cycle and in response to DNA damage. Curr Biol 9:975–978

Luo K, Vega-Palas MA, Grunstein M 2002 Rap1-Sir4 binding independent of other Sir, yKu, or histone interactions initiates the assembly of telomeric heterochromatin in yeast. Genes Dev 12:1528–1539

Maillet L, Boscheron C, Gotta M, Marcand S, Gilson E, Gasser SM 1996 Evidence for silencing compartments within the yeast nucleus:a role for telomere proximity and Sir protein concentration in silencer-mediated repression. Genes Dev 10:1796–1811

Maillet L, Gaden F, Brevet V et al 2001 Ku deficient strains exhibit alternative states of silencing competence. EMBO Rep 2:203–210

Marcand S, Buck SW, Moretti P, Gilson E, Shore D 1996 Silencing of genes at nontelomeric sites in yeast is controlled by sequestration of silencing factors at telomeres by Rap 1 protein. Genes Dev 10:1297–1309

Marshall WF 2002 Order and disorder in the nucleus. Curr Biol 12:R185–192

Marshall WF, Dernburg AF, Harmon B, Agard DA, Sedat JW 1996 Specific interactions of chromatin with the nuclear envelope:positional determination within the nucleus in Drosophila melanogaster. Mol Biol Cell 7:825–842

Marshall WF, Straight A, Marko JF et al 1997 Interphase chromosomes undergo constrained diffusional motion in living cells. Curr Biol 7:930–939

Martin SG, Laroche T, Suka N, Grunstein M, Gasser SM 1999 Relocalization of telomeric Ku and SIR proteins in response to DNA strand breaks in yeast. Cell 97:621–633

Mishra K, Shore D 1999 Yeast Ku protein plays a direct role in telomeric silencing and counteracts inhibition by rif proteins. Curr Biol 9:1123–1126

Pryde FE, Louis EJ 1999 Limitations of silencing at native yeast telomeres. EMBO J 18:2538–2550

Renauld H, Aparicio OM, Zierath PD, Billington BL, Chhablani SK, Gottschling DE 1993 Silent domains are assembled continuously from the telomere and are defined by promoter distance and strength, and by SIR3 dosage. Genes Dev 7:1133–1145

Rusche LN, Kirchmaier AL, Rine J 2002 Ordered nucleation and spreading of silenced chromatin in Saccharomyces cerevisiae. Mol Biol Cell 7:2207–2222

Shei GJ, Broach JR 1995 Yeast silencers can act as orientation-dependent gene inactivation centers that respond to environmental signals. Mol Cell Biol 15:3496–3506

Spann TP, Goldman AE, Wang C, Huang S, Goldman RD 2002 Alteration of nuclear lamin organization inhibits RNA polymerase II-dependent transcription. J Cell Biol 156:603–608

Taddei A, Hediger F, Neumann FR, Bauer C, Gasser SM 2004 Separation of silencing from perinuclear anchoring functions in yeast Ku80 Sir4 and Esc1 proteins. EMBO J 25:1301–1312

Tham WH Wyithe JS Ferrigno PK Silver PA Zakian VA 2001 Localization of yeast telomeres to the nuclear periphery is separable from transcriptional repression and telomere stability functions. Mol Cell 8:189–199

Thompson JS, Johnson LM, Grunstein M 1994 Specific repression of the yeast silent mating locus HMR by an adjacent telomere. Mol Cell Biol 14:446–455

Vinciguerra P, Stutz F 2004 mRNA expert: an assembly line from genes to nuclear pores. Curr Opin Cell Biol 16:285–292

DISCUSSION

Wilson: Can you describe Esc1?

Gasser: It is a large protein of about 200 kDa and has three short coiled coil domains, each about 30–40 amino acids long. It is highly acidic and has no obvious transmembrane domain. It has multiple potential lipidation sites, but they are not at the C- or N-terminus, so modification may involve some kind of processing.

Wilson: Do you know that Esc1 is actually lipidated?

Gasser: Esc1p is modified. It migrates aberrantly in a gel, making a smear from 200 kDa upwards. This size of protein is difficult to work with.

Gruenbaum: Are there any Ku-minus mammalian cells in which you can look to see if there is dynamic change in the spots?

Gasser: There are Ku-deficient mice. They age early and are immunodeficient, but there are telomere defects and telomere–telomere fusions occur very frequently. Whereas mammalian telomeres are not anchored at the nuclear periphery, they do seem to be anchored internally: they move less than other parts of the chromosome.

Gruenbaum: In meiosis they seem to be anchored.

Gasser: Yes, they cluster at the nuclear periphery during the bouquet stage. Is this something unique to meiosis? I don't know. One cannot expect our studies of yeast to identify the precise molecules relevant in mammalian cells, although the general mechanisms may be relevant.

Starr: Can you use a LexA fusion library to screen for the proteins that act at the nuclear envelope? You could do this perhaps by putting different LexA fusion proteins in a genomewide screen and then inducing some sort of repression or a GFP screen to see what is localized to the envelope.

Gasser: We have performed a two-hybrid screen to identify the Ku anchor. One of the problems when you do this kind of screen is that Ku binds so many things. It binds machinery of the end-joining apparatus and it binds the Sir protein, so that targeting Ku to a reporter will induce repression. We have screened with a mutant that does not nucleate repression and now have a good two-hybrid screen yielding candidate anchors.

Starr: You could take the Lac repressor with the LexA sites, and put in random LexA proteins to see whether you pull out the Lac repressor. You wouldn't have to worry about Ku in this situation: you might get a nuclear envelope protein binding there.

Gasser: I am all for microscopy-based screens, but not a single student or postdoc in my lab shares my enthusiasm! The problem is that even tethered sites are in constant motion. You can't just look at something once under the microscope: it has to be quantitated. After visual inspection alone you can't say anything.

Gerace: I wonder how much the movement of these telomeric chromatin spots is affected by the size. The telomeres are really small in yeast, so if you made a super-sized telomere, like 10 tandem repeats, would this move the same way?

Gasser: We haven't done that, so I can't say. Extra-long telomeres are still at the periphery, so length alone doesn't disrupt binding. Your question is, would they move less? I don't know. Actually telomeres move more than individual pores, and we can readily quantify the diffusion of a pore. Moreover, the telomere movement is more than that of the spindle pole body, which is embedded in the envelope. In other words, the telomere movement does not simply reflect movement of its anchor in the membrane.

Goldman: Do all those movements require ATP?

Gasser: Not all: the very small movements do not, while chromatin movements larger than 200–300 nm do. The movement of the pore in the envelope is not energy dependent. If we add CCCP, we see that the chromatin stops moving but the pores keep shifting around.

Goldman: But you don't know what the motor is.

Gasser: That is a completely different study. It is probably linked to ATP-dependent chromatin remodellers.

Wilson: There might be a link between telomeres and lamins. Even though telomeres are not anchored at the nuclear envelope in interphase mammalian cells, nevertheless for a few minutes during telophase, early in nuclear envelope assembly, BAF, emerin, LAP2α and A-type lamins co-localize in an area that might correspond to telomeres, according to Roland Foisner. After this brief time, LAP2α and the other proteins become evenly distributed. Something special might happen to telomeres, at this early stage of nuclear envelope assembly.

Gasser: I agree that this could well be the case. There are intranuclear lamins, of course.

Goldman: In mammalian cells heterochromatin is not restricted to the nucleolus.

Gasser: That's true. In many differentiated cells, a lot of the heterochromatin is peripheral. It is also perinucleolar. Sometimes it just aggregates with itself to form a chromocentre. The fact is, there are lots of different kinds of heterochromatin in mammalian cells, and in budding yeast we basically have one type, which is Sir-dependent heterochromatin.

Goldman: It is always at the perimeter of the nucleus.

Gasser: Yes. Sir4 is an integral component of yeast heterochromatin, and it itself has an affinity for Esc1 at the nuclear periphery. In experiments I didn't show you we have popped out a ring of chromatin that moves throughout the nucleoplasm when it is actively transcribed. If it is silent and bound by Sir proteins, the ring moves straight to the periphery and binds Esc1. The heterochromatin that we have in yeast itself has an address: it can target itself to the periphery through its chromatin proteins.

Oshima: Can you comment on human Sir proteins?

Gasser: As far as we know, only Sir2 is conserved in mammalian cells. The Sir complex is made of three unrelated proteins, Sir2, Sir3 and Sir4. Sir2 is an NAD-dependent histone deacetylase (HDAC). There are seven Sir2 family members in human cells. Moreover, there is a Sir2 mutant in *Caenorhabditis elegans* which has an ageing phenotype, as in yeast. But Sir2 proteins aren't always involved in gene repression and many are not even nuclear. Finally, Sir3 has domains which are closely related to a universally conserved protein called ORC1 (origin recognition complex). The core of Sir3 is related to the family of AAA ATPases, but it is not an ATPase. As for Sir4, it has a neurofilament-like N-terminus, and its C-terminus long ago claimed to be lamin-like due to a coiled-coil structure.

Goldman: I just want to mention one thing. Yeast does have an intermediate filament (IF)-like system. These papers were published just over a decade ago by Michael Yaffe who discovered a mutant yeast in which the nuclei and mitochondria could not move into the daughter cells. When he isolated, expressed and cloned this protein it made beautiful IFs in the test-tube. As I recall it had 27% homology to mammalian vimentin. Antibodies to vimentin cross-reacted fully with this, so his antibody would stain beautiful vimentin networks in cultured fibroblasts. About three weeks ago on the front cover of *Cell* there was a report that in *Colibacter* there is an IF-like protein that also has 25% homology to vimentin. It is critical in determining the spiral-form shape of *Colibacter*. I would therefore say that IF-like proteins might exist in the nucleus. It is fascinating to hear that you have something that has a neurofilament-like N-terminus and a lamin-like C-terminus.

Gasser: Obviously, these are proteins that have coiled-coiled domains that form protein–protein interactions and influence chromosome positioning. It is risky, however, to push the homologies so far that you are deducing function based on a common structure. After all, we cannot conclude anything about Sir2 function in mammalian cells on the basis of what we know in yeast, except that it is a NAD-dependent protein deacetylase. I am very conservative about drawing conclusions about function based on structural homology.

Gruenbaum: You described an effect in G1 and S but not in G2. What is the explanation?

Gasser: This is very nice. Telomeres are released in mitosis. They have to be released or the chromosomes can't segregate. It is interesting that Ku doesn't anchor in G2 and M in the basic relocalization anchor. We don't know whether it is the Ku or its target X that gets phosphorylated and modified. We assume it's one or the other.

Warner: Is there any evidence that the effect of Sir2 on yeast lifespan is related to acetylation of histones or other proteins?

Gasser: There is good evidence that the effect on lifespan has to do with the instability of repetitive genomic regions. Not telomeres, but the rDNA. Sir2 is the only one of the Sir proteins that has a secondary function, which is to suppress recombination in the rDNA. In yeast the rDNA is a series of direct 9 kb repeats. Together with another protein in the rDNA Sir2 suppresses recombination of these direct repeats. When they are excised they accumulate as rDNA circles: that is, rDNA repeats pop out as circles. Their accumulation correlates well with ageing and the restriction of lifespan in what we call a mother cell. rDNA instability is regulated by other factors such as SGS1, which is the Werner's helicase homologue in yeast. Still we have to be very careful about these homologies. Because both Sir2 and Werner's homologues affect rDNA stability and ageing in yeast, people assume there must be a link between DNA instability and cellular senescence in mammalian cells. In fact, it is not clear that ageing in human cells has anything to do with rDNA stability. In addition, I think that none of the mammalian Sir2 homologues are localized to the nucleolus, so it is unlikely that the mechanism is conserved.

Hutchison: Is there any evidence that Sir4 mutants will cause accelerated ageing in yeast?

Gasser: No, in fact they do the opposite. Sir2 has two functions. It is bound to telomeres and to large stretches of rDNA. If you mutate Sir4 you release the Sir2 which is bound to subtelomeric regions. As a result, more Sir2 goes to the rDNA, which leads to hyperstabilization of the rDNA repeats. This allows cells to go through more divisions. In fact, this argues that rDNA in stability is truly a natural cause of ageing; it is not just the pathology of a mutant. This is why I stressed a strong link with repeat stability. This may be the aspect for which analogies can be drawn between yeast and mammalian cells. Still one should be careful about inferring a Sir2-dependant mechanism. The molecular events can be very different although the analogous outcome suggests they are the same.

Gerace: So to knock the telomeres off the nuclear envelope you would have to mutate both Sir and Ku. When you do that, what do you get?

Gasser: You get increased chromosome loss in mitosis. In vegetatively growing yeast in lab conditions this has little effect on cell growth, but in an evolutionary sense the effects are disastrous, because these cells can't mate. If you lose mating type silencing, you no longer have a and α cells. Without opposite mating types the species cannot form a diploid zygote, and it therefore dies out quickly.

Gruenbaum: Has anyone tried to do the reverse experiment? This would involve trying to express lamins in yeast and analyzing their effects on telomere movement.

Gasser: There have been two reports. Both were done by Erich Nigg (Enoch et al 1991), partly in collaboration with Paul Nurse and partly with my lab. The overexpression of chicken lamin B2, in either *Schizosaccharomyces pombe* or *Saccharomyces cerevisiae*, leads to a ring of lamin around the nucleus. There was a big debate as to whether this is on the inside or outside of the envelope, and it remains unresolved because no one performed electron microscopy. My prejudice is that at least in budding yeast it was on the outside of the nuclear envelope, and that the lamins polymerized around the envelope. For the *S. pombe* case they argued strongly that it was intranuclear because they could demonstrate Cdk kinase modification of lamin (Enoch et al 1991). Indeed, in fission yeast Cdc2 is activated at mitosis and the lamins are depolymerised. Their presence did not seem to affect growth or gene expression and the cells survived unless lamins were expressed at such high levels that they became toxic.

Goldman: It would be interesting to put a mutant lamin in.

Gasser: I think there are many potentially interesting experiments to do.

Burke: Chromosome 6 isn't always looped, is it?

Gasser: Chromosome 6 is a small chromosome and the telomeres stay together much of the time, but not all the time. Certainly in mitosis the telomeres come apart. A good part of the time the telomeres are close together.

Burke: That may just be a reflection of its size, not any special zipper proteins.

Gasser: We have looked in mutants. In Ku mutants the two telomeres are not so close together. So I think this is an active interaction. Still the mechanism is probably not simple: if it were simply Ku holding telomeres together, then any telomere could interact with every other telomere. Addressing the question of chromosome territories, we have looked at two small chromosomes in yeast, three and six. The two telomeres of each chromosome are close together: their separation is below the level of resolution of the light microscope. However, chromosome 5 is a bigger chromosome, as is chromosome 14, and in these cases right and left telomeres are apart.

Goldman: Are they always in the same position?

Gasser: They tend to have a zone, yes. We use the spindle-pole body as our 'North pole', because it has a fixed position in the nuclear envelope. The nucleolus is also in a fixed position. Then we ask whether the telomeres are in the same place relative to these points. They are in the same general positions, even though they move in a Brownian fashion around a given site.

Wilson: Tom Misteli has evidence that different pairs of chromosomes tend to be close neighbours in different human tissues. Two chromosomes tend to be close neighbours in liver but not in other tissues, for example. These neighbours correlate with translocations that frequently cause cancer in that particular tissue. Do specific telomeres associate with each other in yeast?

Gasser: For a long time we have been trying to identify which yeast telomeres cluster together. This is how we found that 6R and 6L, and 3R and 3L interact. We haven't deduced the rule yet that determines what interacts with what. It sounds as if it should be trivial, but it is not.

Gruenbaum: This was actually done many years go in *Drosophila* salivary gland nuclei by John Sedat's group where they showed no specific association between the arms of different chromosomes.

Gasser: Apart from Tel 6R-6L and Tel 3R-3L, it could be random.

Capeau: You said that there is a $Ku^{-/-}$ mouse which shows advanced ageing. Can you comment on the mechanisms responsible?

Gasser: This is not really understood, although there are models. One is that increased DNA damage causes ageing, because end-joining is the most frequently used repair mechanism for a double-stranded break in mammals. Moreover, Ku-mediated end-joining repairs accurately: Ku binds the double strand break and ensures that there is no degradation of the ends and no accumulation of mutations. There are other mechanisms that will repair double-strand breaks, yet with less precision. Actually, there is a third pathway that repairs with precision (homologous recombination) but it requires contact with the homologous partner sequence. The other end-joining mechanism is inaccurate: endonucleases chew backwards until they encounter a sequence that favours an inaccurate end-joining or recombinational repair event. In brief, one theory is that general genomic instability is part of the ageing process. The other theory is that telomeres shorten more rapidly in Ku mutants, as they do in Werner's syndrome patients, and that this limits cell division capacity. The problem with this model is that mice have between 50 and 100 kb of telomere repeats, making it hard to imagine why a slight reduction in this long repetitive DNA should have any effect on chromosome stability. Some scientists have proposed a mechanism based on the release of heterochromatin proteins bound at subtelomeric regions. The idea is that subtelomeric repetitive DNA is a reservoir for general repressor proteins such as HP1. If one shortens the amount of TG, there is a release of heterochromatin proteins, which then relocalize to repress other genes. These remain hypotheses — how either event really triggers ageing is unclear.

Capeau: Is there some specific tissue activation in these mice?

Gasser: They are immunodeficient, of course. They are grey and wrinkled like old mice, and otherwise I don't know much else about them.

Warner: They get premature osteoporosis.

Gasser: Is that a laminopathy?

Goldman: It has been seen in some of the laminopathies.

Hutchison: Could you speculate generally about why chromosomes occupy territories? I know Wendy Bickmore has put forward this idea that it is to do with silencing and gene activation. Is it possible that chromosomes occupy territory for the purposes of inheritance?

Gasser: I think that could be the case. First, let me say that I think it may be different for lower eukaryotes and for species that differentiate. In differentiating tissues it makes sense that a spatial pattern could be established and inherited as part of the differentiation process. In lower eukaryotes, such as yeast and perhaps flies, I think the spatial positioning of chromosomes has more to do with efficient mitotic segregation. In yeast centromeres are clustered at one position within the nucleus and the telomeres are anchored. The phenotype of spatial disorganization is chromosome loss. Thus, in very simple organisms the positioning may primarily ensure efficient chromosome segregation. In higher eukaryotes it is possibly more interesting. The model is that yeast chromosomal domains might assume fixed positions once they are stably repressed, such as through Polycomb-type repression. These zones then tend to interact with each other, or interact with the nuclear periphery or a peri-nucleolar region. This position often ensures that they replicate late, and this could help propagate their chromatin structure onto newly replicated DNA. For instance, the DNA could be replicated in a zone that is enriched for HDACs or methylases which could recreate a specific structure. Position and repression could cooperate to establish a pattern that could then be inherited. It is unclear whether the propagation or re-establishment of a repressed chromatin state after replication is a result or a cause of its subnuclear position: in yeast both are true. There is some indirect evidence favouring a correlation, but the idea that 3D chromosomal or domain positioning is a reflection of heritable gene patterns must be tested rigorously in a genetically modifiable organism.

Wilson: The concept of chromosome territories can be misleading or confusing, because it sounds like there is something solid, like a wall. Right now, the Wilson territory is wherever I happen to be. Chromosomes are so big their territory is wherever they are. There is no wall that separates these 'territories'. In terms of thinking about inheritance, what you have said is very clear: territories may just reflect where chromosomes end up after each cell division.

Goldman: It would be interesting to know whether the anchorage sites to the nuclear surface were relatively fixed in terms of their position.

Wilson: This is an important question. We don't know how chromosomes, which are huge in mammalian cells, attach to the lamina and/or nuclear envelope.

Gruenbaum: Territories were also found in *Drosophila*. In addition, there are only limited sites that almost always associate with the nuclear envelope including centromeres, telomeres and intercalary heterochromatin.

Gasser: Here I see a bit of confusion. It is true that some chromosomal sites are frequently at the periphery, but it is unclear whether a given locus will always have as its neighbour chromosome X, Y, or Z. This is a different question. I like to think of general subnuclear position as stemming from the heterochromatin up, from certain classes of binding sites, or tissue specific kinds of chromatin. These may interact with binding sites at the periphery — generic binding sites. We have shown in yeast that active chromatin can diffuse anywhere in a nucleus, and this is probably also true in mammals, unless impediments are imposed by nucleoli or other dense structures or so-called nuclear 'bodies'. We are setting up to do this kind of pop-out of the chromatin ring in the mammalian nucleus. My bet is that it will diffuse all over the nucleus.

Goldman: It would be interesting to look at this as a function of the cell cycle. My guess is that in late prophase prior to nuclear envelope breakdown you will be able to see chromosomes inside the interphase nucleus, and they all look like they have attachments. Getting ready for mitosis may be a different story from the rest of interphase.

Gasser: There are clearly attachment sites. This is what we are saying. This is different from saying that a whole chromosome has an address, like a grid on a city map.

Goldman: So we are saying that the attachment sites might be relatively fixed and stable.

Gasser: I don't know whether these sites are lamins or emerin or other LEM proteins, but they are probably dispersed at many different sites around the nuclear periphery. It is not as if chromosome 11R has to go to one unique site. Nonetheless, you can ask whether one detects reproducible positioning. Yes, we do see it in that the ends of chromosome 6 will tend to be in one subregion of the nucleus. In yeast, I think this is because the centromeres are clearly attached to the spindle pole body, which is positioned relative to the site of bud emergence. There is a limited distance a telomere can go from that depending on the length of the chromosome arm and the average compaction rate. This may occur without having a telomere-specific attachment site. With respect to mammalian nuclei, when you compare chromosome positions within nuclei from differentiated tissues, you do not see nearest neighbour patterns that are reproducible.

Wilson: Tom Misteli is starting to find such patterns (Roix et al 2003).

Gasser: In all tissues?

Wilson: No. In one tissue there is a certain pair of chromosomes typically next to each other. I don't know whether he has addressed the other chromosomes, but if you look at that same pair they are not next to each other in other tissues. This might add new information about how chromosomes are organized in three dimensions.

Gruenbaum: Although chromosomes do not tend to be specifically organized with respect to one another, we cannot rule out that they have a limited number of positions. This makes the analysis much more complicated because it is hard to get statistically meaningful data.

Goldman: The chromosome painting technique is interesting. You have to remember that 25–30 years ago the nucleus was thought to be a bowl of spaghetti and that chromatin was thought to be disorganized. Now at least we are saying that there is some order and semblance of organization. Perhaps it is the anchorage to the envelope that is the important organizing principle, as has been suggested for many years.

Gruenbaum: In *Drosophila* salivary glands, both arms of the same chromosome are preferentially positioned next to each other. This is also true for mammalian cells.

Wilson: That is the Rabl configuration: how chromosomes are positioned after they've been dragged apart during mitosis.

Gasser: This can change as G1 phase progresses.

Goldman: It is quite obvious that the chromosome territories would have to be different in these laminopathies because they show gross abnormalities in nuclear shape. Anything inside the nucleus is also going to be altered. What we're really interested in is whether these alterations are critical with respect to changing patterns of gene expression.

Gasser: The interesting question is, are things disorganized because the lamins and emerins are absent there and the proper heterochromatin structure is not setup? This is basically deregulation of genes first and then misorganization. Or is it the direct effect of silent or active genes from the periphery?

Goldman: There are many point mutations in lamins, and these could affect the structure of the intermediate filament polymer which in turn could alter the attachments that are critical in organizing chromatin structure.

Ellis: This work has actually been done by Wendy Bickmore. She has taken the X-linked and the autosomal dominant mutations for EDMD and shown that there are no territory changes for some of them. But the mobility for a few mutations is definitely altered.

Gasser: In other words, she is seeing differences in gene expression. That is a question of the mobility of individual genes and domains, not chromosomes.

Ellis: Yes, but it was mutation specific. She has looked at three so far.

Wilson: It sounds like GFP-fused lamin might behave differently.

Gasser: Has anyone done a whole microarray on the genes deregulated in any of these specific mutations?

Goldman: It has been done with progeria.

Stewart: We have done a microarray analysis on the soleus muscle from lamin-deficient mice. The changes in gene expression were very modest and those genes in which expression was increased or decreased didn't tell us much. The soleus muscle is already showing signs of dystrophy and weakness: we don't know whether the changes in gene expression are really a consequence of the pathology occurring in the muscle. Intriguingly, there were a group of genes that we picked up in the soleus that were co-ordinately deregulated in embryonic fibroblasts made from the same mice. I'm not sure what this means. In terms of overall changes, if there are, they are very subtle and may be in some specific genes that we haven't yet zeroed in on.

Gasser: Are you doing this in tissues, or are you doing it in cultured cells?

Stewart: We have done both. In fact, we did an analysis of embryonic fibroblasts from our progeric line of mice and wild-type fibroblasts but failed to find a single statistically significant difference.

References

Enoch T, Peter M, Nurse P, Nigg EA 1991 p34cdc2 acts as a lamin kinase in fission yeast. J Cell Biol 112:797–807

Roix JJ, McQueen PG, Munson PJ, Parada LA, Misteli T 2003 Spatial proximity of translocation-prone gene loci in human lymphomas. Nat Genet 34:287–291

A-type lamin-linked lipodystrophies

Corinne Vigouroux and Jacqueline Capeau[1]

INSERM U.402, Faculty of Medicine Saint-Antoine and Department of Biochemistry, Tenon Hospital, Pierre and Marie Curie University, 27 rue Chaligny, 75571 Paris Cedex 12, France

Abstract. Lipodystrophies represent a group of diseases characterized by altered body fat repartition and major metabolic alterations with insulin resistance. Genetic forms of partial lipodystrophy are currently recognized as two syndromes with subcutaneous lipoatrophy but preserved or increased fat at the level of face and neck (Dunnigan syndrome or FPLD due to *LMNA* mutations) and/or abdomen (PPARγ-linked forms) and are both transmitted as dominant diseases. FPLD is further characterized by muscular hypertrophy, hyperandrogenism, acanthosis nigricans, hepatomegaly with steatosis and at the biological level, marked hypertriglyceridaemia, low HDL cholesterol, insulin resistance and altered glucose tolerance or diabetes. These signs occur after puberty and their prevalence and severity are more marked in female than in male patients. At the genetic level, *LMNA* mutations concern in most cases the type-A lamin C-terminal domain and more than 80% are heterozygous substitutions located at position 482 (R482W/Q/L). The other locations are G465D, K486N, R582H and R584H. The presence of signs evocative of limb-girdle muscular dystrophy has been reported in patients with typical forms of FPLD. In addition, forms presenting with lipodystrophy and myopathy have been reported for patients with mutations at position R28W, R60G, R62G or R527P. In addition, lipodystrophy, either partial or generalized, can be associated with syndromes of premature ageing like Hutchinson-Gilford progeria or acromandibular dysplasia, but also with other phenotypes, as we described in a patient bearing the *LMNA* R133L heterozygous substitution.

2005 Nuclear organization in development and disease. Wiley, Chichester (Novartis Foundation Symposium 264) p 166–182

Adipose tissue biology

Lipodystrophies are characterized by an altered distribution and/or amount of body fat. They are associated with alterations in metabolic parameters associated with insulin resistance. Such an association of clinical and biological alterations was reported not only in human diseases but also in different models of transgenic animals and points to an important role for adipose tissue in metabolism.

---

[1]This paper was presented at the symposium by Jacqueline Capeau to whom all correspondence should be addressed.

Adipocyte differentiation requires the chronological activation of a network of transcription factors. Factors from the C/EBP family are required to initiate the differentiation process and to maintain the differentiated phenotype in the later stages. The factor SREBP1 participates in the differentiation process by activating the central factor peroxisome proliferator-activated receptor (PPAR)γ directly and indirectly, and is involved in the response to insulin. PPARγ is required to acquire differentiation and increase insulin sensitivity. Thiazolidinediones, pharmacological agonists of PPARγ, are used in diabetic patients to decrease insulin resistance. Adipocytes play a major role in energy metabolism since they store triglycerides after meals under the control of insulin and release free fatty acids and glycerol long after meals in response to catecholamines. Fatty acids are used by most tissues and in particular by the liver and muscles to recover energy, while glucose is mainly produced by the liver to feed the brain and glucose-dependent tissues. In addition to its metabolic role, adipose tissue is a major endocrine tissue which secretes numerous cytokines and hormones (named adipokines) and various proteins (Rajala & Scherer 2003). In particular, adipocytes secrete leptin which controls the body energy level and adiponectin which acts at the level of the liver and muscles. These adipokines are able to increase fatty acid oxidation in these tissues, decrease hepatic glucose production and increase muscle glucose utilization by activating the energy sensitive enzyme AMP-kinase, thus acting as an insulin-sensitizer. In the states of lipodystrophy and insulin resistance, the pattern of secreted adipokines appears to be altered: adiponectin secretion is decreased while that of interleukin 6 (IL6) and tumour necrosis factor α (TNFα) is increased. These two latter cytokines act mainly through autocrine/paracrine mechanisms and induce fat insulin-resistance resulting in an increased production of free fatty acids (lipolysis being a pathway inhibited by insulin). The increased levels of fatty acids together with the decreased levels of adiponectin and possibly leptin are probably responsible for an accumulation of fatty acid derivatives (probably as acyl-CoA) together with triglyceride storage in the liver and muscles, which is toxic and induces insulin resistance (i.e. lipotoxicity). In addition, liver synthesis and secretion of triglycerides as very low density lipoprotein (VLDL) are increased while triglyceride clearance by adipose tissue is decreased leading to hypertriglyceridaemia. The pathophysiology of most human lipodystrophies is still unknown. However, murine models of lipoatrophic diabetes have revealed that primary genetic alterations in fat development result in diabetes and dyslipidaemia and that diabetes could be reversed by fat transplantation in one model (Reitman et al 2000). Leptin and adiponectin deficiencies, due to the absence of adipose tissue, are important determinants of metabolic abnormalities since exogenous administration or transgenic overexpression of leptin or adiponectin has been shown to markedly improve insulin sensitivity, glycaemic control, dyslipidaemia and hepatic steatosis in these mice with a complete reversion of the metabolic disorders when the two adipokines were added together further indicating the major role of adipokines in the control of insulin sensitivity (for a review see Rajala & Scherer 2003). These pathophysiological mechanisms help to understand why these clinical and biological features are associated in most cases in the syndromes of lipodystrophy whatever their origin (Reitman et al 2000).

Clinical forms of lipodystrophies

Lipodystrophies represent a heterogeneous group of diseases characterized by generalized or partial alterations in body fat development or distribution and insulin resistance. The other cardinal clinical signs of these syndromes are acanthosis nigricans, which is a skin disorder associated with insulin resistance, frequent hyperandrogenism in females, muscular hypertrophy and liver steatosis. Insulin resistance is associated with a progressive altered glucose tolerance leading to diabetes, and with hypertriglyceridaemia (Garg 2000a, Reitman et al 2000). Diabetes is generally difficult to treat due to insulin resistance and rapidly leads to diabetic complications. Acute pancreatitis due to severe hypertriglyceridaemia, and liver cirrhosis arising from the frequent nonalcoholic steatohepatitis are also responsible for the morbidity and mortality of these diseases. However, there is a wide range of severity among the patients from the rare and serious congenital generalized form to the milder acquired partial one.

The main forms of lipodystrophies are classified according to their origin, either genetic or acquired, and to the clinical pattern of the lipoatrophy, either generalized or partial (Table 1) (Reitman et al 2000, Hegele 2003). Genetic forms with generalized lipoatrophy are transmitted according to a recessive pattern. In the typical Berardinelli-Seip congenital lipodystrophy (BSCL), mutations in two genes are responsible for most cases (Magré et al 2003): BSCL1 is caused by mutations in the *AGPAT2* gene which encodes the enzyme 1-acylglycerol-3-phosphate-acyl transferase 2 responsible for the synthesis of phosphatidic acid from lysophosphatidic acid in the pathway of triglyceride synthesis (Agarwal et al 2002). *BSCL2* encodes the protein seipin (of which the function is unknown) expressed in numerous tissues, in particular in the brain and testes (Magré et al 2001). There might be other loci for BSCL but they would account only for a minority of cases (Magré et al 2003).

In addition, a complete lipoatrophy was found in some patients with mandibuloacral dysplasia (MAD) and a mutation in the *ZMPSTE24* gene, encoding a zinc metalloproteinase probably involved in prelamin processing, was discovered in one of these patients (Agarwal et al 2003). Otherwise, in some patients with the R133L *LMNA* mutation, a complete lipoatrophy was

TABLE 1 Classification of the main syndromes of lipodystrophy (LD)

	Genetic forms					Acquired forms	
	Generalized lipoatrophy Berardinelli-Seip synd. BSCL	Dunnigan-type LD FPLD	Kobberling-type LD	LIRLLC	PPARG-linked LD	Generalized lipoatrophy Lawrence syndrome	HIV-linked LD
	Complete lack of fat	Lack of fat in the limbs, buttocks and trunk increased fat in the face and neck	Lack of fat in the limbs and buttocks increased fat in trunk	Generalized lipoatrophy insulin-resistant diabetes leuko-melanodermic papules liver steatosis cardiomyopathy premature aging	Lack of fat in the limbs and buttocks preserved fat in trunk	Complete lack of fat	Lipoatrophy and/or visceral fat accumulation
	Onset at birth or early infancy	Clinical signs after puberty	After puberty	Young adulthood	Clinical signs in adults hyperinsulinaemia in children	Infancy or adulthood	Related to treatment
	Autosomal recessive	Autosomal dominant	Generally unknown	Probably autosomal dominant	Autosomal dominant	Acquired	Acquired
	Severe insulin resistance	Severe insulin resistance	Severe insulin resistance	Severe insulin resistance	Severe insulin resistance	Severe insulin resistance	Variable insulin resistance
	Severe hyperTG diabetes	Severe hyperTG diabetes	Severe hyperTG diabetes	HyperTG diabetes	HyperTG diabetes severe hypertension	Severe hyperTG diabetes	VariableTG and glycemia
	Mutations in BSCL1 encoding AGPAT2 and BSCL2 encoding seipin	Mutations in LMNA exons 8 and 11 R482W/Q/L, K486N G465D, R582H, R584H	In general no mutations in LMNA or PPARγ 1 mutation in ZMPSTE241 mutation in LMNA	Mutations in LMNA R133L	Mutations in PPARγ V290M, F388L, R425C, P467L	Unknown	Role of drugs PI and NRTI

observed together with other alterations leading to the identification of a new laminopathy (Caux et al 2003, see below).

In genetic syndromes of partial lipodystrophy, mutations in two genes have been reported up to now. LMNA is responsible for the familial partial lipodystrophy of the Dunnigan-type (FPLD). Otherwise, in a few patients, dominant negative mutations in PPARγ result in a phenotype of 'metabolic syndrome' with partial lipodystrophy affecting limbs and buttocks but sparing abdominal subcutaneous fat, severe insulin resistance, diabetes, and severe hypertension. Four heterozygous mutations have been reported: V290M, F388L, R425C and P467L (see Hegele 2003 review).

In a family with partial lipodystrophy and a different phenotype from FPLD, resembling the Kobberling-type of partial lipodystrophy with preserved fat at the truncal level, a R582H LMNA mutation was observed (Garg et al 2001).

Some forms of acquired lipodystrophies could have an immunological basis: autoimmune diseases are sometimes associated with sporadic cases of generalized lipoatrophies (Lawrence syndrome). The C3 nephritic factor, an IgG antibody against complement components, is demonstrable in some cases of partial lipodystrophy of the Barraquer-Simons type (lipoatrophy of face and trunk, with excess accumulation of fat in the lower part of the body) (Reitman et al 2000, Garg 2000a).

Finally, probably the most frequent form of lipodystrophy is the redistribution of fat that occurs in HIV-infected patients under antiretroviral therapy. These patients frequently present a peripheral lipoatrophy and/or increased visceral adipose tissue, and could develop hypertriglyceridaemia and insulin resistance (Chen et al 2002). Concerning the HIV-linked lipodystrophy syndrome, several authors pointed out the deleterious effects of some antiretroviral treatments on adipogenesis (Caron et al 2001, Chen et al 2002). We have found that *in vitro* the antiretroviral class of protease inhibitors is able to inhibit prelamin maturation in cultured adipocytes resulting in abnormal lamina structure and altered nuclear stability. These data suggest that protease inhibitors could induce an acquired laminopathy (Caron et al 2003).

It is interesting to note that the very common syndrome of insulin resistance or metabolic syndrome, largely prevalent in developed countries, could represent a mild form of acquired lipodystrophy since it involves android repartition of fat, glucose intolerance, hypertension, dyslipidemia and liver steatosis. This common condition represents a strong risk factor for cardiovascular disease (Hegele 2003).

LMNA mutations resulting in lipodystrophy

Mutations causing typical FPLD cluster only in exons 8 and 11 of *LMNA*, coding for the globular C-terminal domain of type A-lamins (Cao & Hegele 2000,

Shackleton et al 2000, Speckman et al 2000, Vigouroux et al 2000, Hegele et al 2000, Garg et al 2001). The most frequent mutation substitutes a basic amino acid at position 482 (arginine) for a neutral one (tryptophan, glutamine, leucine). All the patients are heterozygous for these mutations. They occur in several haplotypes in the different families, suggesting that codon 482 is a hot spot for mutations. Other mutations also induce a complete or partial loss of a positive charge (from lysine for K486N substitution, or from arginine in the mutations R582H or R584H which affect exon 11) or the appearance of a negative charge (aspartate in the G465D mutation). The structure of the C-terminal domain of lamin A was characterized and found to contain an Ig-like structure: all the mutations observed in FPLD patients were found to be localized to the domain surface while mutations responsible for cardiac and skeletal muscle abnormalities were located inside the domain (Dhe-Paganon et al 2002, Krimm et al 2002).

Alterations in exon 11 are rare, and there are only two reported families for the R584H alteration (Hegele et al 2000, Vigouroux et al 2000) and only one for the R582H substitution (Speckman et al 2000). They specifically affect the A isoform of lamin.

In patients with the A-type phenotype of MAD with a pattern of partial lipodystrophy, loss of subcutaneous fat from the extremities and normal or slight excess fat in the neck and truncal regions, the homozygous R527H mutation in *LMNA* was discovered in all cases (Simha et al 2003, Novelli & d'Apice 2003).

In a patient with premature ageing and complete lipoatrophy together with insulin resistance classified as LIRLLC, a heterozygous R133L *LMNA* mutation was found (Caux et al 2003).

Studies of *LMNA* in other metabolic disorders have been performed. We excluded *LMNA* mutations as responsible for generalized lipodystrophy (Vigouroux et al 2000). No mutations in exon 8 of *LMNA* have been found in subjects with HIV therapy-associated lipodystrophy (Behrens et al 2000). Hegele et al (2001) suggested that a common variation in *LMNA* could be associated with obesity-related phenotypes. However, there is at present no evidence for an association of this variant with type 2 diabetes (Wolford et al 2001).

Clinical alterations observed in A-type lamin-linked lipodystrophies

FPLD, dominantly inherited, is a rare disease characterized by the disappearance, after puberty, of subcutaneous adipose tissue in the limbs, buttocks, abdomen and trunk. This progressive lipoatrophy spares the neck and face, where adipose tissue can accumulate, frequently causing a cushingoid appearance. However, in lean patients, this latter feature can be lacking, and differential diagnosis with total acquired lipoatrophy can be difficult if the familial dominant transmission of the disease is not evident. Prominence of muscles and superficial veins is partly due to lipoatrophy, but, as in the other syndromes of insulin resistance, a real muscular hypertrophy is present. The android aspect of patients is particularly striking in females, but does not systematically draw attention in males, in which this condition is frequently unrecognized. Garg et al (1999) performed magnetic resonance imaging studies in four affected patients, three females and one male and confirmed the near-absence of fat in extremities and gluteal regions, its reduction in truncal area, and increase in neck, face and in the visceral area. In females, breasts have markedly reduced subcutaneous fat whereas adipose tissue accumulates in labia majora (Garg et al 1999). The partial adipose tissue loss in FPLD is associated with reduced plasma leptin and adiponectin levels (Hegele 2003).

Metabolic alterations associated with FPLD are responsible for the severity of the disease with insulin resistance, generally attested by hyperinsulinaemia with concomitant normal or elevated glycaemia. Clinically, the presence of acanthosis nigricans, brownish hyperkeratotic skin localized to the axillary and inguinal regions, is associated with the insulin resistant state. When hyperinsulinaemia no longer compensates for insulin resistance and glucose intolerance, then diabetes occurs.

Dyslipidaemia is frequent among FPLD patients. Hypertriglyceridaemia, due to an elevated level of VLDL, is the more prevalent feature. When severe, it can lead to a life-threatening acute pancreatitis. Other perturbations of the lipid profile can also be associated with FPLD, for example decreased high-density lipoprotein (HDL) with or without elevated total cholesterol (Schmidt et al 2001). These patients accumulate numerous cardiovascular risks: visceral fat accumulation, hypertension, diabetes, hyperinsulinaemia, high triglycerides, low HDL, hypoadiponectinaemia, elevated free fatty acids (FFAs), elevated C-reactive protein (CRP), particularly in women, leading to early coronary heart disease (Hegele et al 2003). Like in other lipodystrophic syndromes, liver steatosis, with its risk of evolution towards cirrhosis, is often observed in these patients.

Although this disease affects males and females, both clinical traits and metabolic complications are more severe in women (Vigouroux et al 2000, Garg 2000b). In addition, women affected by FPLD frequently complain of hirsutism, and a polycystic ovary syndrome with ovarian dysfunction and hyperandrogenism is usual.

Two sisters affected by the R582H substitution have a less severe loss of subcutaneous adipose tissue and milder metabolic abnormalities reminiscent of the Kobberling-type partial lipodystrophy (Garg et al 2001). However, such an attenuated phenotype was not observed in patients harbouring the very similar R584H mutation (Vigouroux et al 2000, Hegele et al 2000). Environmental factors, such as diet and physical activity, are probably important determinants of the severity of metabolic complications (Hegele et al 2000).

The treatment of FPLD is difficult. Appropriate diet and physical training are important to minimize metabolic alterations. However, diabetes mellitus, which appears secondarily during the disease, usually requires large doses of insulin. Insulin sensitizers, like metformin, could improve the control of glycaemia. PPARγ agonists, such as thiazolidinediones, which promote both adipocyte differentiation and insulin sensitivity, seem a promising treatment for lipodystrophic syndromes. A patient was treated with thiazolidinedione roziglitazone and an increase in subcutaneous fat was observed together with a decrease in insulin resistance. However, glycaemic control did not improve and lipid profile even worsened (Owen et al 2003). Some patients received a leptin treatment which was able to improve metabolic parameters but not lipodystrophy (Oral et al 2002).

Mandibuloacral dysplasia (MAD) is characterized by mandibular and clavicular hypoplasia, acro-osteolysis, joint contractures and skin alterations together with partial lipodystrophy sparing the trunk. Mutations in *LMNA* are always homozygous R527H (Novelli et al 2003). Patients with MAD and a more generalized loss of subcutaneous fat involving the trunk have no mutation in *LMNA* but one patient had a ZMPSTE24 mutation (Agarwal et al 2003).

A new clinical condition, involving generalized lipoatrophy, insulin resistant diabetes, hypertriglyceridaemia, hepatic steatosis, hypertrophic cardiomyopathy with valvular involvement and disseminated leukomelanodermic papules, (LIRLLC) was described in a patient with heterozygous R133L *LMNA* mutation (Caux et al 2003). The same mutation was reported in two other patients considered to have atypical Werner syndrome (Chen et al 2003, Vigouroux et al 2003).

Skeletal and cardiac muscle symptoms
in lamin-linked lipodystrophies

In most cases, there is no association between skeletal or cardiac muscular signs and lipodystrophy or metabolic abnormalities in patients bearing *LMNA* mutations. However, in some patients with FPLD due to the *LMNA* R482W alteration, muscular signs compatible with LGMD1B were present (Vanthygem et al 2004, personal communication). A cardiac septal hypertrophy can be observed in FPLD patients but could be secondary to diabetes and hypertension.

Four *LMNA* mutations associating with lipodystrophy and muscular alterations have been published. One patient with R60G *LMNA* mutation presented with typical partial lipodystrophy and metabolic alterations together with heart failure due to cardiomyopathy. Patients from two unrelated families with a heterozygous R527P *LMNA* mutation had features of partial lipodystrophy and minor metabolic alterations together with features of EDMD and cardiac rhythm disturbances (van der Kooi et al 2002). Patients from one family with R28W *LMNA* mutation had signs of myopathy and cardiac manifestations together with signs of FPLD. In another pedigree with signs of FPLD and conduction system defects and cardiomyopathy, a R62G *LMNA* mutations was observed (Garg et al 2002).

Cellular alterations observed in cells from patients with lamin-linked lipodystrophies

We performed a study of skin fibroblasts of FPLD patients with R482W and R482Q *LMNA* mutations (Vigouroux et al 2001). Protein expression of type A and type B lamins, LAP2β and emerin was normal in the whole population of fibroblasts. However, about 5–25% of these cells had abnormal blebbing nuclei with A-type lamins forming a peripheral meshwork, which was frequently disorganized. Emerin strictly colocalized with this abnormal lamin A/C meshwork. Cells from lipodystrophic patients often had other nuclear envelope defects, mainly consisting of nuclear envelope herniations that were deficient in B-type lamins, and eventually in nuclear pore complexes and LAP2β. Furthermore, the heterogeneous staining by DAPI suggests that chromatin was decondensed in nuclear areas flanking disorganized nuclear envelope domains. The mechanical properties of nuclear envelopes were altered, as judged from the low stringency of extraction of nuclear envelope proteins. These structural nuclear alterations were caused by the lamin A/C mutations, since the same changes were introduced in human control fibroblasts by expression of R482W mutated lamin A. However, despite these abnormalities, we showed that the fibroblasts from FPLD patients were euploid and able to cycle and divide. In a recent study, Capanni et al (2003) reported that fibroblasts from a FPLD patient carrying the R482L mutation presented lamin A/C aggregates close to the nuclear lamina and that emerin did not colocalize with these aggregates. Interestingly, emerin failed to interact with lamin A in R482L mutated fibroblats *in vivo* while the interaction with lamin C was preserved *in vitro*.

Transfection studies of mutant *LMNA* alleles have been performed in several cell types. In C2C12 myoblasts, there were no gross abnormalities in the localization or stability of lamin A-containing mutations found in FPLD (Östlund et al 2001). However, structural abnormalities in the nuclear envelope and chromatin similar to those observed in patients' skin fibroblasts were generated in fibroblasts, myoblasts and preadipocytes mouse cell-lines overexpressing lamin A harbouring missense mutations at codon 482 and 453 (responsible for EDMD). The occurrence of nuclear abnormalities was reduced when lamin B1 was coexpressed with mutant lamin A (Favreau et al 2003).

The observation of heterogeneous, but very similar nuclear alterations in cells from FPLD or EDMD patients, from $Lmna^{-/-}$ mice and from patients with other laminopathies affecting lamin A, confirms that lamin A is a major determinant of the nuclear architecture. However, the pathophysiology of these diseases remains unknown.

The mutational hot-spot found in the FPLD-linked phenotype suggests a more specific alteration of a lamin A function that could only be expressed at the adipocyte level. The systematic alteration of the charge of an amino acid in the C terminal domain of the protein could alter binding properties of lamin A/C. The recent crystallization of the C-terminal end of lamin A showed that it is spatially organized as an immunoglobulin-like domain and that mutations in amino acids at positions 465, 482 and 486, found in FPLD, are localized at the external surface of the structure (Dhe-Paganon et al 2002, Krimm et al 2002). Immunoglobulin folds are known to mediate protein–protein, protein–lipid and protein–DNA interactions. Interestingly, an interaction between lamin A/C and the adipogenic transcription factor SREBP1 has been reported *in vitro* (Lloyd et al 2002). It has been recently found that the C-terminal domain of lamin A/C contains a DNA binding domain which interacts non-specifically with DNA (Stierlé et al 2003). When lamin A/C is mutated at position 482 the interaction with SREBP1 is reduced as is the interaction with DNA. This could suggest that lamin A/C is required for the proper localization of some DNA binding proteins on chromatin in cell lineages issued from mesenchymal stem cells.

References

Agarwal AK, Arioglu E, De Almeida S et al 2002 AGPAT2 is mutated in congenital generalized lipodystrophy linked to chromosome 9q34. Nat Genet 31:21–23

Agarwal AK, Fryns JP, Auchus RJ, Garg A 2003 Zinc metalloproteinase, ZMPSTE24, is mutated in mandibuloacral dysplasia. Hum Mol Genet 12:1995–2001

Behrens GM, Lloyd D, Schmidt HH et al 2000 Lessons from lipodystrophy: *LMNA*, encoding lamin A/C, in HIV therapy-associated lipodystrophy. AIDS 14:1854–1855

Cao H, Hegele RA 2000 Nuclear lamin A/C R482Q mutation in Canadian kindreds with Dunnigan-type familial partial lipodystrophy. Hum Mol Genet 9: 109–112

Capanni C, Cenni V, Mattioli E et al 2003 Failure of lamin A/C to functionally assemble in R482L mutated familial partial lipodystrophy fibroblasts: altered intermolecular interaction with emerin and implications for gene transcription. Exp Cell Res 291:122–134

Caron M, Auclair M, Vigouroux C et al 2001 The HIV protease inhibitor indinavir impairs sterol regulatory element- binding protein-1 intranuclear localization, inhibits preadipocyte differentiation, and induces insulin resistance. Diabetes 50:1378–1388

Caron M, Auclair M, Sterlingot H et al 2003 Some HIV protease inhibitors alter lamin A/C maturation and stability, SREBP-1 nuclear localization and adipocyte differentiation. AIDS 17:2437–2444

Caux F, Dubosclard E, Lascols O et al 2003 A new clinical condition linked to a novel mutation in lamins A and C with generalized lipoatrophy, insulin-resistant diabetes, disseminated leukomelanodermic papules, liver steatosis, and cardiomyopathy. J Clin Endocrinol Metab 88:1006–1013

Chen D, Misra A, Garg A 2002 Lipodystrophy in human immunodeficiency virus-infected patients. J Clin Endocrinol Metab 87:4845–4485

Dhe-Paganon S, Werner ED, Chi YI, Shoelson SE 2002 Structure of the globular tail of nuclear lamin. J Biol Chem 277:17381–17384

Chen L, Lee L, Kudlow BA et al 2003 LMNA mutations in atypical Werner's syndrome. Lancet 362:440–445

Favreau C, Dubosclard E, Ostlund C et al 2003 Expression of lamin A mutated in the carboxyl-terminal tail generates an aberrant nuclear phenotype similar to that observed in cells from patients with Dunnigan-type partial lipodystrophy and Emery-Dreifuss muscular dystrophy. Exp Cell Res 282:14–23

Garg A 2000a Lipodystrophies. Am J Med 108:143–152

Garg A 2000b Gender differences in the prevalence of metabolic complications in familial partial lipodystrophy (Dunnigan variety). J Clin Endocrinol Metab 85:1776–1782

Garg A, Peshock RM, Fleckenstein JL 1999 Adipose tissue distribution pattern in patients with familial partial lipodystrophy (Dunnigan variety). J Clin Endocrinol Metab 84:170–174

Garg A, Vinaitheerthan M, Weatherall PT, Bowcock AM 2001 Phenotypic heterogeneity in patients with familial partial lipodystrophy (Dunnigan variety) related to the site of missense mutations in lamin A/C gene. J Clin Endocrinol Metab 86:59–65

Garg A, Speckman RA, Bowcock AM 2002 Multisystem dystrophy syndrome due to novel missense mutations in the amino-terminal head and alpha-helical rod domains of the lamin A/C gene. Am J Med 112:549–555

Hegele RA 2003 Monogenic forms of insulin resistance: apertures that expose the common metabolic syndrome. Trends Endocrinol Metab 14:371–377

Hegele RA, Cao H, Anderson CM et al 2000 Heterogeneity of nuclear lamin A mutations in Dunnigan-type familial partial lipodystrophy. J Clin Endocrinol Metab 85:3431–3435

Hegele RA, Huff MW, Young TK 2001 Common genomic variation in *LMNA* modulates indexes of obesity in Inuit. J Clin Endocrinol Metab 86:2747–2751

Hegele RA, Kraw ME, Ban MR, Miskie BA, Huff MW, Cao H 2003 Elevated serum C-reactive protein and free fatty acids among nondiabetic carriers of missense mutations in the gene encoding lamin A/C (LMNA) with partial lipodystrophy. Arterioscler Thromb Vasc Biol 23:111–116

Krimm I, Ostlund C, Gilquin B et al 2002 The Ig-like structure of the C-terminal domain of lamin A/C, mutated in muscular dystrophies, cardiomyopathy, and partial lipodystrophy. Structure (Camb) 10:811–823

Lloyd DJ, Trembath RC, Shackleton S 2002 A novel interaction between lamin A and SREBP1: implications for partial lipodystrophy and other laminopathies. Hum Mol Genet 2002 11:769–777

Magré J, Delépine M, Khallouf E et al 2001 Identification of the gene altered in Berardinelli-Seip congenital lipodystrophy on chromosome 11q13. Nat Genet 28:365–370

Magré J, Delepine M, Van Maldergem L et al 2003 Prevalence of mutations in AGPAT2 among human lipodystrophies. Diabetes 52:1573–1578

Novelli G, d'Apice MR 2003 The strange case of the "lumper" lamin A/C gene and human premature ageing. Trends Mol Med 9:370–375

Oral EA, Simha V, Ruiz E et al 2002 Leptin-replacement therapy for lipodystrophy. N Engl J Med 346:570–578

Östlund C, Bonne G, Schwartz K et al 2001 Properties of lamin A mutants found in Emery-Dreifuss muscular dystrophy, cardiomyopathy and Dunnigan-type partial lipodystrophy. J Cell Sci 114:4435–4445

Owen KR, Donohoe M, Ellard S, Hattersley AT 2003 Response to treatment with rosiglitazone in familial partial lipodystrophy due to a mutation in the LMNA gene. Diabet Med 20:823–827

Rajala MW, Scherer PE 2003 Minireview: the adipocyte — at the crossroads of energy homeostasis, inflammation, and atherosclerosis. Endocrinology 144:3765–3773

Reitman ML, Arioglu E, Gavrilova O et al 2000 Lipoatrophy revisited. Trends Endocrinol Metab 11:410–416

Schmidt HH, Genschel J, Baier P et al 2001 Dyslipemia in familial partial lipodystrophy caused by an R482W mutation in the *LMNA* gene. J Clin Endocrinol Metab 86: 2289–2295

Simha V, Agarwal AK, Oral EA et al 2003 Genetic and phenotypic heterogeneity in patients with mandibuloacral dysplasia-associated lipodystrophy. J Clin Endocrinol Metab 88:2821–2824

Shackleton S, Lloyd DJ, Jackson SN et al 2000 *LMNA*, encoding lamin A/C, is mutated in partial lipodystrophy. Nat Genet 24:153–156

Speckman RA, Garg A, Du F et al 2000 Mutational and haplotype analyses of families with familial partial lipodystrophy (Dunnigan variety) reveal recurrent missense mutations in the globular C-terminal domain of lamin A/C. Hum Genet 66:1192–1198

Stierle V, Couprie J, Ostlund C et al 2003 The carboxyl-terminal region common to lamins A and C contains a DNA binding domain. Biochemistry 42:4819–4828

van der Kooi AJ, Bonne G, Eymard B et al 2002 Lamin A/C mutations with lipodystrophy, cardiac abnormalities, and muscular dystrophy. Neurology 59:620–623

Vantyghem MC, Pigny P, Maurage CA et al 2004 Patients with familial partial lipodystrophy of the dunnigan type due to a *LMNA* R482W mutation show muscular and cardiac abnormalities. J Clin Endocrinol Metab, in press

Vigouroux C, Magré J, Vantyghem MC et al 2000 Lamin A/C gene: sex-determined expression of mutations in Dunnigan-type familial partial lipodystrophy and absence of coding mutations in congenital and acquired generalized lipoatrophy. Diabetes 49:1958–1962

Vigouroux C, Auclair M, Dubosclard E et al 2001 Nuclear envelope disorganization in fibroblasts from lipodystrophic patients with heterozygous R482Q/W mutations in lamin A/C gene. J Cell Sci 114:4459–4468

Vigouroux C, Caux F, Capeau J et al 2003 LMNA mutations in atypical Werner's syndrome. Lancet 362:1585

Wolford JK, Hanson RL, Bogardus C et al 2001 Analysis of the lamin A/C gene as a candidate for type II diabetes susceptibility in Pima Indians. Diabetologia 44:779–782

DISCUSSION

Young: In your Western blots on the protease inhibitor, did you use an N-terminal antibody? It didn't look like you saw a doublet band of lamin A and pre-lamin A. You saw the accumulation with the C-terminal antibody, but did you see it also with the N-terminal antibody?

Capeau: To perform these experiments we used an antibody directed against an internal part of lamin A which recognizes lamin A and C as pre-lamin A. We do not see a doublet for lamin A and pre-lamin A but a broad band in accordance with the results obtained in the *Zmpste24* deficient mice (Pendas et al 2002). To test the pre-lamin A presence we used the Santa Cruz antibody used by Pendas (see above) which appeared to be specific for pre-lamin A.

Young: So do you see a doublet band consisting of both mature lamin A and pre-lamin A, or is all of it pre-lamin A?

Capeau: No there is no doublet. However, both lamin A and pre-lamin A are present in these cells as indicated by the experiments of sequential extraction of the nuclear proteins.

Goldman: Do you use the pre-lamin A-specific antibody raised against the C-terminal 18 residues?

Capeau: We used the antibody to the C-terminal region from Santa Cruz (C-20) used by Pendas, and it works.

Goldman: When you used that, in your controls there was almost no detectable pre-lamin A. This means that the process must be taking place very rapidly. You didn't even see a trace of pre-lamin A.

Wilson: Is SREBP1 normally processed from a precursor to a mature form?

Capeau: Yes. Regarding prelamin A in the control cells, we indeed observed a faint band.

Wilson: Does SREBP1 still depend on lamins for stability? It is a direct effect of proteolytic processes.

Capeau: It is quite complicated. SREBP1 is formed as a precursor in the ER. It requires a double proteolytic step for it to leave the ER. The N-terminal part, which is the active transcription factor is released into the cytosol and then enters the nucleus through a pore. During this process there is probably an interaction with the lamin, which is required for activity.

Wilson: Are the CPK elevations seen in lipodystrophy patients evidence of muscle breakdown? Is CPK elevated in EDMD patients?

Bonne: In EDMD the CPK level is not highly elevated compared to other muscular dystrophies. It is either normal or mildly elevated.

Gruenbaum: What is the current evidence that SREBP1 requires lamin *in vivo* for its normal activity?

Capeau: That is just a hypothesis. The important thing is that there is an interaction.

Shackleton: You are right; there are no functional data. But I thought you produced some data a while back showing that SREBP1 doesn't get into the nucleus when cells were treated with protease inhibitors?

Capeau: That's correct. In about half of the cells treated with the protease inhibitor indinavir, SREBP1 couldn't enter the nucleus. It seemed to be maintained at the nuclear periphery with the lamina network looking abnormal. The cells where SREBP1 was unable to enter had nuclear herniations.

Gruenbaum: Herniations are found in all kinds of laminopathies.

Shackleton: I'm now developing embryonic stem (ES) cell lines with lamin mutations. I hope to then get them to differentiate into adipocytes and look at what is happening to SREBP1.

Burke: Do you know of any obvious differences between adipocytes in the face and neck versus those in the trunk and limbs? Are they of different embryological origin? Do they develop out of something else?

Capeau: No. What is striking is that the phenotype observed in FPLD patients is quite similar to that observed in patients with Cushing syndrome. As you know, those patients have increased fat in the neck and face, with reduced peripheral fat in the limbs and abdomen but increased visceral fat. Probably, this has something to do with cortisol. It has been shown that adipocytes express the enzyme 11-β-hydroxysteroid dehydrogenase type 1 (11βHSD1) that can transform inactive cortisone to active cortisol. It has been shown that adipocytes in certain parts of the body, such as visceral adipocytes, express this enzyme at a high level. When there is one reason or another to activate this enzyme these adipocytes could make a lot of cortisol which can result in fat hypertrophy. This activation of 11βHSD1 could be due to increased levels of cytokines such as TNFα. It has become apparent that obesity and diabetes are inflammatory diseases of adipose tissue with TNFα probably playing a major role. There could be a local inflammation, inducing release of TNFα which could activate the hypertrophy of adipocytes from some locations and not from others. This is speculative, of course.

Goldman: I have a question about your protease inhibitor experiments. What happens to the lamins under those conditions? What do the nuclei look like?

Capeau: They are dysmorphic. The stability of the nuclear envelope is reduced.

Goldman: If you take cells in culture and treat them with a protease inhibitor, is this what you see?

Capeau: Yes.

Young: We have fibroblasts from the *Zmpste24* knockout mice. This might be an interesting control for you to use to show that the nuclear morphology doesn't change. Nuclear morphology is a little bit abnormal in the *Zmpste24* knockout fibroblasts, but it is not terribly abnormal. We find $\sim 19\%$ abnormal nuclei in *Zmpste24*-deficient cells *versus* about 8% in our control cells. Similarly for the cells lacking the methyltransferase *Icmt*: they have higher levels of pre-lamin A, and they have a higher frequency of nuclear blebbing.

Goldman: Could it be that compensatory proteolytic cleavage is occurring? How specific are these proteases for cleaving lamins?

Young: There is no mature lamin A in the *Zmpste24* knockouts. With the methyltransferase *Icmt*, it is a little different. We initially reported that this was definitely required for the formation of lamin A. We would now back off from this and say that it is important but not required. Clearly, the processing is reduced in the *Icmt* knockout cells, but it is not completely eliminated. *Zmpste24* is essential, as is the farnesyltransferase. The methyltransferase *Icmt* is important but not essential.

Goldman: What is the phenotype of the *Zmp* knockouts?

Young: They have reduced subcutaneous fat. There is another group from Spain that has made the same mice, and they have reported that their *Zmpste24* knockout mice had lipodystrophy. They might have been correct. Importantly, however, they didn't notice that all the *Zmpste24*-deficient mice have osteolytic lesions in the zygomatic arch by 8 weeks of age, along with osteolytic lesions in other bones. The zygomatic arch is the origin of the masseter muscle, the main muscle for moving the jaw. Thus, the destruction of the zygomatic arch could easily cause a nutritional problem. Thus, the *Zmpste24*-deficient mice don't have much fat, but they aren't eating very well, so it is not completely clear that they truly have a lamin-induced lipodystrophy.

Worman: Do they have insulin resistance or diabetes?

Young: No. They have low levels of cholesterol and triglycerides, which I think reflects starvation.

Capeau: In mouse models it is very difficult to induce insulin resistance if animals are eating a normal kind of food. You have to feed mice with a high fat diet to see the insulin resistant phenotype. If the mice have trouble eating you probably won't see any insulin resistance.

Goldman: Do you think that lipodystrophy in AIDS patients is due to protease inhibitors?

Capeau: Yes, in part, but it is much more complicated than that.

Goldman: There are nuclear lobulations due to the AIDS virus itself. This has been shown by Werner Greene (de Noronha et al 2001).

Capeau: The herniations were seen in cultured cell-lines in the absence of any HIV. We think that the herniation is the ultimate sign of nuclear alteration. It is seen in a small percentage of cells (about 10%). I would think that some nuclear alterations are present in the nucleus before the herniation is seen. We have shown the increased fragility of the nuclear envelope. Even cells that look normal could have altered nuclei.

Goldman: In the Greene paper (de Noronha et al 2001) they did say that the blebs or herniations are transient, and then they break open and the nucleus reseals. Because they are transient it is hard to capture enough of them in any given sequence of observations. Why do AIDS patients show lipodystrophy when they are on these PI drugs?

Capeau: The pathophysiology of HIV-related lipodystrophy is complex and probably related to the two main classes of antiretroviral molecules: protease inhibitors but also nucleoside analogues, inhibitors of the viral reverse transcriptase. In these two classes, some individual molecules act in synergy to alter adipose tissue.

Young: ZMPSTE24 and the site 2 protease (a cleavage enzyme for SREBP) are structurally similar. They are both zinc metalloproteinases with multiple membrane-spanning helices. For SREBP it would be very interesting to look at the site 1 and site 2 cleavage steps specifically, rather than just looking at the SREBP nuclear localization.

Capeau: I agree. We saw that the SREBP band was a little larger in the presence of PI. This could suggest that the S2P protease, which is a zinc metalloproteinase close to ZMPSTE24 implied in lamin maturation could be inhibited. This could result in altered maturation by protease inhibitors of both pre-lamin and SREBP.

Wilson: I know of one autosomal dominant Emery-Dreifuss muscular dystrophy (EDMD) patient who was very distressed by her extreme thinness. She approached doctors who treat HIV patients to alleviate the side-effects of protease inhibitor regimens. She asked for the same hormonal therapy used on HIV patients to overcome their weight loss problems. For her this treatment appears to have successfully augmented subcutaneous fat above the waist, but it had no effect below the waist.

Shackleton: Did you look in your cultured cells to see whether other proteins were prevented from entering the nucleus apart from SREBP1?

Capeau: We looked at PPARγ level and found it to be decreased. But since this transcription factor is located downstream of SREBP1 in the transcription cascade, this result was expected.

Shackleton: Were the cells in which SREBP couldn't get into the nucleus able to differentiate in your system?

Capeau: No.

Fatkin: We had some interesting observations in the lamin knockout mouse. We looked at SREBP1 localization and levels in nuclear and cytoplasmic extracts from heart tissue. At a very early stage before the mice exhibited severe dilated cardiomyopathy, distribution of SREBP1 appeared normal. Later in the disease when there were marked nuclear morphological changes and severe dilated cardiomyopathy, we saw perinuclear accumulation of SREBP1 and with reduced PPARγ levels. This suggests that there might be an acquired defect in the nuclear translocation of the active domain of SREBP1.

Goldman: This is a good point. We haven't discussed the potential impact of nuclear lamin changes on nuclear transport. They might be quite subtle but they could have profound effects on cell physiology and pathology.

Stewart: We did a follow-up study on the lamin knockout mice with Mark Reitman and Dedra Cutler and the mice showed no sign of lipodystrophy (Cutler et al 2002). They were not insulin resistant nor did they have elevated triglyceride levels. The lack of fat, which we initially described, was probably due to their problems with eating, due to severe dystrophy in their pharyngeal and tongue muscles.

Fatkin: It is going to be hard to reproduce the metabolic phenotype that we see in patients since the homozygous mice are very unwell and have multiple factors that may contribute to changes in metabolic parameters.

References

Cutler DA, Sullivan T, Marcus-Samuels B, Stewart CL, Reitman ML 2002 Characterization of adiposity and metabolism in Lmna-deficient mice. Biochem Biophys Res Commun 291:522–527

de Noronha CMC, Sherman M, Lin HW et al 2001 Dynamic disruptions in nuclear envelope architecture and integrity induced by HIV-1 Vpr. Science 294:1105–1108

Pendas AM, Zhou Z, Cadinanos J et al 2002 Defective prelamin A processing and muscular and adipocyte alterations in Zmpste24 metalloproteinase-deficient mice. Nat Genet 31:94–99

Cytoskeletal defects in amyotrophic lateral sclerosis (motor neuron disease)

Jean-Pierre Julien, Stephanie Millecamps and Jasna Kriz

Research Center of CHUL and Department of Anatomy and Physiology of Laval University, 2705 Boulevard Laurier, Quebec, QC, G1V 4G2 Canada

Abstract. There is growing evidence for the involvement of cytoskeletal defects in the pathogenesis of motor neuron disease and especially in components of the microtubule-based transport system. Here we will review our recent work aiming to elucidate the role of peripherin in amyotrophic lateral sclerosis (ALS) and to address the mechanism of disease caused by deletions in the *ALS2* gene that cause recessive forms of juvenile ALS and primary lateral sclerosis (PLS). Peripherin is an intermediate filament protein detected in spheroids, a hallmark of ALS, and increased levels of peripherin mRNA have been found in some ALS cases. Our transgenic mouse and cell culture studies support the view of a peripherin involvement in ALS. However, a gene knockout approach demonstrated that peripherin is not a key contributor of motor neuron disease caused by mutant superoxide dismutase linked to familial ALS. A recent breakthrough in the field of ALS came with the discovery of frameshift deletions in the *ALS2* gene coding for Alsin. Our transfection experiments in cultured cells suggest that Alsin is a cytoskeletal protein with dual endosomal and centrosomal localizations. We have generated a mouse knockout for Alsin that develops progressive motor dysfunction during ageing. Thus, it is anticipated that this mouse model will be useful to investigate the pathogenic pathways linked to Alsin gene mutations.

2005 Nuclear organization in development and disease. Wiley, Chichester (Novartis Foundation Symposium 264) p 183–196

Amyotrophic lateral sclerosis (ALS) is a neurodegenerative disease characterized by the progressive loss of upper and lower motor neurons resulting in paralysis and death. Little is known about the causes of ALS and most cases are sporadic. Numerous loci have been identified for ALS but only few causative genes are known. Most ALS studies in the past decade have been focused on mutations in the Cu/Zn superoxide dismutase (*SOD1*) gene that are linked to autosomal dominant forms with adult onset (ALS1) corresponding to $\sim 20\%$ of the familial cases of the disease (FALS) (Rosen et al 1993). Yet, the mechanism of disease caused by *SOD1* mutations is not fully understood.

There is growing evidence for the involvement of cytoskeletal components in ALS. Spheroids composed of neurofilament and peripherin proteins are a hallmark of degenerating spinal motor neurons occurring in most familial and sporadic ALS cases (Carpenter 1968, Corbo & Hays 1992, Migheli et al 1993, Wong et al 2000). Transgenic mouse studies carried out by us and others demonstrated that certain types of intermediate filament (IF) accumulations can provoke motor neuron dysfunction (Larivière & Julien 2004). Furthermore, in a small number of sporadic ALS patients (~1% of cases), codon deletions or insertions have been discovered in the phosphorylation repeat motif of the heavy neurofilament NF-H gene (Figlewicz et al 1994, Al-Chalabi et al 1999).

More recently, there have been many reports demonstrating that cytoskeletal alterations in the microtubule-based transport machinery can also provoke motor neuron disease in humans and mice (Table 1). The overexpression in mice of dynamitin disrupts dynein/dynactin function and induces motor neuron disease with neurofilamentous swellings in motor axons (LaMonte et al 2002). Mutant dynactin gene was also discovered in human cases of motor neuron disease (Puls et al 2003). Further proof that defects in axonal transport can provoke neurodegeneration came from the analysis of a mouse knockout for the kinesin *KIF1B* gene (Zhao et al 2001) and from the identification of mutations in the motor domain of the *KIF1B* gene in some cases of Charcot-Marie-Tooth disease type 2A (Zhao et al 2001). Moreover, two groups recently identified that the progressive motor neuronopathy (*Pmn*) mutation in the mouse is due to a mutation in the tubulin-specific chaperone (Tbce) protein (Bommel et al 2002, Martin et al 2002). *Tbce* is essential for proper tubulin assembly and for the maintenance of microtubules in motor axons. The overexpression of tau, a stabilizer of microtubules, can also provoke motor neuron disease in mice (Ishihara et al 1999). This suggests that altered function of tubulin cofactors might be implicated in human motor neuron diseases.

Another breakthrough in the field of motor neuron disease came with the discovery of deletion mutations in coding exons of a new gene mapping to chromosome 2q33, *ALS2* coding for Alsin, from patients with an autosomal recessive form of juvenile ALS (JALS), primary lateral sclerosis (PLS) and infantile-onset ascending hereditary spastic paralysis (IAHSP) (Yang et al 2001, Hadano et al 2001, Eymard-Pierre et al 2002, Gros-Louis et al 2003, Devon et al 2003). The Alsin protein contains guanine nucleotide exchange factor (GEF) homology domains that are known to activate small guanosine triphosphatase (GTPase) belonging to the Ras superfamily.

Here, this review will focus on motor neuron disease caused by peripherin abnormalities and by Alsin gene mutations with emphasis on studies based on genetically modified mice.

Peripherin in motor neuron disease

The neuronal cytoskeleton is composed of three interconnected filaments: the actin microfilaments, microtubules and intermediate filaments (IFs). Five major types of

TABLE 1   Cytoskeletal defects in motor neuron degeneration

Gene	Feature	References
Human gene mutations		
NEFH	Codon deletions in KSP domain of ~ 1% sporadic ALS cases	(Figlewicz et al 1994, Al-Chalabi et al 1999)
NEFL	Charcot-Marie-Tooth neuropathy type 2	(Georgiou et al 2002, Jordanova et al 2003, De Jonghe et al 2001, Mersiyanova et al 2000)
Dynactin	Motor neuron disease	Puls et al 2003
KIF1Bβ	Charcot-Marie-Tooth neuropathy type 2	Zhao et al 2001
Alsin	Juvenile ALS and PLS	Yang et al 2001, Hadano et al 2001
Mouse models		
Disorganization of IFs		
Mutant NF-L	Early-onset loss of motor neurons	Lee et al 1994
Age-dependent IF aggregates	Loss of motor neurons at ~ 2 years old	Beaulieu et al 1999
Peripherin;NF-L$^{-/-}$	Loss of motor neurons at ~ 4 months old	Beaulieu et al 1999
Defects of microtubule-based transport		
Dynamitin overexpressor	Disruption of dynein function causing motor neuron degeneration	LaMonte et al 2002
Pmn gene	Missense mutation in tubulin-specific chaperone E	Bommel et al 2001, Martin et al 2002
Dynein	Missense point mutations causing motor neuron disease	Hafezparast et al 2003
KIF1B	Heterozygous knockout causing late-onset motor neuron degeneration	Zhao et al 2001
Tau overexpressor	Perikaryal IF accumulations, degeneration of motor axons	Ishihara et al 1999

IF proteins are expressed in adult neurons: the three neurofilament proteins, peripherin and α-internexin. Spheroids, composed of peripherin and neurofilament proteins are a hallmark of degenerating spinal motor neurons occurring in most familial and sporadic ALS cases (Carpenter 1968, Corbo & Hays 1992, Migheli et al 1993, Wong et al 2000). Many factors can potentially lead to accumulation of IF proteins including deregulation of IF protein synthesis, defective axonal transport, abnormal phosphorylation and proteolysis. Until recently, it was widely believed that such IF abnormalities were simply the consequence of neuronal dysfunction. However, studies with transgenic mice suggest that IF alterations can sometimes produce deleterious effects and even cause neuronal death.

Toxicity of peripherin overexpression

An example of neurotoxicity caused by IF disorganization was provided by studies using transgenic mice overexpressing wild-type peripherin (Beaulieu et al 1999, 2000). Mice overexpressing by four- to sevenfold a wild-type peripherin transgene developed a late-onset and progressive motor neuron disease characterized by the presence of IF accumulations resembling axonal spheroids found in ALS patients. Unlike transgenic mice overexpressing wild-type neurofilament proteins, the peripherin transgenic mice exhibited late-onset degeneration of motor neurons. Interestingly, the disease was exacerbated by NF-L deficiency (Beaulieu et al 1999). Our interpretation is that in absence of NF-L, the NF-M and NF-H subunits exert deleterious effects upon the assembly of peripherin leading to the formation of protein aggregates. In this regard, it may be of relevance that the mRNA levels for NF-L are selectively down regulated in motor neurons of ALS patients (Bergeron et al 1994, Wong et al 2000) and that an up-regulation of peripherin mRNA has been detected in a familial ALS case (Robertson et al 2003).

Surprisingly, the overexpression of an *NF-H* transgene completely rescued the peripherin-mediated degeneration of motor neurons *in vivo* (Beaulieu & Julien 2003). The overexpression of *NF-H* shifted the intracellular distribution of peripherin from the axonal to the perikaryal compartment of spinal motor neurons. This suggests that the protective effect of extra NF-H proteins in this situation may be related in part to the sequestration of peripherin into the perikaryon of motor neurons thereby abolishing the development of axonal IF inclusions that might block transport. These findings again illustrate the importance of IF protein stoichiometry in formation, localization and toxicity of neuronal inclusion bodies.

The deleterious effect of excess peripherin has also been demonstrated using *in vitro* cultured motor neurons that were microinjected with an expression vector coding for the peripherin gene (Robertson et al 2001). The overexpression of peripherin induced apoptotic death of motor neurons and the toxicity was attenuated by co-expression of NF-L, illustrating the importance of IF protein stoichiometry. Moreover, dorsal root ganglion (DRG) neurons derived from peripherin transgenic embryos exhibited apoptotic death when cultured in a proinflammatory CNS environment rich in activated microglia (Robertson et al 2001). The apoptotic death of DRG neurons overexpressing peripherin was mediated by tumour necrosis factor (TNF)α secreted by activated microglia. This led us to suggest that peripherin aggregates might predispose neurons to the detrimental effects of a proinflammatory environment.

In addition to the normal 58 kDa peripherin, we recently reported the existence of an abnormal neurotoxic splicing variant of peripherin in ALS cases (Robertson et al 2003). This peripherin variant of 61 kDa contains a 32 amino acid insertion within the α-helical rod domain derived from retention of intron four sequences. The 61 kDa peripherin was assembly-incompetent and very toxic when expressed in cultured neuronal cells (Robertson et al 2003). Whereas this 61 kDa peripherin was undetected in normal mice, it was detectable in motor neurons of mice expressing mutant SOD1 linked to ALS (Robertson et al 2003). Remarkably, specific antibodies against the 61 kDa isoform were also able to detect motor neurons and axonal spheroids in two cases of familial ALS (Robertson et al 2003). Considering that neurodegenerative diseases are often associated with defects in alternative splicing, it seems reasonable that in some ALS cases the production of such an abnormal 61 kDa peripherin variant could be a factor contributing to disease.

To further examine the role of peripherin in disease caused by *SOD1* mutations, we generated Sod^{G37R} mice with altered levels of peripherin. Surprisingly, neither the up-regulation nor suppression of peripherin expression affected disease onset, mortality and loss of motor neurons in this mouse model (Larivière et al 2003). From this study, we conclude that peripherin is not a key contributor to motor neuron degeneration in disease caused by *SOD1* mutations. Nonetheless, mutations in *SOD1* are responsible for only $\sim 2\%$ of total ALS cases and further research is needed to investigate the role of peripherin in human ALS of other aetiologies.

Motor neuron disease caused by Alsin gene mutations

A recent breakthrough in the field of ALS research came with the discovery of deletion mutations in coding exons of a new gene mapping to chromosome 2q33—*ALS2* coding for Alsin—from patients with an autosomal recessive form of juvenile ALS (JALS), juvenile PLS or IAHSP (Yang et al 2001, Hadano et al 2001, Eymard-Pierre et al 2002, Gros-Louis et al 2003, Devon et al 2003). All

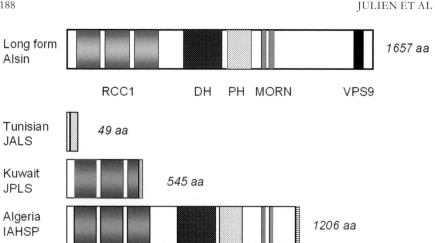

FIG. 1. Schematic representation of the long form of human Alsin protein and predicted truncated forms due to frameshift mutations in cases of human disease. Alsin has in the N-terminus multiple motifs with homology to the regulator of chromosome condensation (RCC1). The mid-portion contains domains homologous to the DBl homology (DH), to pleckstrin homology (PH) domains, and membrane occupation and recognition motifs (MORN). At the C-terminus there is a vacuolar protein sorting 9 (VPS9) domain.

reported mutations in *ALS2* predict premature truncation of Alsin protein. The wide variability in size of the truncated Alsin mutants together with the recessive inheritance of motor neuron disorders suggests a loss of function of Alsin. The human *ALS2* gene spans ~80 kb and it contains 36 exons. It produces two transcripts resulting from alternative splicing. The large Alsin mRNA of 6394 nucleotides is predicted to code for a 184 kDa protein of 1657 amino acids whereas the short transcript of 2651 nucleotides has a single open reading frame of 396 amino acids.

The *ALS2* gene is ubiquitously expressed. It encodes a protein having motifs with GEF homology domains that are known to activate GTPase belonging to the Ras superfamily (Fig. 1).

The N-terminal has multiple domains of the regulator of chromosome condensation (RCC1). The mid-portion has domains homologous to the DBl homology (DH) and pleckstrin homology (PH) domains, and membrane occupation and recognition motifs (MORN). At the C-terminal there is a vacuolar protein sorting 9 (VPS9) domain. The RCC1-like, DH/PH and VPS9 domains are GEFs for the small GTPase Ran (Ras-related nuclear), Rho (Ras-homologous member) and Rab5 (Ras-related in brain 5), respectively (Yang et al 2001, Hadano et al 2001). Ran promotes the nucleation and stability of spindle microtubules, the protein transport through the nuclear pore complex, the

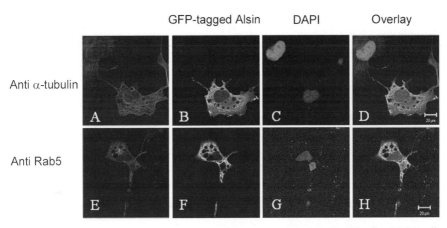

FIG. 2.   Overexpression of alsin induces endosome accumulation in COS-7 cells. COS-7 cells were transfected with EGFP-tagged Alsin cDNA expression vector. Transfected cells (B and F) exhibited accumulation of vesicles in cytoplasm that are positive for early endosome marker Rab5 (E). Dapi positive nucleus (C,G).

chromosome partitioning and cell cycle regulation. Rho is a key regulatory molecule for the organization of actin cytoskeleton controlling cell morphology and motility in response to extracellular signals whereas Rab5 promotes the association of endosomes with microtubules and endosome trafficking toward the microtubule organizing centre (MTOC) (Nielsen et al 1999).

Dual endosomal membrane and centrosomal location

In order to investigate the cellular distribution of Alsin and the role of its protein domains, we generated cDNA expression vectors for the full length and truncated forms of the mouse protein. Overexpression of intact Alsin resulted in the enlargement of endosomes in cultured COS-7 cells. Otomo et al (Otomo et al 2003) reported a similar endosomal activation after transfection of full length human Alsin cDNA in cultured neuronal cells. As shown in Fig. 2, COS-7 cells overexpressing the full length Alsin with a fusion EGFP at the N-terminus developed multiple vacuoles of different size in the cytoplasm. These vesicles were detected with antibodies against the early endosomal marker Rab5. The overexpression of full length Alsin also induced a disorganization of the microtubule network and disappearance of tubulin asters in COS-7 cells. Transfection of truncated mutants of Alsin showed that the VPS9 domain was required for induction of the phenotypes.

Transfection of Alsin cDNA in SW13 cells yielded different results than in COS-7 cells (data not shown). There was also formation of vesicles in transfected SW13 cells but they were of smaller sizes than those occurring in COS-7 cells. Moreover, the majority of SW13 cells expressing full length Alsin exhibited a centrosomal localization of the protein. Alsin colocalized with γ-tubulin, a centrosome marker. Again, the C-terminal VPS9 domain of the protein was found to be a mediator of targeting into this compartment.

The finding of dual endosomal membrane and centrosomal localizations for Alsin suggest research avenues to unravel the mechanism of disease caused by Alsin gene mutations. For example, defects in endosomal trafficking might result in alterations in neurotrophin signalling. Furthermore, the centrosome is a crucial organelle that produces the microtubules required for the intracellular transport. In neurons, the centrosome is also a microtubule-nucleating structure. After nucleation in the pericentriolar material, microtubules are released and actively transported by available motor proteins in axons or dendrites (Baas 1999). Accordingly, alterations of centrosome function through a loss of Alsin might put motor neurons at risk because of their long axons that can reach a metre in length. Hopefully, the availability of knockout mice for Alsin will help to address the mechanism by which Alsin mutations cause selective motor neuron degeneration.

A mouse knockout

Following the discovery of the Alsin gene, our group quickly decided to generate a knockout mouse for Alsin. As mentioned above, the deletion mutations linked to juvenile ALS or PLS result in frameshifts that generate premature stop codons suggesting a disease resulting from a loss-of-function mutation. Therefore, we hypothesized that the targeted disruption of the *Alsin* gene in the mouse could provide a suitable model of the human disease. Sequence analysis of the mouse *Alsin* cDNA showed 87% sequence identity with human Alsin. The mouse *Alsin* cDNA has 6349 nucleotides and codes for a 183 kDa protein of 1651 amino acids. Using the technique of gene targeting in embryonic stem (ES) cells, we succeeded in generating mice homozygous for the deletion of exon 3 and part of exon 4 of the Alsin gene. Immunoblotting confirmed the absence of Alsin proteins in the homozygous knockout mice. The *Alsin* knockout mice are born with no obvious developmental defects and they are of normal size. However, during their first year of life, the Alsin knockout mice developed slow and progressive motor dysfunction as determined by rotarod tests. Substantial loss of motor axons was also detected in L5 ventral roots of knockout mice at 6 months old. Thus, it is anticipated that the *Alsin* knockout mouse will provide a useful model to investigate the function of Alsin and to elucidate the molecular mechanism responsible for the selective degeneration of motor neurons in juvenile ALS and PLS.

Acknowledgements

This work was supported by the Canadian Institutes of Health Research (CIHR) and the National Institute of Neurological Disorders and Stroke (NINDS, USA). J.-P. Julien holds a Canada Research Chair on mechanisms of neurodegeneration. S.M. was supported by the Fondation pour la Recherche Médicale and was awarded (by the Fondation Liliane) the Bettencourt-Schueller prize.

References

Al-Chalabi A, Andersen PM, Nilsson P et al 1999 Deletions of the heavy neurofilament subunit tail in amyotrophic lateral sclerosis. Hum Mol Genet 8:157–164

Baas PW 1999 Microtubules and neuronal polarity: lessons from mitosis. Neuron 22:23–31

Beaulieu JM, Julien JP 2003 Peripherin-mediated death of motor neurons rescued by overexpression of neurofilament NF-H proteins. J Neurochem 85:248–256

Beaulieu JM, Nguyen MD, Julien JP 1999 Late onset death of motor neurons in mice overexpressing wild-type peripherin. J Cell Biol 147:531–544

Beaulieu JM, Jacomy H, Julien JP 2000 Formation of intermediate filament protein aggregates with disparate effects in two transgenic mouse models lacking the neurofilament light subunit. J Neurosci 20:5321–5328

Bergeron C, Beric-Maskarel K, Muntasser S et al 1994 Neurofilament light and polyadenylated mRNA levels are decreased in amyotrophic lateral sclerosis motor neurons. J Neuropathol Exp Neurol 53:221–230

Bommel H, Xie G, Rossoll W et al 2002 Missense mutation in the tubulin-specific chaperone E (Tbce) gene in the mouse mutant progressive motor neuronopathy, a model of human motoneuron disease. J Cell Biol 159:563–569

Carpenter S 1968 Proximal axonal enlargement in motor neuron disease. Neurology 18:841–851

Corbo M, Hays AP 1992 Peripherin and neurofilament protein coexist in spinal spheroids of motor neuron disease. J Neuropathol Exp Neurol 51:531–537

De Jonghe P, Mersiyanova IV, Nelis E et al 2001 Further evidence that neurofilament light chain mutations can cause Charcot-Marie Tooth disease type 2E. Ann Neurol 49:245–249

Devon RS, Helm JR, Rouleau GA et al 2003 The first nonsense mutation in alsin results in a homogeneous phenotype of infantile-onset ascending spastic paralysis with bulbar involvement in two siblings. Clin Genet 64:210–215

Eymard-Pierre E, Lesca G, Dollet S et al 2002 Infantile-onset ascending hereditary spastic paralysis is associated with mutations in the alsin gene. Am J Hum Genet 71:518–527

Figlewicz DA, Krizus A, Martinoli MG et al 1994 Variants of the heavy neurofilament subunit are associated with the development of amyotrophic lateral sclerosis. Hum Mol Genet 3:1757–1761

Georgiou DM, Zidar J, Korosec M et al 2002 A novel NF-L mutation Pro22Ser is associated with CMT2 in a large Slovenian family. Neurogenetics 4:93–96

Gros-Louis F, Meijer IA, Hand CK et al 2003 An ALS2 gene mutation causes hereditary spastic paraplegia in a Pakistani kindred. Ann Neurol 53:144–145

Hadano S, Hand CK, Osuga H et al 2001 A gene encoding a putative GTPase regulator is mutated in familial amyotrophic lateral sclerosis 2. Nat Genet 29:166–173

Ishihara T, Hong M, Zhang B et al 1999 Age-dependent emergence and progression of a tauopathy in transgenic mice overexpressing the shortest human tau isoform. Neuron 24:751–762

Jordanova A, De Jonghe P, Boerkoel CF et al 2003 Mutations in the neurofilament light chain gene (NEFL) cause early onset severe Charcot-Marie-Tooth disease. Brain 126:590–597

LaMonte BH, Wallace KE, Holloway BA et al 2002 Disruption of dynein/dynactin inhibits axonal transport in motor neurons causing late-onset progressive degeneration. Neuron 34:715–727

Larivière RC, Julien JP 2004 Functions of intermediate filaments in neuronal development and disease. J Neurobiol 58:131–148

Larivière RC, Beaulieu JM, Nguyen MD, Julien JP 2003 Peripherin is not a contributing factor to motor neuron disease in a mouse model of amyotrophic lateral sclerosis caused by mutant superoxide dismutase. Neurobiol Dis 13:158–166

Martin N, Jaubert J, Gounon P et al 2002 A missense mutation in Tbce causes progressive motor neuronopathy in mice. Nat Genet 32:443–447

Mersiyanova IV, Perepelov AV, Polyakov AV et al 2000 A new variant of Charcot-Marie-Tooth disease type 2 is probably the result of a mutation in the neurofilament-light gene. Am J Hum Genet 67:37–46

Migheli A, Pezzulo T, Attanasio A, Schiffer D 1993 Peripherin immunoreactive structures in amyotrophic lateral sclerosis. Lab Invest 68:185–191

Nielsen E, Severin F, Backer JM, Hyman AA, Zerial M 1999 Rab5 regulates motility of early endosomes on microtubules. Nat Cell Biol 1:376–382

Otomo A, Hadano S, Okada T et al 2003 ALS2, a novel guanine nucleotide exchange factor for the small GTPase Rab5, is implicated in endosomal dynamics. Hum Mol Genet 12:1671–1687

Puls I, Jonnakuty C, LaMonte BH et al 2003 Mutant dynactin in motor neuron disease. Nat Genet 33:455–456

Robertson J, Beaulieu JM, Doroudchi MM et al 2001 Apoptotic death of neurons exhibiting peripherin aggregates is mediated by the proinflammatory cytokine tumor necrosis factor-alpha. J Cell Biol 155:217–226

Robertson J, Doroudchi MM, Nguyen MD et al 2003 A neurotoxic peripherin splice variant in a mouse model of ALS. J Cell Biol 160:939–949

Rosen DR, Siddique T, Patterson D et al 1993 Mutations in Cu/Zn superoxide dismutase gene are associated with familial amyotrophic lateral sclerosis. Nature 362:59–62

Wong NK, He BP, Strong MJ 2000 Characterization of neuronal intermediate filament protein expression in cervical spinal motor neurons in sporadic amyotrophic lateral sclerosis (ALS). J Neuropathol Exp Neurol 59:972–982

Yang Y, Hentati A, Deng HX et al 2001 The gene encoding alsin, a protein with three guanine-nucleotide exchange factor domains, is mutated in a form of recessive amyotrophic lateral sclerosis. Nat Genet 29:160–165

Zhao C, Takita J, Tanaka Y et al 2001 Charcot-Marie-Tooth disease type 2A caused by mutation in a microtubule motor KIF1Bbeta. Cell 105:587–597

DISCUSSION

Goldman: The mutation appeared to be in the 2B region of the central rod domain of peripherin. Does it alter the heptad repeats?

Julien: I don't know the structure of the abnormal human peripherin species detected with antibodies against the peptide sequence of the mouse peripherin intron 4. This will await the cloning and sequence of this variant from this ALS patient. It is a good suggestion to look more closely at the effect of the 32 residue insert in the mouse peripherin species. At this time, we know that the effect of cDNA expression is to disrupt the IF assembly in cultured cells. We haven't done this *in vitro* with the purified variant protein.

Goldman: You might find that it can interact in some way *in vitro* at certain levels and become integrated into the polymer. This would be interesting to know.

Julien: It can be rescued if you overexpress NFL in cultured cells.

Goldman: What exactly does Alsin do and where is it found? It wasn't clear to me. If you look at Alsin in normal nerve cells, where is it localized?

Julien: In the LAN2 human cells that we tested, we detected it in the centrosome and in the tips of the neurites. Otomo et al (2003) have overexpressed it in primary cortical neurons which undergo endosomal enlargement similar to that seen in COS cells.

Goldman: Overexpression could introduce artefacts. What about normal cells? Is it just found in the centrosomes and neurite tips?

Julien: You get a different picture depending on which cell type you transfect.

Goldman: I am more interested in untransfected nerve cells.

Julien: We can't detect Alsin in untransfected cells of mouse origin with our antibodies. However, recently we have been able to detect the centrosome with anti-Alsin antibodies in cells of human origin including Hela cells, SKNSH neuroblastoma, and 293 cells

Morris: Did you detect it by Western blotting?

Julien: You need to do immunoprecipitation of brain samples followed by Western blotting. Alsin is a protein of extremely low abundance. In non-transfected mouse cells we can't detect it.

Goldman: Is it specific to the nervous system?

Julien: No, the mRNA is expressed in various cell types.

Bonne: If you aren't able to detect the protein, what is the level of the mRNA?

Julien: We didn't do Northern blots. Hadano et al (2001) have and they can see the mRNA clearly.

Bonne: Has anyone looked at this protein in tissues other than the nervous system?

Julien: Alsin has been detected in lymphoblasts from normal individual. However, the alsin mutant proteins are very unstable and were undetectable in lymphoblasts from ALS2 patients (Yamanaka et al 2003).

Starr: Are there homologues in flies or worms?

Davies: No. Not that this means anything, of course!

Julien: I am not aware of any.

Shumaker: Overexpression of Alsin causes disruption of tubulin organization. What kind of interaction is there between the two?

Julien: I don't know. If it is localized in the centrosome region, it could be very important for nerve cells. This is where the microtubules assemble.

Goldman: You also normally find it at the ends of neurons. This could imply that it would be both at the plus and minus ends of microtubules, which would be interesting and unusual. At plus ends you think about capping and the minus end is associated with the centrosome.

Julien: It is a bit mysterious at this stage. There was a recent paper by Don Cleveland and colleagues (Yamanaka et al 2003) that the truncated mutant proteins are very unstable and degraded rapidly.

Shumaker: I recall some literature saying that Rab5 is often associated with compartments associated with the centrosome. Perhaps in those tissues you will see more Alsin with your antibodies.

Wilson: Mutations in *LMNA* also cause one form of CMT.

Julien: There is a form of CMT caused by *NFL* mutations. To date I think there are about five families.

Wilson: With different genetic causes of CMT, do you see exactly the same axonal or neuronal traffic phenotype in each case? In other words, do you get aggregations in every case?

Julien: I don't think anyone has looked. They know it is axonal.

Levy: All these disorders linked to NFL and lamin are axonal neuropathies. This means that there is loss of fibres. Essentially, the large diameter fibres are totally lost but the myelin sheath is preserved.

Wilson: There are multiple underlying causes, but do you think they all ultimately disrupt axonal transport?

Julien: The cell type specificity is something to look at. If axonal transport is involved in the mechanism, then long nerve cells with long axons are more at risk than the small ones.

Wilson: Is it true that MTs in axons have defined lengths and are organized by neurofilaments?

Julien: We don't know to what extent they are organized by neurofilaments, but you are right that they are not one length. In some of the NFL knockout mice we made, when we cut transverse sections from the axons we see a lot of microtubules. The density of microtubules has increased a lot while there are no neurofilaments. Yet the transport is probably faster. If we do double knockout of NFM and NFH, then most NFL remains in the cell body. It is not carried down the axon. What migrates goes very fast, but transport is frequently interrupted. If you look at the mutation in CMT and NFL you'll see they are not in the highly conserved NFs such as in the keratin disease. In keratin diseases most mutations are in the rod domain, in 1A and 2B, which are the helix initiation and termination sites. They are very important in coiled-coil formation. In NFs we don't see the mutations there. It would probably be too disruptive and would be embryonic lethal. For CMT which is later onset there is a mild mutation that would have an effect over the long-term. So the NFH mutation in the C-terminal phosphorylation may just affect assembly slightly, but enough over 40–60 years to make a difference. This is very difficult to study in mice.

Goldman: It has been shown that phosphorylation of the neurofilament heavy chain tail, which is one of the longest C-terminal domains on any IF protein, accumulates as a function of time as NF protein is transported anterograde along the axon. It could be that these kinds of changes lead to alterations in the association of motors and transport in the long term.

Julien: We also need to think of protein aggregation and accumulation as not always being toxic. If you overexpress NFH protein it is not that good for the mice — they develop a kind of neuropathy — but it doesn't kill the motor neurons. If you breed these mice with mice expressing mutant SOD you can extend their life by five months. It is the best rescue so far of the SOD mutant mouse. We believe it is acting as a phosphorylation sink. There could be other mechanisms, though.

Goldman: Is SOD found in the aggregates also?

Julien: It depends on the mutation. The G85R forms large aggregates. They all form high molecular weight complexes. These are probably the toxic form.

Wilson: Among all the laminopathies, CMT is the most confusing to me. There are so many subtle ways that neurons and axons can be harmed. It seems like CMT mutations linked to lamin A might cause very subtle defects, because the axon is such a sensitive assay for things that go wrong.

Julien: That's an interesting concept. To have a large axon that requires a lot of protein synthesis puts the cell at risk.

Goldman: The nerve cell is unique in having the bulk of its protein synthetic machinery in the cell body, having to deliver proteins along axons that can be several feet in length. When you have a SOD mutation why is the major phenotype in the nervous system?

Julien: All the potential explanations of this have so far failed. At first it was thought that it was a loss of function activity. This is not the case because knockout SOD mice don't develop motor neuron disease. Then there was a hypothesis that perhaps it has an abnormal enzymatic activity. It uses NO and forms peroxynitrite, which is toxic. The problem here is the *in vivo* result: if you overexpress wild-type SOD or remove endogenous SOD it doesn't affect disease progression. Removing the capacity to bind copper doesn't affect the disease (this is the catalytic site). If you express it selectively in motor neurons in mice they don't develop the disease. We have made chimeric mice where 30% of the nerve cells express the mutant SOD. We can label them with antibodies, and we have a marker for the wild-type also. We found no change at all five months after the time when they should have been dead. Expressing SOD only in motor neurons is not enough. Bad environment is needed, and we think that inflammation is part of the problem.

Goldman: Why isn't the liver affected? Is it true that these patients have only neurological disorders with no other obvious problems.

Gasser: My understanding of SOD is that there are both mitchondrial and cytoplasmic SOD enzymes. The liver cell is full of mitochondria, so perhaps there is compensation potential between cytoplasm, mitochondria and nuclei. In the axon the cytoplasm is physically separated from the nucleus.

Julien: If you cut the sciatic nerve distally, for example, it will regrow. But if you cut it very close to the spinal cord the cell will die. You need trophic factors, which probably come from Schwann cells. It is not only the distance, but also the cells that are in the environment.

Goldman: Maybe it is the complex transport machinery, which is unique to the nerve cell that causes this pathological change.

Julien: That is an interesting idea. Sensory neurons in these mice are often long but they are not affected as much.

Davies: If you cross some of these mutants with the SOD mutant you should see higher additive motor neuron phenotypes. This might give you a clue as to the common pathways.

Julien: Hafezparast et al (2003) have recently shown a slowing down of disease in mutant SOD1 mice with perturbations in dynein function. This is very surprising

Davies: The argument was that dynein is not a good model for ALS but it does have an effect on this process.

Goldman: If the accumulation of filaments is a sign of the pathology in ALS, then you can think not only of kinesin but also dynein and even myosin V.

Julien: There is a down-regulation of dynein in this mouse, and when it is crossed with the SOD1 mutant mouse you would think that this would accelerate the disease. It doesn't, though. It rescues it — the mouse lived as much as 6 weeks longer.

Davies: That is a remarkable result.

References

Hadano S, Hand CK, Osuga H et al 2001 A gene encoding a putative GTPase regulator is mutated in familial amyotrophic lateral sclerosis 2. Nat Genet 29:166–173

Hafezparast M, Klocke R, Ruhrberg C et al 2003 Mutations in dynein link motor neuron degeneration to defects in retrograde transport. Science 300:808–812

Otomo A, Hadano S, Okada T et al 2003 ALS2, a novel guanine nucleotide exchange factor for the small GTPase Rab5, is implicated in endosomal dynamics. Hum Mol Genet 12:1671–1687

Yamanaka K, Vande Velde C, Eymard-Pierre E, Bertini E, Boespflug-Tanguy O, Cleveland DW 2003 Unstable mutants in the peripheral endosomal membrane component ALS2 cause early-onset motor neuron disease. Proc Natl Acad Sci USA 100:16041–16046

LMNA mutations in progeroid syndromes

Shurong Huang, Brian K. Kennedy* and Junko Oshima[1]

*Department of Pathology, University of Washington, Box 357470, Seattle, WA 98195 and *Department of Biochemistry, University of Washington, Seattle, WA 98195, USA*

Abstract. Segmental progeroid syndromes are disorders in which affected individuals present various features that suggest accelerated ageing. The two best-known examples are Hutchinson-Gilford progeria syndrome (HGPS, 'Progeria of childhood') and Werner syndrome (WS, 'Progeria of the adult'). A novel, recurrent *de novo* mutation in the *LMNA* gene, responsible for the majority of HGPS cases, results in an in-frame deletion of 50 amino acids, including endoproteolytic sites required for processing of prelamin A to mature lamin A protein. Another mutation results in a 35 amino acid in-frame deletion with a milder HGPS phenotype. *WRN*, the gene responsible for the majority of WS cases, encodes a multifunctional nuclear protein with exonuclease and helicase activities and may participate in optimizing DNA repair/recombination. A subset of WS patients do not show mutations at the *WRN* locus (atypical WS), but show heterozygous amino acid substitutions in the heptad repeat region of lamin A. Structural analysis suggests that mutations in atypical WS may interfere with protein–protein interactions. When compared to *WRN*-mutant WS, *LMNA*-mutant atypical WS patients appear to show earlier onset and possibly more severe ageing-related symptoms.

2005 Nuclear organization in development and disease. Wiley, Chichester (Novartis Foundation Symposium 264) p 197–207

Werner syndrome (WS, 'Progeria of Adults') is a progeroid syndrome characterized by the early onset of common age-related disorders combined with an aged appearance (Epstein et al 1966). Common age-related disorders include diabetes mellitus type 2, ocular cataracts, osteoporosis, hypogonadism, atherosclerosis and malignancies. Classical WS is caused by mutation of the *WRN* gene that encodes a multifunctional nuclear protein possessing both helicase and exonuclease functions (Chen & Oshima 2002). The International Registry of Werner Syndrome at the University of Washington (Seattle, WA,

---

[1]This paper was presented at the symposium by Junko Oshima to whom correspondence should be addressed.

FIG. 1.   Atypical Werner syndrome patient with L140R *LMNA* mutation. (Reproduced with permission from Oshima et al 2003.)

USA) (*http://www.wernersyndrome.org*) has enrolled over 100 suspected WS cases from all over the world. Approximately 80% of these cases do not have mutations at the *WRN* locus, though they satisfy the clinical criteria used to warrant positional cloning of the *WRN* gene. We tentatively categorized these non-*WRN* WS cases as 'atypical' WS. Among atypical WS cases, we identified two heterozygous *LMNA* mutations, R133L and L140R, in three cases with Caucasian and African American origin (Fig. 1) (Chen et al 2003). We also identified a new heterozygous mutation (A57P) in an Iranian case exhibiting features of mandibuloacral dysplasia and cardiomyopathy.

All three cases with the R133L or L140R mutations had an overall aged appearance (scleroderma-like skin and graying/thinning of hair) that contributed to the clinical diagnosis of WS. Relatively short stature and type II diabetes mellitus were reported near age 20 in both R133L mutant cases. Ocular cataracts were not seen in either case, but the patients may have been too young for the tests to be conclusive. One patient with the L140R mutation died of aortic stenosis/ insufficiency at age 36 and had a history of cataracts, osteoporosis, soft tissue calcification, and premature atherosclerosis all of which are common in classical WS. Caux (Caux et al 2003) reported a 27 year old with the R133L mutation who exhibited an overall aged-appearance and a wide range of systemic abnormalities including lipodystrophy, diabetes mellitus type 2 and cardiomyopathy. In 1985, long before *WRN* or *LMNA* mutations were identified, Imura (Imura et al 1985) categorized WS patients according to similar clusters of clinical features and inferred that there were at least three distinct clinical types of WS. Among the three groups, type 2 WS was the least common, had the earliest age for onset, and most resembled the current category of *LMNA*-type WS. As a group, patients with R133L and L140R *LMNA* mutations appear to have an earlier initial onset of symptoms when compared to classical mutant-*WRN* WS (13 vs. 18 years of age, respectively). Once established, the symptoms may progress faster in the atypical cases.

R133L and L140L mutations reside in the heptad repeat region of the alpha-helical coiled coil segment 1B that is unique to nuclear intermediate filaments (Fig. 2). Structural analysis predicts that R133L and L140R mutations may not alter the dimerization of lamins but may affect their interaction with other proteins (Chen et al 2003). The E145K mutation that was reported in unusual cases of HGPS also resides in this heptad repeat region (Eriksson et al 2003) (Fig. 2). The E145K cases are unusual in that the patients retain their scalp hair as well as ample subcutaneous tissues over arms and legs.

Hutchinson-Gilford progeria syndrome (HGPS) is the definitive child onset progeroid syndrome (Brown 1979). Affected individuals typically exhibit growth retardation during infancy along with accelerated degenerative changes of the cutaneous, musculoskeletal and cardiovascular systems. Baldness and a characteristic 'plucked-bird appearance' generally develop by age 2. The median age of death of HGPS patients is 13.5 years of age. Over 90% of the deaths are due to myocardial infarction or congestive cardiac failure. Ogihara et al (1986) reported an otherwise clinically typical HGPS patient who died of myocardial infarction at age 45, an exceptionally long-lived HGPS patient.

To date, there are six different *LMNA* mutations reported as causing HGPS, all of which are amino acid substitutions (Cao & Hegele 2003, De Sandre-Giovannoli

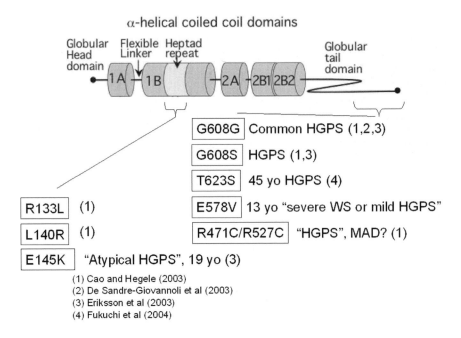

FIG. 2. Functional domains of lamin A and *LMNA* mutations identified in HGPS and atypical WS. The boxes indicate the site of mutation, along with the clinical diagnosis and references. Only mutations identified in atypical WS and HGPS are shown.

et al 2003, Eriksson et al 2003). The G608G (1824C > T) and G608S (1833G > A) mutations create cryptic splicing sites in exon 11 that result in 50 amino acid in-frame deletions in the C-terminus of the lamin A protein. G608G is the most common HGPS mutation while G608S has been reported in only one case. The 50 amino acid in-frame deletion retains a CAAX-box motif for farnesylations but eliminates the endoproteolytic site necessary for the conversion from pre-lamin A to mature lamin A (Sinensky et al 1994). The T623S (1868C > G) mutation reported in the 45 year-old HGPS case resulted in a 35 amino acid in-frame deletion that completely overlaps with the more common 50 amino acid deletion (Fukuchi et al 2004), further supporting the hypothesis that incomplete processing of lamin A may be the underlying cause of typical HGPS.

The characteristics of cells isolated from mutant-*LMNA* progeroid syndromes include abnormal nuclear morphology and abnormal subnuclear lamin A/C distribution. In primary skin fibroblasts isolated from the patient with a L140R mutation, a majority but not all nuclei showed various degrees of herniation

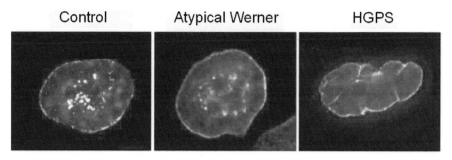

FIG. 3.   Nuclear morphology of control, atypical WS and HGPS fibroblasts. Cells are stained for anti-lamin A/C. Shape of the nuclei can be seen with the DAPI staining.

instead of a normal smooth, ovular shape. Immunocytostaining of lamin A/C showed smooth, thin staining along the perimeter of the nuclei as well as intranuclear foci co-localized with nucleolar fibrilarin marker. Detailed studies previously showed the latter to be perinucleolar staining (Kennedy et al 2000). In L140R mutant cells, perinucleolar staining was absent in nuclei exhibiting abnormal nucleolar morphology. G608G HGPS cells showed even more dramatic changes: the nuclei were generally larger and showed severe denting or blebbing. Though HGPS fibroblasts retained perinuclear lamin A/C staining patterns, perinucleolar staining patterns were absent in virtually all the affected cells (Fig. 3).

Previous literature suggests at least three non-exclusive models for the pathogenesis of laminopathies such as HGPS and atypical WS:

- Increased nuclear fragility. General alterations in nuclear structure resulting from *LMNA* mutations would lead to cell instability and ultimately tissue atrophy
- Altered chromatin silencing. Many lamina-associated proteins bind to BAF, an abundant chromatin-associated protein. Though the function of BAF is not well understood, abnormal lamin protein could alter global transcriptional regulation
- Mislocalization of nuclear proteins. Lamins A and C act to regulate a number of critical nuclear components whose activities are important to prevent onset of disease. One such candidate may be the WRN protein

At least five other disorders have been linked to *LMNA* mutations: Emery-Dreifuss muscular dystrophy (EDMD), dilated cardiomyopathy type 1A (DCM1A), limb-girdle muscular dystrophy type 1B (LGMD1B), Charcot-Marie-Tooth disease type 2 (CMT2), Dunnigan-type familial partial lipodystrophy (FPLD), and mandibuloacral dysplasia (MAD) (Burke & Stewart 2002). It is not yet known why mutations in *LMNA* are able to cause such a large variety of disorders. Nuclear fragility that eventually leads to cell instability cannot explain the phenotypic differences between *LMNA*-associated disorders, although abrogated regulation of other nuclear components of gene expression may account for such differences.

Acknowledgments

The authors thank Mr Joel Janes for his editorial assistance. This work was supported by funding from the NIH HD44782.

References

Brown WT 1979 Human mutation affecting aging — a review. Mech Ageing Dev 9:325–336

Burke B, Stewart CL 2002 Life at the edge: the nuclear envelope and human disease. Nat Rev Mol Cell Biol 3:575–585

Cao H, Hegele RA 2003 LMNA is mutated in Hutchinson-Gilford progeria (MIM 176670) but not in Wiedemann-Rautenstrauch progeroid syndrome (MIM 264090). J Hum Genet 48:271–274

Caux F, Dubosclard E, Lascols O et al 2003 A new clinical condition linked to a novel mutation in lamins A and C with generalized lipoatrophy, insulin-resistant diabetes, disseminated leukomelanodermic papules, liver steatosis, and cardiomyopathy. J Clin Endocrinol Metab 88:1006–1013

Chen L, Oshima J 2002 Werner syndrome. J Biomed Biotechnol 2:46–54

Chen L, Lee L, Kudlow BA et al 2003 LMNA mutations in atypical Werner's syndrome. Lancet 362:440–445

De Sandre-Giovannoli A, Bernard R, Cau P et al 2003 Lamin a truncation in Hutchinson-Gilford progeria. Science 300:2055

Epstein CJ, Martin GM, Schultz AL, Motulsky AG 1966 Werner's syndrome: a review of its symptomatology, natural history, pathologic features, genetics and relationships to the natural aging process. Medicine 45:172–221

Eriksson M, Brown WT, Gordon LB et al 2003 Recurrent de novo point mutations in lamin A cause Hutchinson-Gilford progeria syndrome. Nature 423:293–298

Fukuchi K-I, Katsuya T, Sugimoto K et al 2004 LMNA mutation in a 45-year old Japanese with Hutchinson-Gilford progeria syndrome. J Med Genet 41:e67

Imura H, Nakao, Kuzuya H, Okamoto M, Okamoto M, Yamada Y 1985 Clinical, endocrine and metabolic aspects of the Werner syndrome compared with normal aging. In: Salk D, Fujiwara Y, Martin GM (eds) Werner's syndrome and aging. Vol 190. Plenum Press, New York, p171–186

Kennedy BK, Barbie DA, Classon M, Dyson N, Harlow E 2000 Nuclear organization of DNA replication in primary mammalian cells. Genes Dev 14:2855–2868

Ogihara T, Hata T, Tanaka K et al 1986 Hutchinson-Gilford progeria syndrome in a 45-year-old man. Am J Med 81:135–138

Oshima A, Garg A, Martin GM, Kennedy BK 2003 *LMNA* mutations in atypical Werner's syndrome. (authors' reply) Lancet 362:1586

Sinensky M, Fantle K, Trujillo M et al 1994 The processing pathway of prelamin A. J Cell Sci 107:61–67

DISCUSSION

Gasser: In these cases of spontaneous mutations, how do you know that there aren't other mutations that are causing the phenotypes?

Oshima: We don't. The only evidence we have is genetic. If we sequence one of these cases and we don't see the mutation in 100 or so normal controls, then this is considered sufficient genetic proof.

Gasser: With your two cases of spontaneous progeria, have you introduced the mutations to mice, to see whether they get the syndrome? Is this something that can be done?

Goldman: I don't think there is a progeria mouse which is a definitive model for the human disease.

Stewart: We published what started off to be an attempt to make an autosomal dominant Emery-Dreifuss muscular dystrophy (AD-EDMD) mouse (Mounkes et al 2003). For reasons we still don't understand, our attempt to make the AD-EDMD mouse resulted in the introduction of a rather bizarre splicing defect at the 3′ end of exon 9. This splicing defect significantly disrupted the C terminal globular domain in the lamin A and C proteins, with a substantial portion of the globular domain being deleted. It may have also affected polyadenylation of the lamin C transcript, making the transcript unstable. The deletion in the C-terminal globular domain in the lamin A protein may also partially resemble the deletion resulting from the most common progeria mutation in the human lamin A gene. We however feel that our mice strongly resemble human Hutchinson–Gilford progeria in that both their phenotypes are remarkably similar. In terms of introducing these other specific missense mutations, I know that this is being tried by Steve Young and Francis Collins, but we haven't heard any results yet.

Young: I would add that the *Zmpste24*-deficient mice have a lot of the phenotypes that resemble progeria, too. Their phenotypes are similar to the ones that Colin Stewart identified in his 'progeria' mice. We have tried to put the Hutchinson-Gilford deletion into a mouse, and we were successful in doing so. We sustained a bit of setback, though. In our mice, we deleted the last 50 amino acids from exon 11, and we deleted intron 11. (We did this rather than introducing the point mutation.) Our targeted mutation was perfect and we did get a tiny bit of the Hutchinson-Gilford transcript. However, the new exon 11/12 fusion in our mutant allele was not recognized efficiently by the mRNA splicing machinery. Lamin A and lamin C formation is controlled by alternative splicing in exon 10. The completely unanticipated consequence of deleting intron 11 is that we ended up with nearly all lamin C production. Thus, we actually made a 'lamin C-only' mouse rather than a Hutchinson-Gilford model. So far the lamin C-only mice look fairly healthy. Now we will go back and make additional gene targeted mice to make 'progerin-only' mice. For this, we are going to try to introduce the same mutation (the exon 11 deletion and the intron 11 deletion). However, we will also delete *intron 10*, so the mutant allele cannot possibly make lamin C.

Burke: Have you attempted modifying the cleavage site to make a non-cleaver?

Young: We are doing this, but we don't yet have any information.

Warner: Dr Oshima, since you have worked a long time on Werner's, what is the usual explanation for why these patients have short stature?

Oshima: They don't have a growth spurt because they have a problem with their bones.

Goldman: You say you don't detect Werner's syndrome until later. HGPS can be detected quite early, at the age of two.

Oshima: The Werner short stature is not obvious immediately. It is more the lack of the first growth spurt, but this is usually noted retrospectively.

Gerace: Are there any neuroendocrine disorders that cause premature ageing?

Warner: I can't think of any.

Young: 80% of your patients have Werner's mutations and 5% of them have lamin A mutations. Have you looked in the remainder to see whether any of them have an accumulation of prelamin A?

Oshima: No, I haven't looked at them. I haven't looked in the classical Werner's cases, either.

Young: Are those fibroblast cell lines available?

Oshima: Yes.

Young: It would be interesting to see whether any of these had *ZMPSTE24* deficiency.

Oshima: I sent the remaining cells to Dr Abhimanyu Garg at the University of Texas for this purpose. He didn't find anything.

Levy: In our patients classified as having progeria or progeria-like syndromes we have looked at both LMA and STE24 mutations. We haven't found any mutations in STE24, but when we looked at the nuclear shape and localization of lamin A/C, in some cases there were exactly the same aspects as in typical HGPS. It seems that in some cases of premature ageing protein processing is identical but gene defects are different.

Young: But you see an accumulation of pre-lamin A with some of the point mutations.

Levy: Yes.

Goldman: Is this by Western blotting or immunolocalization?

Levy: Immunolocalization, using the same Santa Cruz antibody which recognizes both pre-lamin and lamin A.

Hutchison: Could you clarify the classification of these diseases? Using the name 'atypical Werners' implies that there is a convergent mechanism between the helicase and the lamin. Is it atypical Werner's simply because of the age of onset, or is this atypical Hutchinson-Gilford?

Oshima: The clinical diagnosis of the disease was devised when we knew nothing. Then classifications were based on pathology. Now, the current trend is to use the causative genes.

Levy: As a clinical geneticist, I feel that you have to talk about Werner's syndrome if there is an associated neoplasm. If there isn't, it doesn't enter Werner's classification. I feel that the lamin-associated mutations are involved in a continuum of disorders ranging from mild muscular disorders to very severe systemic disorders such as progeria or progeric disorders. I feel that the classification of Werner's should only be used for patients with a mutation in the Werner's protein. In your Western blots, is there a decrease of the Werner protein?

Oshima: No, they are all the same. I sequenced it and there is no mutation.

Goldman: I want to raise a cautionary note about looking at Westerns and talking about amounts of defective and normal proteins. This requires quantitative Western blot analysis. Also, intermediate filaments, including the lamins, are in general susceptible to proteolysis in cell free preparations. Determining what pre-lamin A is requires very specific antibodies.

Young: I think that the Santa Cruz antibody is actually a good antibody, but you have to do an immunoprecipitation first with an N-terminal lamin A/C antibody and then do a Western blot with the C-terminal pre-lamin A-specific antibody. It doesn't work consistently or very well when you just do a straight Western blot with the C-terminal pre-lamin A-specific antibody. You get very nice results if you first do an immunoprecipitation and then a Western.

Goldman: We know that the lamins are some of the most insoluble proteins in cells. It is known that immunoprecipitation does not quantitatively remove all of the lamin protein. I am not arguing about the utility of this technique, but if you are going to try to distinguish differences in amounts of wild-type and mutant protein, which is important for understanding the effects of dominant negative mutations, then quantitative Westerns need to be done very carefully.

Young: I agree. For quantitative work, there is a lot of doubt.

Gruenbaum: You showed the same lamin A staining in Werner's syndrome and HGPS. Is it true for the different passage numbers?

Oshima: I just look at one time point, so I can't answer this.

Bonne: I would like to return to the question of how we can prove the pathogenicity of any missense mutation. One of the mutations found by Junko Oshima, the L140R mutation was previously found in EDMD heterozygote patients. The genetics of the lamins can confuse and complicate things. Perhaps one way to dissect out the various syndromes is as follows. On one hand, there is a progeria mutation that affects only lamin A and leads to deficiency of lamin A and an accumulation of pre-lamin A. This is quite homogeneous. It is severe and there is early death. On the other hand there is a large batch of missense and nonsense mutations all along the protein that lead to various syndromes, but not as severe as progeria. The patient Jacqueline Capeau mentioned earlier shares the same mutation R133, and looks quite similar from the clinical point of view. I wonder whether we are looking at two separate issues. First there is the mechanism at the molecular level. We have a whole batch of overlapping syndromes. At the other side there is the accumulation of pre-lamin A that may be even worse. In one of the progeria papers, there is a compound heterozygous mutation reported. Have people checked whether the mutations are on the same chromosome or not? In our experience with EDMD patients we have several patients with two mutations. When we cloned these mutations we found that they were on the same gene, which is not recessive but dominant. The gene has two mutations. The protein may be mutated twice but 50% of normal protein is still there. This is something that we have to keep in mind.

Shumaker: It was on two separate chromosomes.

Julien: Are there any mutations in lamin A that affect its phosphorylation?

Goldman: I thought that there was only one known phosphorylation site, the MPF site.

Burke: No, there are two.

Goldman: Has the C-kinase site been mapped?

Shumaker: It is at 359.

Gerace: Alan Fields mapped it on lamin B1.

Goldman: We have done some preliminary experiments with John Eriksson looking at interphase phosphorylation sites. We find many sites using comparative 2D phosphopeptide mapping.

Burke: It is very unlikely that the Cdc2 sites are modified. This would be the kiss of death. If you alter either one of them you create a mitotic catastrophe.

Goldman: Those experiments were done a long time ago. We can find similar aggregates and changes in lamins just by overexpressing the wild-type. They saw deep nuclear inclusion bodies and said that the cells couldn't go into mitosis. Cells become abnormal when you overexpress wild-type A or B. These experiments should be re-done by controlling mutant protein expression levels. I don't think that this has been carefully analysed even in mitotic cells.

Gasser: This work was done in Erich Nigg's laboratory, just next to mine. They did careful phosphotryptic peptide maps, comparing both *in vivo* and *in vitro* patterns.

Goldman: But did they sequence the peptides to find out exactly which residue was involved?

Gasser: They knew how many mitosis-specific modifications there were, because they could analyse the charge of the spots on the 2D separation by both charge and size.

Goldman: In the case of the intermediate protein, vimentin, it is known that *in vitro* it is a very good substrate for many kinases. However, if you label cells with ^{32}P for 10, 20 or 30 min you see virtually no vimentin phosphorylation. If you treat cells with calyculin A, a phosphatase inhibitor, it increases vimentin phosphorylation rapidly and dramatically within a few minutes. In preliminary experiments with John Eriksson, we see similar changes in the lamins. This tells us that the exchange, the phosphatase–kinase equilibrium, is rapid. It is also possible that the lamins are similarly dynamic.

Wilkins: I have a question about the basic biology of these two progeroid syndromes. They must be different fundamentally in how they start. Is it known whether the defect in either or both is due to replication defects in stem cells or cells capable of renewal, or does the pathogenesis in either or both cases involve differentiated cells that go wrong? Is there anything known about the fundamental course of what happens in these progeroid syndromes? To put it another way, is the basic defect in either the lamin defect or Werner's syndrome due to a failure of cell renewal?

Oshima: I would say yes. There is more apoptosis in Werner's cells.

Wilkins: So you may be preferentially losing cells capable of self-renewal, like stem cells?

Oshima: Yes. But Werner's mutant cells tend to senesce with the telomeres still long. The cells die because they accumulate DNA and are susceptible to some genotoxic agent.

Reference

Mounkes LC, Kozlov S, Hernandez L, Sullivan T Stewart CL 2003 A progeroid syndrome in mice is caused by defects in A-type lamins. Nature 423:298–301

A genetic approach to study the role of nuclear envelope components in nuclear positioning

Daniel A. Starr and Min Han*

*Center for Genetics and Development and Section of Molecular Biology, 313 Briggs Hall, University of California, Davis, CA, 95616, and *Department of Molecular, Cellular and Developmental Biology, HHMI and University of Colorado, Boulder, CO, 80309, USA*

Abstract. In many cell types, the nucleus is positioned to a specific location. Our work and that of others has demonstrated that several integral nuclear envelope proteins function to move the nucleus and to anchor it in place. Our forward genetics approach has identified three components of the nuclear envelope involved in nuclear positioning. ANC-1 consists of two actin-binding calponin domains, a huge central coiled domain, and a nuclear envelope targeting a domain termed the KASH domain. ANC-1 functions to physically tether the actin cytoskeleton to the outer nuclear membrane. UNC-83 is a novel protein that functions in an unknown manner during nuclear migration. UNC-83 contains a domain with weak homology to the KASH domain of ANC-1. UNC-84 is a SUN protein that is required for both nuclear migration and anchorage. UNC-84 recruits both UNC-83 and ANC-1 to the nuclear envelope. We propose a model where UNC-84 is an integral component of the inner nuclear membrane, with its SUN domain in the perinuclear space. The SUN domain then recruits ANC-1 and UNC-83, through interactions with their KASH domains, to the outer nuclear envelope. Together these proteins function to bridge the two membranes of the nuclear envelope, connecting the nuclear matrix to the cytoskeleton.

2005 Nuclear organization in development and disease. Wiley, Chichester (Novartis Foundation Symposium 264) p 208–226

The nuclear envelope (NE) is a unique structure that performs multiple essential functions. It consists of two membranes; the inner nuclear membrane (INM) is closely associated with the nuclear lamina and is contiguous with the outer nuclear membrane (ONM) through the nuclear pores. The ONM is contiguous with the endoplasmic reticulum (ER). In addition to separating the nucleoplasm from the cytoplasm, the NE also controls the movement of molecules between these two compartments, gives physical structure to the nucleus, organizes chromatin, and functions in multiple other roles. Despite all these functions, fewer than 15 integral membrane nuclear envelope components have been studied in any detail. Other chapters of this book go into great detail on many of these proteins and functions. Here we discuss the role of the NE in respect to nuclear positioning within the cell. Specifically, how does the NE function in the regulation of its own movement from one location to another in the cytoplasm? And, after the establishment of a specific location, how does the NE function to anchor the nucleus in place?

We have taken a genetic approach using the relatively simple model organism *Caenorhabditis elegans* to study nuclear positioning. Three integral components of the NE, UNC-83, UNC-84, and ANC-1 are essential for proper nuclear positioning within the cell. At least two of these proteins are conserved in mammals. Our data suggest a model where these proteins function to bridge both membranes of the nuclear envelope, effectively connecting the nuclear matrix to the cytoskeleton.

Results and discussion

Genetic screens for mutations disrupting nuclear migration and anchorage

C. elegans is an excellent choice for a non-biased genetic approach to study nuclear positioning. Since the entire cell lineage for the development of *C. elegans* and the normal position for all nuclei have been precisely documented (Sulston & Horvitz 1977, Sulston et al 1983), genetic screens for defects in the position of nuclei are simple to carry out. Horvitz and Sulston isolated the first mutants that disrupted nuclear migrations more than 20 years ago (Horvitz & Sulston 1980, Sulston & Horvitz 1981). They screened for mutants that altered the normally invariant cell lineage of *C. elegans* and found a range of phenotypic classes. One class that is relevant to our work consists of two complementation groups, *unc-83* and *unc-84*, which disrupt nuclear migration in three different cell types (Fig. 1A–C). Defects in the nuclear migration of P cells, which normally migrate from a lateral position to the ventral cord during the first larval stage, lead to cell death. Therefore, the normal descendants of P cells are missing, which results in egg laying defects and uncoordinated movement (unc) in *unc-83* and *unc-84* mutants (Sulston & Horvitz 1981). Since these initial studies, nearly 20 alleles of both *unc-83* and *unc-84* have been isolated from various screens, but no third complementation class has been identified (Malone et al 1999, Starr et al 2001).

Hedgecock and Thomson identified the first mutants disrupting nuclear anchorage in a screen for defects of nuclear positioning (1982). Five alleles of a single complementation group, *anc-1* (for anchorage defective), caused nuclei of the syncytial hypodermis to float freely through the cytoplasm, often forming large clumps of nuclei (Fig. 1D). In fact, *anc-1* mutants probably disrupt the anchorage of nuclei in all somatic, post-embryonic cells (Hedgecock & Thomson 1982). *unc-84* mutants were found to have similar defects in nuclear anchorage

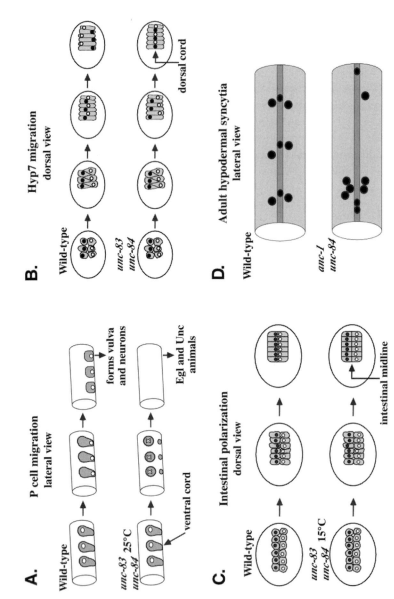

A. P cell migration lateral view

Wild-type

unc-83 25°C
unc-84

ventral cord

forms vulva and neurons

Egl and Unc animals

B. Hyp7 migration dorsal view

Wild-type

unc-83
unc-84

dorsal cord

C. Intestinal polarization dorsal view

Wild-type

unc-83 15°C
unc-84

intestinal midline

D. Adult hypodermal syncytia lateral view

Wild-type

anc-1
unc-84

FIG. 1. *unc-83*, *unc-84*, and *anc-1* mutant phenotypes. *unc-83* and *unc-84* mutations disrupt the nuclear migrations of three cell types. (A) Left lateral view of P-cell nuclear migration in wild-type and *unc-83* or *unc-84* mutant larvae at 25 °C (nuclear migration is nearly normal at 15 °C). P-cell cytoplasm is gray and nuclei are white. White X marks dying nuclei where nuclear migration has failed. (B) The dorsal surface of a pre-elongation embryo illustrating intercalation and nuclear migration of hyp7 precursors in wild-type and *unc-83* or *unc-84* mutant embryos. Cytoplasm of the hyp7 precursors is gray, nuclei that migrate from right to left are black, and nuclei that migrate from left to right are white. Anterior towards left, right is upwards. Nuclei that fail to migrate are abnormally located in the dorsal cord. (C) A dorsal view, through the middle of a pre-elongation embryo, of nuclear migration during intestinal polarization during the E16 stage in wild-type and *unc-83* or *unc-84* mutant embryos raised at 15 °C. Cytoplasm of embryonic intestinal cells is gray, nuclei right of the midline are black and nuclei left of the midline are white. Anterior is leftwards. In *unc-83* or *unc-84* mutants, the nuclei do not localize to the midline. (D) *anc-1* and *unc-84* mutations disrupt nuclear anchorage. A lateral view of the adult syncytial hypodermis in wild-type and *anc-1* or *unc-84* mutant animals. Cytoplasm of the four syncytia that cover the entire mid-body of an adult animal is gray, and nuclei are white. Normally nuclei are evenly spaced, but in the mutants, they are unanchored, and often cluster. Parts of this figure (A–C) have been reproduced with permission from Starr et al 2001.

(Malone et al 1999). Mutations in *anc-1* also disrupt the anchorage of mitochondria (Hedgecock & Thomson 1982, Starr & Han 2002). We have isolated five additional alleles of *anc-1* in screens for un-anchored nuclei, but none of *unc-84* (unpublished results). Due to the large target size of *anc-1* (see below) it is not known if this screen has been saturated.

UNC-84, UNC-83, and ANC-1 are components of the NE

unc-84 encodes a 1111 residue protein with a predicted transmembrane domain in the middle and a conserved C-terminal SUN domain (for Sad1p, UNC-84 homology; Malone et al 1999). Both the C-terminal SUN domain and the unique N-terminal domain are required for nuclear migration (Malone et al 1999). Both GFP-tagged UNC-84 and antibodies against UNC-84 localize to the nuclear envelope of nearly all somatic cell nuclei (Fig. 2A,B; Lee et al 2002, Malone et al 1999). *unc-83* encodes a completely novel protein; the only identified motif is a predicted transmembrane region 17 residues from the C-terminus (Starr et al 2001). Monoclonal antibodies against UNC-83 localize to the nuclear envelope, where they co-localize with antibodies against UNC-84. Unlike UNC-84, which is found at the nuclear envelope of nearly all somatic cell nuclei from the 24-cell stage through to adulthood, UNC-83 is found on only a subset of nuclei. UNC-83 is first observed at the NE of migrating hyp7 nuclei during embryogenesis (Fig. 2C,D; Starr et al 2001). Later it is also localized to the NE of P cells and intestinal cells; both populations of these nuclei fail to migrate in *unc-83* and *unc-84* mutations. UNC-83 was also observed in a limited number of other somatic cells (Starr et al 2001).

anc-1 encodes a huge protein of 8546 residues (Starr & Han 2002). The bulk of the protein is repetitive and is predicted to form a long helical fibrous domain. Antibodies against this domain localize to the cytoplasm, are excluded from the nucleus, and are enriched at the NE (Fig 2E-F; Starr & Han 2002). The N-terminus of ANC-1 contains two ∼100 amino acid domains with calponin homology (Gimona et al 2002); it binds F-actin *in vitro* and localizes with actin *in vivo* (Starr & Han 2002). The C-terminal 60 residues of ANC-1 are highly conserved (40-60% identity) with the C-termini of *Drosophila* Klarsicht and human Syne-1 and Syne-2. We term this the KASH domain (Klarsicht, ANC-1, Syne homology); it consists of a predicted transmembrane domain followed by 39 residues. When ectopically expressed, the C-terminal 350 residues of ANC-1, including the KASH domain, are sufficient for NE localization and cause a dominant negative nuclear anchorage phenotype (Starr & Han 2002). Our model for ANC-1 function is that the N-terminus binds to the actin cytoskeleton, while the C-terminus binds to the ONM. The long fibrous middle then extends between these two structures, effectively tethering the nucleus to the actin cytoskeleton (Starr & Han 2002, 2003).

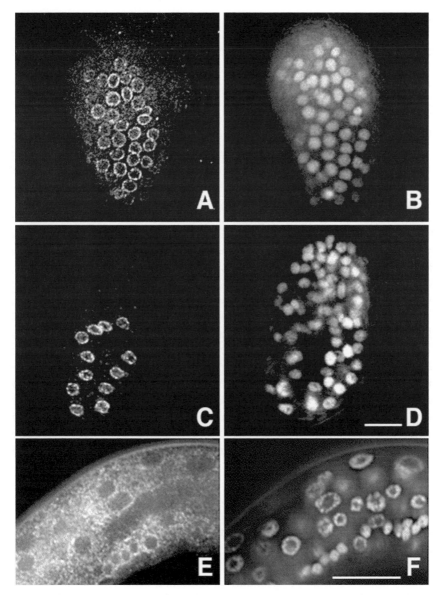

FIG. 2. UNC-84, UNC-83, and ANC-1 localize to the NE. (A–B) An embryo stained with anti-UNC-84 antibodies showing UNC-84 (A) at the NE of all DAPI-stained nuclei (B). (C–D) A similar stage embryo stained with anti-UNC-83 antibodies showing UNC-83 (C) at the NE of only a few (the migrating hyp7 cells) DAPI-stained nuclei (D). Scale bar for A-D is 10 microns. (E–F) The midbody of an L4 larvae showing localization of anti-ANC-1 antibodies (E) and DAPI stained nuclei (F). Scale bar is 10 μm. Parts of this figure (C–F) have been reproduced with permission from Starr et al (2001) and Starr & Han (2002).

The SUN domain of UNC-84 recruits UNC-83 and ANC-1 to the NE

unc-83(lf), *unc-84(lf)*, or double mutant animals have the same nuclear migration phenotype, suggesting that these two proteins function in a single pathway. We therefore determined whether they require each other for localization to the NE. UNC-83 fails to localize to the NE in an *unc-84(null)* mutant embryo. Moreover, missense mutations in the SUN domain of UNC-84 disrupt UNC-83 localization, while missense mutations in the N-terminus of UNC-84 do not, despite their defect in nuclear migration (Starr et al 2001). Mutations in *unc-83* do not disrupt UNC-84 localization (Lee et al 2002). Additionally, point mutations in the N-terminal or SUN domains of UNC-84, which disrupt nuclear migration, do not disrupt the localization of UNC-84 to the NE (Lee et al 2002). These data suggest that UNC-84 recruits UNC-83 to the NE through a genetic interaction with the SUN domain. Only when both UNC-84 and UNC-83 are at the NE, can nuclear migration proceed normally. This suggests that UNC-83 then recruits or controls additional factors required for migration. The molecular mechanisms of UNC-83 are under investigation.

unc-84(lf) and *anc-1(lf)* mutants have indistinguishable nuclear anchorage phenotypes. Given that the SUN domain of UNC-84 recruits UNC-83, we tested whether UNC-84 might act as a docking site for ANC-1. ANC-1 failed to localize to the NE in an *unc-84(null)* mutant and in *unc-84* alleles with missense mutations in the SUN domain (Starr & Han 2002). Thus, the SUN domain of UNC-84 is required for the localization of both ANC-1 and UNC-83 to the NE. It is therefore likely that UNC-84 helps control the switch between migration and anchorage of nuclei. A simple model is that UNC-83 and ANC-1 compete with one another for limited numbers of UNC-84 docking sites. This is unlikely because overexpression of ANC-1 does not lead to mislocalization of UNC-83, nor does the overexpression of UNC-83 lead to nuclear anchorage defects (unpublished data). We propose that other unidentified proteins participate in this important developmental switch between migration and anchorage.

SUN domains have been found in *C. elegans* UNC-84, *S. pombe* Sad1p, *Drosophila* predicted protein CG18584, and two human proteins, SUN1 and SUN2; these SUN domains are between 34 and 47% identical to one another (Hagan & Yanmagida 1995, Malone et al 1999). Sad1p, which is required for setting up the mitotic spindle, localizes to the spindle pole body and, when overexpressed, to the NE (Hagan & Yanmagida 1995). Recently, a divergent SUN domain (26% identity to UNC-84) was identified in *C. elegans* SUN-1. SUN-1 is required to recruit ZYG-12 to the outer nuclear membrane in the early embryo; ZYG-12 then functions to attach the centrosome to the NE (Malone et al 2003). Epitope-tagged versions of human SUN1 localize in transfected tissue culture cells to the nuclear envelope (Dreger et al 2001). Although SUN proteins clearly localize to the NE, the topology of UNC-84/SUN in the NE remains to be determined. Since lamin is required for the localization of UNC-84 (Lee et al 2002), one model suggests that UNC-84 is an integral component of the inner nuclear membrane. In this model, the SUN domain of UNC-84 extends into the perinuclear space where it could interact with UNC-83 and ANC-1, effectively targeting these proteins to the outer nuclear envelope (Starr & Han 2003).

The KASH domain specifies NE localization

KASH domains (for Klarsicht, ANC-1, Syne homology) have been found in a number of proteins that have been shown to localize to the NE. They consist of a predicted transmembrane stretch followed by about 40 amino acids and are found at the C-termini. The founding member of the family is *Drosophila* Klarsicht, which is required for nuclear migration in the developing eye disc and lipid droplet migration in the embryo (Mosley-Bishop et al 1999, Welte et al 1998). ANC-1 is a member of a family of huge proteins that have calponin domains at the N-terminus, a KASH domain at the C-terminus, and a large helical central rod domain. *Drosophila* MSP-300 and mammalian Syne-1 and Syne-2 (also published as myne-1 and -2, nesprin-1 and -2, and NUANCE) are the other identified members of this family. The central rod domains of MSP-300 and the Syne proteins consist of spectrin repeats (Apel et al 2000, Mislow et al 2002a, Starr & Han 2002, Zhang et al 2001, 2002, Zhen et al 2002).

   The KASH domain likely acts as an NE targeting signal. Klarsicht localizes to the NE by a lamin-dependent mechanism and is required for centrosome-to-NE attachment in migrating nuclei in the *Drosophila* eye disc (Mosley-Bishop et al 1999, Patterson et al 2004). As discussed above, the C-terminus of ANC-1 is sufficient for localization to the NE (Starr & Han 2002). Likewise, the 60 residue KASH domains of Syne-1 and Syne-2 have been shown to be necessary and sufficient for localization to the NE (Zhang et al 2001). Whether the localization of KASH domains is to the ONM or the INM remains a point of debate. Our ANC-1 cytoplasmic localization data (Starr & Han 2002), and the digitonin extraction experiments of Zhen et al (2002) strongly suggest that ANC-1 and Syne-1 localize to the ONM. If true, this is the first protein to our knowledge that specifically binds to the ONM but not the ER. However, blot overlay experiments suggest that Syne-1 binds to lamin and emerin, implicating Syne-1 at the INM (Mislow et al 2002b). It is possible that alternatively spliced products of Syne-1 and Syne-2 may be localized to different membranes. However, the exact localization of KASH domain proteins within the NE requires further study.

   We wanted to test if Syne-1 behaves in a mammalian system in a manner similar to ANC-1 in *C. elegans*. Since the overexpression of the C-terminus of ANC-1 leads to a strong dominant negative phenotype (Starr & Han 2002), we are carrying out a

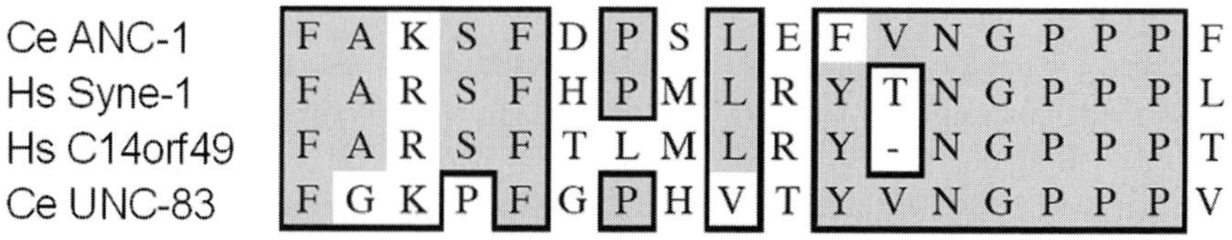

Ce ANC-1	F A K S F D P S L E F V N G P P P F
Hs Syne-1	F A R S F H P M L R Y T N G P P P L
Hs C14orf49	F A R S F T L M L R Y - N G P P P T
Ce UNC-83	F G K P F G P H V T Y V N G P P P V

FIG. 3.   UNC-83 has a C-terminal KASH domain. A clustalW protein alignment of the C-terminal regions of *C. elegans* ANC-1 and UNC-83, and human Syne-1 and C14orf49 is shown. Identical residues are shaded, and similar residues are boxed.

similar experiment in mice. At the mouse neuromuscular junction (NMJ), a cluster of about six muscle nuclei normally forms immediately beneath the NMJ. These nuclei express specific transcripts to respond to signals across the NMJ (Sanes & Lichtman 2001). We have generated a transgenic line that expresses the C-terminal 350 residues of Syne-1 in skeletal muscles and are currently collaborating with R. Grady and J. Sanes (Washington University, St. Louis, MO) to study the role of of Syne-1 in positioning of nuclei at the NMJ.

Since ANC-1 and UNC-83 both require UNC-84 to localize to the NE, it is possible that they are localized by similar mechanisms. We therefore examined the sequence of UNC-83 for a KASH domain. The C-terminus of UNC-83 has a predicted transmembrane domain followed by 18 amino acids; 50% of these residues are conserved in other KASH domains, suggesting that UNC-83 does in fact have a divergent KASH domain (Fig. 3). We are currently testing the hypothesis that the divergent KASH domain of UNC-83 is required for nuclear localization and an interaction with the SUN domain of UNC-84. Our extensive genomic searches identified one other potential KASH domain containing protein in humans, C14orf49 (Fig. 3). Interestingly, this protein was identified as a probable integral membrane component of the NE in a recent proteomic study (Schirmer et al 2003). We are currently testing the hypothesis that C14orf49 is a true NE component and whether C14orf49 might be a homologue of UNC-83.

Summary

A model for the role of the NE in nuclear positioning

We propose a model (Fig. 4) where proteins bridge the two membranes of the nuclear envelope. This bridge can then act as part of a connection between two relatively stable structures: the nuclear matrix and the cytoplasmic cytoskeleton. In this model, the predicted transmembrane region of UNC-84 would be in the inner nuclear membrane. Presumably, UNC-84 would be recruited to the inner nuclear membrane through an interaction with the lamina. In fact, localization of UNC-84 to the NE requires lamin (Lee et al 2002), although it is not known how

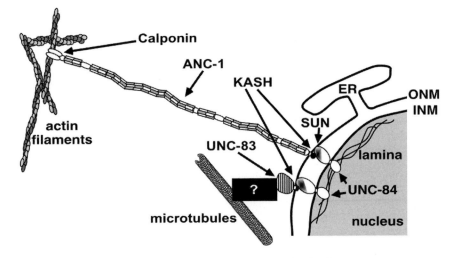

FIG. 4.   A model for nuclear positioning is shown. See text for details.

direct this interaction is. Once at the inner nuclear membrane, we predict that the conserved SUN domain of UNC-84 faces the perinuclear space. From there, the SUN domain can recruit additional proteins to the outer nuclear membrane. The NE localization of both UNC-83 and ANC-1 require UNC-84 (Starr & Han 2002, Starr et al 2001). Specifically ANC-1 localizes to the NE by way of its KASH domain (Starr & Han 2002). Here we show that UNC-83 also has a KASH domain, suggesting that UNC-83 and ANC-1 localize to the outer nuclear membrane through a common mechanism, interaction with the SUN domain. Once recruited to the NE, ANC-1 functions to tether the nucleus to the actin cytoskeleton, while UNC-83 functions through an unknown mechanism to control migration. We propose that this conserved KASH/SUN interaction is a general mechanism to recruit proteins to the outer nuclear membrane, but not the ER.

Acknowledgements

We thank Christian Malone (University of Wisconsin) for comments on the manuscript.

References

Apel ED, Lewis RM, Grady RM, Sanes JR 2000 Syne-1, a dystrophin- and Klarsicht-related protein associated with synaptic nuclei at the neuromuscular junction. J Biol Chem 275:31986–31995

Dreger M, Bengtsson L, Schoneberg T, Otto H, Hucho F 2001 Nuclear envelope proteomics: novel integral membrane proteins of the inner nuclear membrane. Proc Natl Acad Sci USA 98:11943–11948

Gimona M, Djinovic-Carugo K, Kranewitter W J, Winder S J 2002 Functional plasticity of CH domains. FEBS Lett 513:98–106

Hagan I, Yanmagida M 1995 The product of the spindle formation gene sad1+ associates with the fission yeast spindle pole body and is essential for viability. J Cell Biol 129:1033–1047

Hedgecock EM, Thomson JN 1982 A gene required for nuclear and mitochondrial attachment in the nematode C. elegans. Cell 30:321–330

Horvitz HR, Sulston JE 1980 Isolation and genetic characterization of cell-lineage mutants of the nematode Caenorhabditis elegans. Genetics 96:435–454

Lee KK, Starr DA, Cohen M, Liu J, Han M, Wilson KL, Gruenbaum Y 2002 Lamin-dependent localization of UNC-84, a protein required for nuclear migration in C. elegans. Mol Biol Cell 13:892–901

Malone CJ, Fixsen WD, Horvitz HR, Han M 1999 UNC-84 localizes to the nuclear envelope and is required for nuclear migration and anchoring during C. elegans development. Development 126:3171–3181

Malone CJ, Misner L, Le Bot N et al 2003 The C. elegans Hook protein, ZYG-12, mediates the essential attachment between the centrosome and nucleus. Cell 115:825–836

Mislow JM, Kim MS, Davis DB, McNally EM 2002a Myne-1, a spectrin repeat transmembrane protein of the myocyte inner nuclear membrane, interacts with lamin A/C. J Cell Sci 115:61–70

Mislow JM, Holaska JM, Kim MS et al 2002b Nesprin-1alpha self-associates and binds directly to emerin and lamin A in vitro. FEBS Lett 525:135–140

Mosley-Bishop KL, Li Q, Patterson L, Fischer JA 1999 Molecular analysis of the klarsicht gene and its role in nuclear migration within differentiating cells of the Drosophila eye. Curr Biol 9:1211–1220

Patterson K, Molofsky AB, Robinson C, Acosta S, Cater C, Fischer JA 2004 The functions of Klarsicht and nuclear lamin in developmentally regulated nuclear migrations of photoreceptor cells in the Drosophila eye. Mol Biol Cell 15:600–610

Sanes JR, Lichtman JW 2001 Induction, assembly, maturation and maintenance of a postsynaptic apparatus. Nat Rev Neurosci 2:791–805

Schirmer EC, Florens L, Guan T, Yates JR 3rd, Gerace L 2003 Nuclear membrane proteins with potential disease links found by subtractive proteomics. Science 301:1380–1382

Starr DA, Han M 2002 Role of ANC-1 in tethering nuclei to the actin cytoskeleton. Science 298:406–409

Starr DA, Han M 2003 ANChors away: an actin based mechanism of nuclear positioning. J Cell Sci 116:211–216

Starr DA, Hermann GJ, Malone CJ et al 2001 unc-83 encodes a novel component of the nuclear envelope and is essential for proper nuclear migration. Development 128:5039–5050

Sulston JE, Horvitz HR 1977 Post-embryonic cell lineages of the nematode, Caenorhabditis elegans. Dev Biol 56:110–156

Sulston JE, Horvitz HR 1981 Abnormal cell lineages in mutants of the nematode Caenorhabditis elegans. Dev Biol 82:41–55

Sulston JE, Schierenberg E, White JG, Thomson JN 1983 The embryonic cell lineage of the nematode Caenorhabditis elegans. Dev Biol 100:64–119

Welte MA, Gross SP, Postner M, Block SM, Wieschaus EF 1998 Developmental regulation of vesicle transport in Drosophila embryos: forces and kinetics. Cell 92:547–557

Zhang Q, Skepper JN, Yang F et al 2001 Nesprins: a novel family of spectrin-repeat-containing proteins that localize to the nuclear membrane in multiple tissues. J Cell Sci 114:4485–4498

Zhang Q, Ragnauth C, Greener MJ, Shanahan CM, Roberts RG 2002 The nesprins are giant actin-binding proteins, orthologous to Drosophila melanogaster muscle protein MSP-300. Genomics 80:473–481

Zhen YY, Libotte T, Munck M, Noegel AA, Korenbaum E 2002 NUANCE, a giant protein connecting the nucleus and actin cytoskeleton. J Cell Sci 115: 3207–3222

DISCUSSION

Collas: I'm interested in the signalling from the cytoplasm to the nucleus, and possibly the other way round. Howard Worman touched on this with TGFβ, and you touched upon this. Has anyone looked at whether Syne1 has binding sites for kinases and phosphatases?

Starr: It would be a great candidate to be a scaffold in the system, but we haven't looked.

Gruenbaum: I'm very pleased to see that our bridging model is catching up. Also, your data doesn't take into account the possibility that there are shorter forms of nesprin that might be nuclear, as was seen in mammalian cells. Your blot of expressed nesprins showed many lower molecular weight bands. Is it possible that one or some of these bands could represent a nuclear ANC-1 protein?

Starr: The evidence I have that ANC-1 would be on the outer nuclear membrane is that most of my staining is cytoplasmic. But there could be some staining at the inner nuclear envelope that I can't distinguish. One explanation is that there are shorter isoforms, which are inside the nucleus. I can't exclude this. We have not looked in *C. elegans* at what types of smaller isoforms might exist, but given the abundant evidence in mammalian systems that there exist a number of different sizes of transcripts of these proteins, there are likely to be different sized transcripts in *C. elegans*. I did a Western and looked at the lower portion of the gel and I didn't see any obvious bands.

Young: Do you think the situation in your transgenic mice is just due to variegation of transgene expression — so that some nuclei express your transgene and others don't — or do you think there is something more complicated happening, like you suggested?

Starr: This is a good question. One problem is that we only have one transgenic line. The second problem is that it's difficult to imagine that most nuclei in a myotube could be expressing something but that for some reason it cannot be detected on other nuclei in the same syncytium. Perhaps this can be explained if the transgenic protein actually has to be expressed very close to where it ends up. For example, it needs to be expressed by the very nucleus it is localized to. Of course, this is difficult to test. Another possibility is that perhaps these nuclei aren't underneath the NMJ. They could be muscle satellite cells.

Shackleton: Do you have any evidence that UNC-84 is on the inner nuclear membrane as opposed to the outer.

Gruenbaum: When lamin is down-regulated, UNC-84 is displaced to the cytoplasm similar to most known inner nuclear membrane proteins. Although we don't have direct EM evidence for UNC-84 presence in the inner nuclear membrane, we have analysed another protein with a SUN domain, which is present at the inner nuclear membrane.

Wilson: We tested UNC-84 and failed to see any direct interaction between it and lamins.

Bonne: Did your transgenic mice express any visible phenotype? What do the NMJs look like when they don't have any nuclei underneath?

Starr: We saw few (16%) NMJs with no nuclei underneath. This could explain why we don't see a gross phenotype. On average there were 2.3 nuclei under each NMJ. Perhaps this is enough for them to function. Grossly, the bungarotoxin staining looks like it does in wild-type.

Davies: Have you done any electrophysiology?

Starr: No.

Worman: Does anyone know the width of the perinuclear space? Is it small enough that the luminal domains of two proteins can interact with each other?

Starr: Yes.

Goldman: I would argue that from a cell biological viewpoint we can't measure this distance accurately. We can only make an assumption knowing that EMs have the potential for fixation artefacts.

Gerace: In the Fawcett textbook it is about 50–70 nm. This raises the question of the predicted secondary structure for the luminal domain.

Starr: There's nothing obvious.

Malone: Both *C. elegans* SUN proteins have coiled-coil domains that may be long enough to span the luminal domain.

Shumaker: In *Xenopus* nuclei the luminal distance has been measured as 40–130 nm depending on fixation.

Shanahan: Since you found another KLS domain protein in *C. elegans*, UNC-83, does this have any homology with *Drosophila* Klarsicht in any other domain, or with any other proteins?

Starr: The bulk of UNC-83 has no homology with anything. The bulk of Klarsicht has no homology with anything. There's also another protein that I have identified in mammalian systems that has a KASH domain and then a large domain that doesn't have homology to anything. Those three proteins are all predicted to be highly helical. Klarsicht is required to connect the centrosome to the nuclear membrane (Patterson et al 2004), whereas UNC-83 is not.

Gruenbaum: You showed two interesting observations that may shed light on UNC-84 organization. Mutations in the UNC-84 N-terminus can complement mutations in the UNC-84 C-terminus. The other is that the mutations in the

UNC-84 N-terminus do not affect the localization of UNC-83. These observations could imply that UNC-84 self-dimerizes. Has anyone looked at this?

Starr: That is the prediction but we haven't looked.

Shackleton: I have been working with mouse SUN1. This protein seems to be quite different. Perhaps it is a more evolutionarily advanced version of UNC-84, because it seems to have several transmembrane domains. From the work I have done, the SUN domain would appear to be on the exterior of the nucleus. Whether it has evolved to fulfil the function of two proteins, I don't know.

Wilson: So you are modelling the SUN domain as located outside the luminal space.

Goldman: What are your criteria for binding?

Shackleton: We pulled it out in a yeast two-hybrid screen originally, and then we did various GST pull-downs.

Goldman: Which domain of lamin does it bind to?

Shackleton: The globular domain. We didn't even use the helical domain in the two-hybrid screen for lamin.

Goldman: So does it bind to the C- or N-terminus of lamin?

Shackleton: The C-terminus.

Wilson: With mice that overexpress the myc-tagged KASH domain, did you quantify nuclei? Were there any clusters of non-expressing nuclei outside the neuromuscular junction? In other words, were non-expressing nuclei all located near the neuromuscular junction?

Starr: We have tried to quantitate this. It looks like there are some nuclei far from the neuromuscular junction that have no transgenic protein. The problem is, we can't be sure that those nuclei are in the myotube, a satellite cell or associated cells of some sort. We don't have the proper markers.

Gerace: Did you say that you used a muscle-specific promoter to drive the expression of your transgene?

Starr: Yes.

Gerace: So you haven't looked at expression in a more general context.

Starr: Others have looked at Syne1 in other tissues so they can address which tissues it is localized in.

Shanahan: It is in every tissue. You can't even consider it as a single gene. It is basically three overlapping genes and we have identified at least 15 different isoforms, ranging in size from 80 kDa to 1 megaDa. I don't think you can talk about it as a single genetic entity.

Goldman: If you are right, then we can't.

Shanahan: Evolutionarily, there seem to be different genes in *C. elegans* and *Drosophila* that have similar functions to Syne1 and 2 but are slightly different. All the different sized isoforms of the nesprins carry out the functions that perhaps Klarsicht and MSP300 effect. In *Drosophila*, MSP300 was identified as a muscle protein in the Z-lines that, when disrupted, interrupts integrin signalling. The disruption was right up at the N-terminus. This probably corresponds to the larger isoforms of nesprin 1 and 2 that we see in the Z-line and N-line of muscle, and also in mitochondria. We now know that MSP300 does have shorter isoforms from the C-terminus, which would probably be inner nuclear envelope isoforms of nesprins. But they have only just found now that Klarsicht, which was originally identified as a vesicle-moving protein, is actually binding lamin at the nuclear envelope and is involved in nuclear migration.

Starr: That is not a direct interaction. Janice Fischer's (Patterson et al 2004) work shows that there is a genetic interaction, and lamin is required for Klarsicht localization. We would presume that this is an indirect effect.

Worman: Besides the NMJ, if you look at any muscle fibre the nuclei all line up along the side. In dystrophic fibres they are in the middle. Do you think the same things are involved here? Is there a different set of proteins that may be on the outer nuclear membrane that bind to components of the sarcolemma?

Starr: The nuclei do not appear to go into the middle in these mice. That's all I can say.

Davies: Have you tried damaging the muscle to see what happens?

Starr: No.

Davies: What levels of expression do you get from the transgene compared with normal wild-type levels? You could be looking at the limits of expression of the transgene and the distribution could be a result of this.

Starr: We haven't tested that.

Bonne: Howard Worman mentioned that in dystrophic muscle the nuclei are centralized. It is a typical feature, and this is true for every kind of muscular dystrophy, not just laminopathies.

Young: Did *unc-84* mutation also mislocalize the mitochondria?

Starr: No, only mutations in *anc-1* mislocalized mitochondria.

Young: What is it binding to in mitochondria?

Starr: We have no idea. Interesting results have come from Ken Beck's lab (Gough et a 2003) who independently isolated Syne-1. They studied a Golgi-specific protein that turned out to be a Syne-1 isoform. He has dominant negative tissue culture lines in which he overexpresses a piece of the middle of Syne-1 that localizes to and disrupts the structure of the Golgi. Thus, Syne-1 could be acting as a scaffold for a number of different organelles.

Goldman: What does it bind to on the mitochondrial membrane?

Starr: We don't know. We don't know for sure that it directly binds mitochondria, but in the null mutation mitochondria aren't positioned properly. The genetics state that ANC-1 is required for nuclear positioning.

Goldman: That could involve something quite far removed.

Starr: It could be indirect, but it is absolutely required because a null mutation in *anc-1* has severe defects in mitochondrial positioning.

Wilson: The lack of a gross phenotype in the mice is interesting. Do you plan to make a real knockout instead of a dominant disruption?

Starr: Colin Stewart is working on this, as is Min Han. A mouse knockout is not in my plan.

Fatkin: The nuclear position issue has been discussed a lot in skeletal muscle but not cardiac muscle. What would the possible consequences be of abnormal nuclear positioning in cardiac muscle cells?

Starr: Someone else could speculate better on this.

Bonne: Is there an abnormal localization of nuclei in cardiac muscle?

Fatkin: I haven't seen any data on this for cardiac muscle cells.

Bonne: There was a report by Arbustini et al (2002) in Italy. They produced an EM picture of cardiac muscles from patients with mutations. It is the only picture I have seen of cardiac muscle where there was a full disruption of nuclei. In muscle biopsies from patients we have seen very few nuclei with abnormal features — around 5% of the myonuclei. It is quite rare. But we never observed disruption of the membrane. However, the muscle biopsies are not always performed in the most affected muscles of the patients. They are usually made in quadriceps or deltoid muscle, those muscles not being the most affected, so this is not conclusive.

Gasser: What happens if you disrupt lamin in *C. elegans*? Are UNC-83 and UNC-84 properly localized?

Starr: UNC-84 isn't localized in lamin RNAi treated worms but UNC-83 hasn't been looked at. I assume it isn't localized either. Since our genetic experiments show that UNC-84 is required at the nuclear envelope for UNC-83 and ANC-1 localization, I predict that lamin disruption will also mislocalize UNC-83 and ANC-1.

Malone: I study a similar set of proteins, SUN-1 and ZYG-12, earlier in *C. elegans* development. We can clearly detect what the ER looks like in the cell cycle, and it does not look like either of these proteins redistribute to the ER. It is not clear what happens to the ZYG-12 protein when we disrupt localization. I have a question about UNC-84. There is a set of mutations on either side of the protein that both affect nuclear migration, but only one affects UNC-83 localization. Have you any models for how it causes a nuclear migration defect if it doesn't mislocalize UNC-83?

Starr: The model that would fit the best is that N-terminal mutations in UNC-84 would somehow disrupt the interaction between lamin and UNC-84. Then when you start to pull on the outer nuclear membrane, since it is not connected to anything you are not going to move the nucleus.

Gruenbaum: For the first time I think that we can talk about the cytoskeleton and include the nucleoskeleton. Now we have good evidence that every component in the cell is somehow bound to the others and is required to maintain cell integrity. Bob Goldman, I would like to hear your comments on IF localization near the nuclear envelope and whether you think it is physically connected.

Goldman: If you look for cytoskeletal interactions with the nuclear surface, there is some evidence for actin on the outer nuclear surface. Microtubules seem to be concentrated in the centrosomal region, which is close to the nucleus. However, in the axopods of heliozoans Keith Porter and Lew Tilney showed that microtubules looked like they were growing off the nuclear surface. The bulk of cytoskeletal protein on the outside of the nucleus in mammalian cells is invariably IFs. It would be interesting to know whether nesprin can self-associate to form an oligomeric complex. You are showing linkages of nesprin going from the nuclear surface to the cell surface, so can it polymerise into long chains? If you go back to the old literature you will find that there are many ways to nucleate actin *in vitro* which may not reflect normal physiological conditions.

Wilson: Even though the actin binding domain of nesprins is far from the membrane-binding region, there may be additional domains that bind indirectly to actin or stimulate actin events. We can't assume that the middle parts of these long proteins are featureless. They may have interesting functions.

Goldman: It might be interesting to look at the smaller variants of bacterial-expressed nesprin. One might be able to obtain sufficient protein to carry out biochemical experiments.

Shanahan: Using an antibody to the calponin homology (CH) domain region we have found that it is present in foci in the nucleus. We haven't been able to identify what isoform with the CH domain would find its way into the nucleus. But there definitely are foci of CH domain-containing nesprin proteins in the nucleus.

Goldman: How do you know it is nesprin?

Shanahan: There are other proteins in the nucleus with this domain as well. It's not just a nesprin with a CH domain that can get into the nucleus.

Goldman: There are many actin binding proteins. I think we should be excited but cautious. Let's limit our excitement to reality.

Starr: There was an interesting result from Elena Korenbaum's group (Zhen et al 2002). They were the fourth group to identify Syne-1 and they were the first group to identify the longest isoforms, which they named NUANCE. They took the C-terminal KASH domain and bound it to the N-terminal calponin domains. They transfected tissue culture cells and induced an actin cage around the outside of the nuclear envelope.

Goldman: Has the length of nesprin been measured using rotary shadowed preparations for electron microscopy? This should be done. As an aside, we also need to decide what we are going to call this fascinating protein. We need to come to grips with this nomenclature problem.

Shanahan: A lot of the problem is that some of the papers have come out later and the authors have refused to even acknowledge that there are other isoforms. It makes it too complicated for them to discuss their data when they only have one antibody.

Goldman: Few of us can name the person who named tubulin. No one should worry about their reputation on the basis of naming a protein. We call it nesprin because we know Elizabeth McNally and she sent us nesprin antibodies.

Starr: I call it Syne-1 because that is the name that appeared first in the literature — by 14 months (Apel et al 2000).

Gerace: When you did your Western blot to characterize ANC-1, there were a lot of bands recognized. How many alternative splice forms are seen at the mRNA level? Since there are a lot of bands, considering the cytoplasmic localization, is it possible that *in vivo* proteolysis is releasing soluble forms that can move around? Have you used the most rigorous conditions possible to avoid *in vitro* proteolysis?

Starr: We haven't done any of these things. The bands could be because of any of these. This is an important question that needs to be addressed.

Gruenbaum: How many introns are there in *anc-1*?

Starr: Not a lot for *C. elegans* and for a protein of this size. There are about 30. Most of the introns are fairly small. Many other genes that have multiple transcripts in *C. elegans* tend to have larger introns. This one has some larger introns at the 5′ end, but not the middle or at the 3′ end. Also, there is no evidence of SL1 *trans* splicing which is usually associated with multiple start sites.

Hutchison: Can I ask about your model? Your model implies that you need to bring the nucleus to a position in order to help propagate a signal transduction pathway. Are you at the stage where you are prepared to invoke a forced transmission signal transduction mechanism?

Starr: That is completely hypothetical.

Gasser: Is there any evidence that any of this nuclear migration requires actin polymerization?

Starr: It is hard to tell in *C. elegans* whether actin is required or not. The drug studies are hard to do in a multicellular organism.

Goldman: In nuclear migration studies, when the two pronuclei move together, there is literature showing that microtubule-based motors are involved.

Gasser: The same is true in yeast.

Gruenbaum: There were two cases in *Drosophila* where mutations in lamin Dm0 were shown to affect cytoplasmic organization. One was the abnormal cytoplasmic RNA distribution of Gurken RNA, which leads to the dorsalization of the embryo. The other was the disruption of directed outgrowth of cytoplasmic extensions from terminal cells of the tracheal system. We have to keep in mind that there are other examples where nuclear lamins probably affect the cytoplasmic organization.

References

Apel ED, Lewis RM, Grady RM, Sanes JR 2000 Syne-1, a dystrophin- and Klarsicht-related protein associated with synaptic nuclei at the neuromuscular junction. J Biol Chem 275:31986–31995

Arbustini E, Pilotto A, Repetto A et al 2002 Autosomal dominant dilated cardiomyopathy with atrioventricular block: a lamin A/C defect-related disease. J Am Coll Cardiol 39:981–990

Gough LL, Fan J, Chu S, Winnick S, Beck KA 2003 Golgi localization of Syne-1. Mol Biol Cell 14:2410–2424

Patterson K, Molofsky AB, Robinson C et al 2004 The functions of Klarsicht and nuclear lamin in developmentally regulated nuclear migrations of photoreceptor cells in the Drosophila eye. Mol Biol Cell 15:600–610

Zhen YY, Libotte T, Munck M, Noegel AA, Korenbaum E 2002 NUANCE, a giant protein connecting the nucleus and actin cytoskeleton. J Cell Sci 115:3207–3222

General discussion I

Goldman: I want to change the subject a bit to get back to lipodystrophy. It has been shown that vimentin is involved in the formation of lipid droplets. In these lipodystrophies has anyone looked at vimentin?

Capeau: We have done nothing on this. Normally in adipocytes there is one large lipid droplet. This isn't the case in cultured adipocytes, which have 10 or 20 small droplets in the cytosol.

Wilson: I have a general question. It was quite striking that the lipodystrophy patients have muscle hypertrophy, with exaggerated muscle size. This is quite different from the wasting that we expect to see in laminopathies. Can you address that?

Capeau: This is quite a difference. The only comment I can make is that muscular hypertrophy is a common finding in severely insulin-resistant patients. This could be due to the trophic effects of high levels of insulin. They also have skin lesions of acanthosis nigricans that are due to the trophic effects of insulin on keratinocyte proliferation. The muscle is working OK: some of the patients play sport at a high level. I think there is something of a conflict between the atrophic effects and the hypertrophic effects due to insulin resistance.

Young: In the textbook descriptions of lipodystrophy, they say that the muscular phenotype is only apparent because of the loss of fat. You are saying that it is actually an increase in the amount of muscle.

Capeau: Yes. There is a real muscular hypertrophy.

Young: So the textbook chapters need revision.

Fatkin: The cardiac hypertrophy is also an interesting observation. In your patients, ventricular hypertrophy may be caused by hypertension or fatty infiltration associated with a metabolic defect.

Capeau: That is why the hypertrophy is of low specificity. Some of the cardiac alterations seen in these patients could be related to diabetic complications. However, in some of these patients the diabetes is mild, and the hypertension and cardiac disturbances seen were too severe for the diabetes that they had. This would suggest that they were specific defects due to *LMNA* mutations, but this is very difficult to study because they can also have this cardiac problem due to the diabetic complications.

Julien: I have a question concerning therapies for people with laminopathies. Is there much hope? It seems very complicated. If you overexpress lamin it causes disease and it will be hard to down-regulate the specific allele. Is there any potential hope for gene therapy or cell therapy?

Wilson: There is a collective wish!

Julien: For muscular dystrophy caused by dystrophin mutations, the gene has been identified so what progress has been made? Have there been trials of cell transplantation or gene therapy?

Davies: Cell transplantation didn't work because there wasn't fusion of host myoblasts or implanted myoblasts. It works in the mouse but doesn't work in humans. There are a lot of strategies to either replace dystrophin or induce exon skipping. I would say there is a lot of promise, yet a year ago I would have said I was very pessimistic. So maybe you should be more optimistic.

Goldman: It is good to be optimistic because the NIH and other funding agencies should be excited by this!

Davies: We also need to be patient with dystrophies. I am not talking about next year, or even in the next decade potentially.

Gasser: Coming to this fresh, it seems to me that one part of the link is there: you have the mutations which have incredible phenotypes. Whether this is due to gene misregulation or other mechanisms remains unclear. If it had something to do with stabilization of differentiated patterns of gene expression, the hope would be that up-regulating methylation or something else completely unrelated might be an effective therapy. However, you can't even start thinking about therapy if you don't know what is going on downstream from the mutation.

Goldman: On the basis of what we have heard at this meeting it is going to be a long time before we understand all these interactions.

Capeau: I have a general question about the deleterious effect of pre-lamin A. There are two ways by which an increase of pre-lamin A could be deleterious: the accumulation of pre-lamin A or the decrease of normal lamin A which is no longer present. What is the answer?

Young: I don't know, but we have 'lamin C-only' mice that are 10-weeks old and they look healthy. They certainly don't look anything like Colin Stewart's lamin A/C knockout mice that die early. This raises the question as to whether a farnesyltransferase inhibitor which would more-or-less get rid of lamin A formation (this is assuming that non-farnesylated prelamin A wouldn't do anything), would have any effect.

Goldman: You could do dose-dependency experiments.

Young: By the way, when we give farnesyltransferase inhibitors to cells there is a huge accumulation of pre-lamin A. I don't know what this means.

Gasser: It means that it gets degraded when it is at the membrane.

Burke: One of the things you might predict could be occurring with pre-lamin A is that one might expect it to behave differently during mitosis, simply by virtue of retention of this lipid tail. Larry Gerace showed a long time ago that B-type lamins, which are constitutively farnesylated, always partition with membranes in mitotic cells, whereas the A-type lamins behave as soluble proteins. It is possible that one could be subtly screwing up certain aspects of late mitotic progression simply by having lamin A in the wrong place at the wrong time.

Goldman: Junko Oshima, you said that in some cells you didn't see nuclear lamin A inclusions.

Oshima: That is right.

Goldman: The implication of this, if it is Brian Kennedy working with you, is that the initiation or early sites of DNA replication are not present in those nuclei. This is interesting.

Oshima: These are his data so I can't really comment.

Wilson: The Kennedy work is unique and important because it did not use immortalized cells. His findings relate to nuclear organization in primary cells and tissues, which is much more relevant to lamin function in real cells and laminopathy patients.

Gasser: When he looked in transformed cell lines he no longer saw the lamin and Rb correlation with the earliest firing origins. I think the data are solid. There are a lot of artefacts working with transformed cells and particularly in relation to heterochromatin and lamins.

Julien: Do patients with progeria develop pathologies such as Alzheimer's amyloid plaques or Tau protein tangles?

Oshima: No.

Goldman: The incidence is 1 in 8 000 000 and therefore the amount of material available is very small. The nervous system of Hutchinson-Gilford progeria syndrome (HGPS) patients appears to be relatively normal.

Oshima: There is no brain phenotype in mouse models. Progeria patients actually have a slightly higher IQ than the normal population. This comes from an $n = 3$!

Levy: The reason why this syndrome is called a segmental progeroid syndrome is because it doesn't mimic physiological ageing in every respect. There aren't any cognitive impairments, and some processes aren't advanced. If lamins are one of the mechanisms involved in normal ageing, they are just one of the several mechanisms.

Goldman: It has been thought for a long time that if one could identify the gene for progeria, this would reveal some of the mechanisms involved in normal ageing.

Capeau: I would go further and say that this is ageing of the mesenchymal tissues.

Goldman: We heard that there is epithelial involvement early on as indicated by loss of hair.

Capeau: It could be dermal involvement.

Goldman: It is fascinating that mutant lamin As have obvious mesenchymal but not epithelial cell phenotypes.

Stewart: There is a major interaction between the dermis and the epidermis in the hair follicle. Although the hair follicle is largely of epidermal origin, there is a significant contribution, by the dermis, to the dermal papilla. The dermal papilla is important in regulating cycles of hair growth.

Goldman: We have to explain why lamin A in epithelial cells gives no phenotype, and why lamin A defects in mesenchymal cells give a phenotype.

Capeau: Using this protease inhibitor we have also looked at the effect of hepatocytes in culture. We are able to see these dysmorphic nuclei in epithelial cells. There is a phenotype, but it could be silent, and the only expressed phenotype could be mainly in cells of mesenchymal origin.

A lamin-dependent pathway that regulates nuclear organization, cell cycle progression and germ cell development

Ayelet Margalit, Jun Liu*, Alexandra Fridkin, Katherine L. Wilson† and Yosef Gruenbaum[1]

*Department of Genetics, The Institute of Life Sciences, The Hebrew University of Jerusalem, Jerusalem 91904, Israel, *Department of Molecular Biology and Genetics, Cornell University, Ithaca, NY 14853, USA, and †Department of Cell Biology, JHMI, Baltimore, MD, USA*

Abstract. The *C. elegans* genome encodes a single lamin protein (Ce-lamin), three LEM domain proteins (Ce-emerin, Ce-MAN1 and LEM-3) and a single BAF protein (Ce-BAF). Down-regulation of Ce-lamin causes embryonic lethality. Abnormalities include rapid changes in nuclear morphology during interphase, inability of cells to complete mitosis, abnormal condensation of chromatin, clustering of nuclear pore complexes (NPCs), and missing or abnormal germ cells. Ce-emerin and Ce-MAN1 are both embedded in the inner nuclear membrane, and both bind Ce-lamin and Ce-BAF; in addition, both require Ce-lamin for their localization. Mutations in human emerin cause X-linked recessive Emery-Dreifuss muscular dystrophy. In *C. elegans*, loss of Ce-emerin alone has no detectable phenotype, while loss of 90% Ce-MAN1 causes ∼15% embryonic lethality. However in worms that lack Ce-emerin, a ∼90% reduction of Ce-MAN1 is lethal to all embryos by the 100-cell stage, with a phenotype involving chromatin condensation and repeated cycles of anaphase chromosome bridging and cytokinesis. The anaphase-bridged chromatin retained a mitosis-specific phosphohistone H3 epitope, and failed to recruit detectable Ce-lamin or Ce-BAF. Down-regulation of Ce-BAF showed similar phenotypes. These findings suggest that lamin, LEM-domain proteins and BAF are part of a lamina network essential for chromatin organization and cell division, and that Ce-emerin and Ce-MAN1 share at least one and possibly multiple overlapping functions, which may be relevant to Emery-Dreifuss muscular dystrophy.

2005 Nuclear organization in development and disease. Wiley, Chichester (Novartis Foundation Symposium 264) p 231–245

---

[1]This paper was presented at the symposium by Yosef Gruenbaum to whom correspondence should be addressed.

The major components of the nuclear lamina are the lamins. Lamins are nuclear intermediate filament proteins, and are highly conserved in metazoan cells. They are located both at the nuclear periphery and elsewhere in the nucleoplasm. Most or all integral proteins of the inner nuclear membrane interact either directly or indirectly with lamins and are considered to be a part of the nuclear lamina. The nuclear lamina is an essential component of metazoan cells. It is involved in most nuclear activities including DNA replication, RNA transcription, nuclear and chromatin organization, cell cycle regulation, cell development and differentiation, nuclear migration and apoptosis (reviewed in Gruenbaum et al 2003, Holaska et al 2002, Hutchison 2002, Shumaker et al 2003).

Specific mutations in nuclear lamina genes, *LMNA*, *STA/EMR* and *LBR* cause a wide range of heritable human diseases, termed laminopathies. These diseases include Emery Dreifuss muscular dystrophy, limb girdle muscular dystrophy, dilated cardiomyopathy (DCM) with conduction system disease, familial partial lipodystrophy (FPLD), autosomal recessive axonal neuropathy (Chariot-Marie-Tooth disorder type 2, CMT2), mandibuloacral dysplasia (MAD), Hutchinson-Gilford Progeria Syndrome (HGPS), atypical Werner syndrome, Pelger-Huet anomaly (PHA) and Greenberg Skeletal dysplasia (Burke & Stewart 2002, Gruenbaum et al 2003, Wilson et al 2001). Mutations in other nuclear envelope proteins are also expected to cause diseases (Schirmer et al 2003).

The best-studied laminopathy is Emery-Dreifuss muscular dystrophy (EDMD), characterized by early contractures of the Achilles, elbow and neck tendons, progressive muscle wasting, and conduction defects in the heart. The X-linked form of EDMD is caused by loss of emerin, an integral protein of the nuclear inner membrane. The autosomal dominant form of EDMD is caused by missense (and other) mutations in *LMNA*, the gene encoding A-type lamins. Thus, EDMD can result from relatively subtle changes in lamin A filaments, or from the loss of a specific protein (emerin) that binds lamin A (Burke & Stewart 2002, Goldman et al 2002).

Emerin contains a 'LEM-domain', the defining ∼40-residue motif shared by a family of nuclear proteins that includes L̲AP2, e̲merin, M̲AN1, lem-3 and otefin, as well as several uncharacterized proteins (Gruenbaum et al 2003). The LEM-domains of LAP2 and emerin mediate their direct binding to a chromatin protein named Barrier-to-Autointegration Factor (BAF). All 'LEM-proteins' tested (LAP2α, LAP2β, emerin and MAN1) also have a separate domain that confers direct binding to A- or B-type lamins (Mattout-Drubezki & Gruenbaum 2003). The LEM-domain proteins LAP2β and emerin also bind GCL, which is a transcription repressor required for germ line specification in *Drosophila* and for nuclear integrity in mice (Furukawa 1999, Holaska et al 2003, Nili et al 2001).

All LEM-domain proteins probably interact with Barrier-to-Autointegration Factor (BAF). BAF is a small conserved metazoan protein that was first identified for its role in retroviral DNA stability and integration (Cai et al 1998). BAF binds DNA with no sequence specificity. The protein forms dimers and has the ability to bridge dsDNA *in vitro* (Zheng et al 2000). BAF strongly influences chromatin condensation and decondensation during nuclear assembly in *Xenopus* extracts from *in vitro* assembled nuclei (Segura-Totten et al 2002) and is essential in both *C. elegans* (Zheng et al 2000) and *Drosophila* (Furukawa et al 2003). Interestingly, BAF competes with GCL for binding to emerin, suggesting that emerin forms at least two distinct types of nuclear complex *in vivo* (Holaska et al 2003).

Only three LEM-proteins are conserved in *C. elegans*: Ce-emerin and Ce-MAN1, which have transmembrane domains, and LEM-3, with no transmembrane domain (Lee et al 2000). The small number of LEM-proteins in *C. elegans*, and the presence of conserved genes encoding BAF and one B-type lamin, facilitates the study of their functions and interactions *in vivo*. Reducing the level of either lamin or BAF in *C. elegans* causes abnormal nuclear structure, catastrophic exit from mitosis (aberrant chromosome segregation and anaphase chromosome bridging) and early embryonic lethality (Liu et al 2000). In contrast, elimination of emerin has no detectable phenotype in *C. elegans* (Gruenbaum et al 2002). In humans, emerin is expressed in nearly all tissues, but the null phenotype is restricted to skeletal muscles, cardiac function and major tendons, suggesting that the unaffected tissues may express protein(s) that overlap functionally with emerin.

Ce-lamin is an essential protein required for nuclear organization, cell cycle progression, correct spacing of NPCs and germ line development

The *C. elegans* genome contains a single lamin gene, three LEM domain genes, two of which (Ce-emerin and Ce-MAN1) are localized at the inner nuclear membrane, and a single BAF gene (Gruenbaum et al 2003). Ce-lamin, Ce-emerin and Ce-MAN1 are expressed ubiquitously during *C. elegans* development (Gruenbaum et al 2002, Liu et al 2000, 2003), and remain in a spindle envelope until mid-late anaphase during mitosis (Lee et al 2000).

Ce-lamin is present at both the nuclear periphery. RNA interference experiments (RNAi) aimed at down-regulating Ce-lamin (*lmn-1 (RNAi)*) were performed by injecting *lmn-1* dsRNA into the gonads of adult hermaphrodites or by feeding hermaphrodites with bacteria expressing *lmn-1* dsRNA. Most *lmn-1 (RNAi)* embryos were able to form several hundred nuclei before arrest (Liu et al 2000). All *lmn-1 (RNAi)* embryos were abnormal and contained nuclei that varied in size. Later on, the *lmn-1 (RNAi)* embryos degrade and form regions in the embryos that were devoid of nuclei. The reduced *lmn-1* activity and embryonic arrest in the *lmn-1 (RNAi)* embryos indicated that lamins are an essential component of the nuclear envelope.

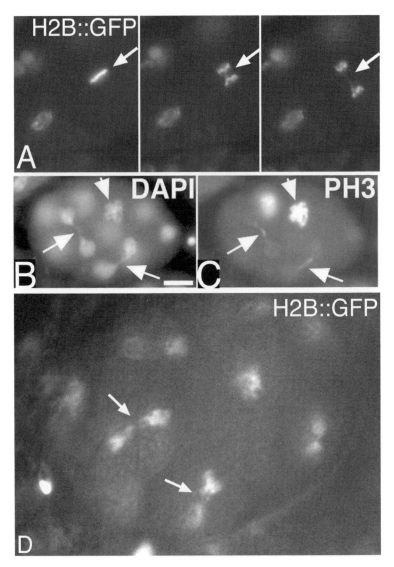

FIG. 1.   Chromatin anaphase bridges are caused by down-regulation of Ce-lamin (A), Ce-emerin plus Ce-MAN1 (B,C) or Ce-BAF (D). (A) An *lmn-1 (RNAi)* embryo expressing H2B::GFP (Praitis et al 2001) was subjected to time-lapse microscopy. The arrow in each of the three images, which are 3 min apart from each other, shows the same nucleus undergoing mitosis. (B) DAPI staining and (C) mitosis-specific anti-phospho-histone H3 staining of an *emr-1(RNAi); lem-2(RNAi)* embryo. Arrows indicate the anaphase-bridged chromatin; arrowhead indicates a late prophase nucleus. (D) GFP fluorescence of a *baf-1 (RNAi)* embryo expressing H2B::GFP (Praitis et al 2001). Arrows indicate the anaphase-bridged chromatin.

The phenotype of *lmn-1 (RNAi)* embryos was examined by 4-D time-lapse microscopy on live embryos. Already at the 2-cell stage the *lmn-1 (RNAi)* nuclei showed rapid changes in shape. Despite the gross defects in nuclear morphology, nuclear divisions still occurred in *lmn-1 (RNAi)* embryos, with these embryos eventually producing several hundred cells before arrest. The *lmn-1 (RNAi)* were unable to complete the cell cycle, had chromatin anaphase bridges (Fig. 1A), and unequal distribution of chromatin in daughter nuclei. These abnormalities were not due to gross defects in microtubule organization, since the pattern of staining of the *lmn-1 (RNAi)* embryos with tubulin antibodies was similar to that of wildtype embryos. In addition, many (but not all) nuclei of the *lmn-1 (RNAi)* embryos showed an abnormal clustering of nuclear pore complexes (Liu et al 2000).

Animals that hatched from eggs laid outside the most potent window of RNAi activity (escaper F1 adults) were sterile (no germ cells) or semi-sterile (reduced amounts and defective germ cells), showing no or dramatic reductions in the number of germ cells. A fraction of these germ cells had multiple nuclei, large nuclei or spermatocyte-like nuclei with condensed chromatin. Some of the F2 embryos from the semi-sterile animals developed into fertile adults. Among the progeny, there was a high incidence of males (12.25%).

The nuclear localization of Ce-emerin and Ce-MAN1 is dependent on Ce-lamin but not on other known nuclear envelope proteins

In *lmn-1 (RNAi)* embryos, in which Ce-lamin could not be detected with Ce-lamin antibodies (>98% down-regulation), both Ce-emerin and Ce-MAN1 were mislocalized in the cytoplasm of embryonic cells (Gruenbaum et al 2002, Liu et al 2003). Ce-emerin was more sensitive than Ce-MAN1 to the level of Ce-lamin down-regulation, since in *lmn-1 (RNAi)* embryos that had higher residual levels of Ce-lamin, Ce-MAN1, but not Ce-emerin, remained localized at the nuclear envelope.

In contrast to Ce-lamin, Ce-emerin remained at the nuclear envelope of embryos with down-regulated Ce-MAN1 or in embryos with a deletion in the gene encoding the inner nuclear membrane protein UNC-84 (Lee et al 2002). Likewise, Ce-MAN1 remained at the nuclear envelope of embryos with an *emr-1* deletion (A. Margalit, unpublished observations) or in embryos with an *unc-84* deletion (Lee et al 2002).

To determine if Ce-lamin or other nuclear envelope proteins depend on Ce-emerin or Ce-MAN1 for their localization, we analyzed the cellular distribution of Ce-lamin, UNC-84, FG-nucleoporins and Ce-BAF in embryos with down-regulated Ce-emerin or Ce-MAN1 or both. While Ce-lamin, UNC-84 or FG-nucleoporins did not require Ce-emerin or Ce-MAN1 for their nuclear envelope localization, Ce-BAF redistributed in the *lem-2 (RNAi); emr-1 (RNAi)* embryos (Liu et al 2003).

These results demonstrate that lamins play a major role in preventing inner nuclear membrane proteins from diffusing back into the ER. However, not every inner nuclear membrane protein requires Ce-lamin for its nuclear envelope localization (A. Fridkin, K. L. Wilson, Y. Gruenbaum, unpublished observations), while other inner nuclear membrane proteins, including UNC-83 (Starr et al 2001), also require partner proteins for their localization. These results indicate that lamins, chromatin and other lamina proteins are all required for proper development of the nuclear envelope.

Ce-MAN1 and Ce-emerin bind both Ce-lamin and BAF

Confocal microscopy showed that both Ce-emerin and Ce-MAN1 colocalized with Ce-lamin and immunogold electron microscopy showed that they were both associated with the peripheral chromatin (Gruenbaum et al 2002, Liu et al 2003). Ce-emerin coimmunoprecipitated with endogenous Ce-lamin from lysates of *C. elegans* (Gruenbaum et al 2002). Blot overlays also showed that Ce-emerin bound directly to both Ce-lamin and Ce-BAF (A. Margalit, unpublished observations). To determine if Ce-MAN1 bound directly to Ce-lamin, we used ^{35}S-Ce-lamin to probe blots of protein lysates from bacteria that expressed either the N-terminal (residues 1-333), or C-terminal (residues 400-500) nucleoplasmic domains of Ce-MAN1. ^{35}S-Ce-lamin bound selectively to the N-terminal fragment, not the C-terminal fragment. Parallel blots were probed with ^{35}S-Ce-BAF, to determine if Ce-BAF binds the LEM-domain of Ce-MAN1. Ce-BAF bound strongly to the N-terminal fragment and weakly to the C-terminal fragment, which has no defined LEM-domain. The presence of a second Ce-BAF-binding region in Ce-MAN1, though not currently understood, strengthened our conclusion that Ce-BAF interaction was important for Ce-MAN1 function. In summary, these biochemical results showed that Ce-MAN1 interacts with two of emerin's conserved partners, Ce-lamin and Ce-BAF, thus supporting the idea that Ce-MAN1 and Ce-emerin have overlapping functions at the inner nuclear membrane.

Ce-emerin is not required for viability in embryos or adults

Animals depleted of Ce-emerin by RNAi (*emr-1 (RNAi)*) or animals with a Ce-emerin deletion (VC237 strain; gk119 homozygous) had no detectable phenotype during development. They developed at normal rates into fertile adult worms. They also displayed normal movement and feeding behaviour, and produced viable fertile offspring with similar brood sizes (Gruenbaum et al 2002, A. Margalit, unpublished observations). These results were not surprising, since emerin loss in humans has no detectable effect until childhood, and then selectively affects a few specific tissues (Emery 1989). In conclusion, Ce-emerin is not essential in *C. elegans*.

Ce-MAN1 is essential for viability

Examination of the *lem-2 (RNAi)* embryos showed that Ce-MAN1 protein was relatively stable. Embryos had significantly reduced (85%–90%), but not eliminated, slightly punctate residual Ce-MAN1 epitopes at the nuclear envelope. The reduction of Ce-MAN1 protein levels caused 15% embryonic lethality, with most dead embryos (>98%) arresting late, after the two fold stage (Liu et al 2003). These results suggested that Ce-MAN1 might be an essential component of the nuclear envelope.

Ce-MAN1 and Ce-emerin have overlapping functions

As mentioned above, Ce-emerin and Ce-MAN1 are the two nuclear membrane-embedded LEM-proteins in *C. elegans*, and Ce-emerin is non-essential. To test the hypothesis that Ce-emerin provides functional backup for Ce-MAN1 in *C. elegans*, we did double-RNAi experiments to reduce or eliminate both proteins. The results were striking, with 100% embryonic lethality by the 100-cell stage in *lem-2 (RNAi); emr-1 (RNAi)* embryos laid 12–36 hours after injection of dsRNA. Thus, in the absence of Ce-emerin, lowering the levels of Ce-MAN1 caused a complete arrest of embryonic development. We concluded that Ce-MAN1 and Ce-emerin perform at least one overlapping essential function in *C. elegans*. DAPI staining of double-*RNAi* embryos at the stages when they arrested (<100 cells) showed that over 50% of the nuclei examined had abnormally condensed chromatin. This condensed chromatin phenotype was probably not the result of anaphase chromatin bridges, since nuclei with condensed chromatin were observed even at the one-cell stage. DIC time-lapse microscopy was used to follow the fate of nuclei and chromatin in *lem-2 (RNAi); emr-1 (RNAi)* embryos. This analysis showed that unlike loss of Ce-lamin, which destabilizes nuclear shape (Liu et al 2000), the down-regulation of both Ce-emerin and Ce-MAN1 did not affect nuclear shape. Thus, at least some lamina functions were still normal. Microtubule patterns as determined by immunofluorescence also appeared normal. The most striking phenotype in *lem-2 (RNAi); emr-1 (RNAi)* embryos was anaphase chromatin bridges, which were present as early as the first nuclear divisions (Fig. 1B). The formation of anaphase bridges apparently delayed, but did not block, cytokinesis.

When cells containing anaphase-bridged chromosomes were immunostained for endogenous Ce-lamin, Ce-lamin was present throughout the two daughter nuclei, but was absent from anaphase-bridged chromatin (Liu et al 2003). In *C. elegans*, Ce-lamin proteins are mitotically disassembled, and therefore completely absent from the chromatin surface during mid-late anaphase. To test the idea that the chromatin found in daughter nuclei had properly exited from mitosis and initiated nuclear envelope formation (whereas anaphase-bridged chromatin remained 'mitotic') cells containing anaphase bridges were immunostained using an antibody specific for phosphorylated serine 10 on histone H3, which is specific for mitotic chromatin. Prophase nuclei stained positively for phosphohistone H3, verifying this marker. The anaphase-bridged chromatin, but not the segregated chromatin, also stained positive for phosphohistone H3 (Fig. 1C). Thus, for cells that lacked both Ce-MAN1 and Ce-emerin, the anaphase-bridged chromatin was fundamentally compromised in its ability to biochemically exit from mitosis and segregate. To our knowledge, this phenotype has not previously been reported. Interestingly, feeding embryos with BAF dsRNA also causes anaphase chromatin bridges (Fig. 1D).

Lamin-based network at the nuclear periphery

We suggest a lamin-based network containing lamins, LEM domain protein and BAF. This network mediates interactions between the nuclear lamina and chromatin and plays a crucial role in regulating chromatin organization and mitosis. The evidence for the existence of such networks is outlined below:

- MAN1 and emerin both depend on lamin for their nuclear lamina localization, while BAF requires these LEM domain proteins for its normal localization
- MAN1 and emerin bind both lamin and BAF
- The phenotypes observed by down-regulating emerin and MAN1 or BAF are similar and include chromatin condensation and specific mitotic defects, including the chromatin anaphase bridges. Down-regulation of lamin also shows these phenotypes
- These phenotypes represent only a subset of those observed in down-regulating Ce-lamin. Elimination of other nuclear envelope proteins that depend on lamin for their nuclear envelope localization (i.e. UNC-84, UNC-83; Gruenbaum et al 2002, Starr et al 2001) or down-regulation of other proteins that interact directly with lamin does not show these phenotypes.

This lamin-based network is probably conserved in evolution, since all these proteins are evolutionarily conserved. However, the larger number of lamin and LEM-domain genes in mammals suggest that this pathway is more complex and involves a larger number of proteins. LEM domain proteins present at the nucleoplasm (i.e. LAP2α, LAP2ζ) can also be a part of this pathway.

Acknowledgments

This work was funded by grants from the Israel Science Foundation (ISF), the Binational Science Foundation Israel-USA, the Jubiläumsfond of the Austrian National Bank (9006), and the National Institutes of Health (GM64535 to K.L.W. and Y.G. and GM66953 to J.L.).

References

Burke B, Stewart CL 2002 Life at the edge: the nuclear envelope and human disease. Nat Rev Mol Cell Biol 3:575–585

Cai M, Huang Y, Zheng R et al 1998 Solution structure of the cellular factor BAF responsible for protecting retroviral DNA from autointegration. Nat Struct Biol 5:903–909

Emery AE 1989 Emery-Dreifuss syndrome. J Med Genet 26:637–641

Furukawa K 1999 LAP2 binding protein 1 (L2BP1/BAF) is a candidate mediator of LAP2-chromatin interaction. J Cell Sci 112:2485–2492

Furukawa K, Sugiyama S, Osouda S et al 2003 Barrier-to-autointegration factor plays crucial roles in cell cycle progression and nuclear organization in Drosophila. J Cell Sci 116:3811–3823

Goldman RD, Gruenbaum Y, Moir RD, Shumaker DK, Spann TP 2002 Nuclear lamins: building blocks of nuclear architecture. Genes Dev 16:533–547

Gruenbaum Y, Lee KK, Liu J, Cohen M, Wilson KL 2002 The expression, lamin-dependent localization and RNAi depletion phenotype for emerin in C. elegans. J Cell Sci 115:923–929

Gruenbaum Y, Goldman RD, Meyuhas R et al 2003 The nuclear lamina and its functions in the nucleus. Int Rev Cytol 226:1–62

Holaska M, Wilson KL, Mansharamani M 2002 The nuclear envelope, lamins and nuclear assembly. Curr Opin Cell Biol 14:357–364

Holaska JM, Lee KK, Kowalski AK, Wilson KL 2003 Transcriptional repressor germ cell-less (GCL) and barrier-to-autointegration factor (BAF) compete for binding to emerin in vitro. J Biol Chem 278:6969–6975

Hutchison CJ 2002 Lamins: building blocks or regulators of gene expression? Nat Rev Mol Cell Biol 3:848–858

Lee KK, Gruenbaum Y, Spann P, Liu J, Wilson KL 2000 C. elegans nuclear envelope proteins emerin, MAN1, lamin, and nucleoporins reveal unique timing of nuclear envelope breakdown during mitosis. Mol Biol Cell 11:3089–3099

Lee KK, Starr D, Liu J et al 2002 Lamin-dependent localization of UNC-84, a protein required for nuclear migration in C. elegans. Mol Biol Cell 13:892–901

Liu J, Rolef-Ben Shahar T, Riemer D et al 2000 Essential roles for Caenorhabditis elegans lamin gene in nuclear organization, cell cycle progression, and spatial organization of nuclear pore complexes. Mol Biol Cell 11:3937–3947

Liu J, Lee KK, Segura-Totten M, Neufeld E, Wilson KL, Gruenbaum Y 2003 MAN1 and emerin have overlapping function(s) essential for chromosome segregation and cell division in C. elegans. Proc Natl Acad Sci USA 100:4598–4603

Mattout-Drubezki A, Gruenbaum Y 2003 Dynamic interactions of nuclear lamina proteins with chromatin and transcriptional machinery. Cell Mol Life Sci 60:2053–2063

Nili E, Cojocaru GS, Kalma Y et al 2001 Nuclear membrane protein, LAP2b, mediates transcriptional repression alone and together with its binding partner GCL (germ-cell-less). J Cell Sci 114:3297–3307

Praitis V, Casey E, Collar D, Austin J 2001 Creation of low-copy integrated transgenic lines in Caenorhabditis elegans. Genetics 157:1217–1226

Schirmer EC, Florens L, Guan T, Yates JR 3rd, Gerace L 2003 Nuclear membrane proteins with potential disease links found by subtractive proteomics. Science 531:1380–1382

Segura-Totten M, Kowalski AK, Craigie R, Wilson KL 2002 Barrier-to-autointegration factor: major roles in chromatin decondensation and nuclear assembly. J Cell Biol 158:475–485

Shumaker DK, Kuczmarski ER, Goldman RD 2003 The nucleoskeleton: lamins and actin are major players in essential nuclear functions. Curr Opin Cell Biol 15:358–366

Starr DA, Hermann GJ, Malone CJ et al 2001 unc-83 encodes a novel component of the nuclear envelope and is essential for proper nuclear migration. Development 128:5039–5050

Wilson KL, Zastrow MS, Lee KK 2001 Lamins and disease: insights into nuclear infrastructure. Cell 104:647–650

Zheng R, Ghirlando R, Lee MS, Mizuuchi K, Krause M, Craigie R 2000 Barrier-to-autointegration factor (BAF) bridges DNA in a discrete, higher-order nucleoprotein complex. Proc Natl Acad Sci USA 97:8997–9002

DISCUSSION

Gasser: It is not clear why one is getting a cut phenotype. Usually this is associated with either lack of segregation — incomplete termination of replication or topoisomerase 2 inactivity — and not interference with transcriptional events due to lack of association with the periphery. Have you ever tried to couple this with topoisomerase 2 down-regulation or inhibition to see whether it is highly synthetic?

Gruenbaum: Indeed, mutations in the yeast Top2 also cause the cut phenotype. In mammalian cells the APC complex of fizzy/fizzy-related also causes the cut phenotype. There are two questions that we are currently addressing. (1) Why do we get the cut phenotype? Is it due to lack of cohesion-degradation? Is it due to cycles of breakage and rejoining of the telomeres, etc.? (2) What is the initial effect of the lamin/lem-domain/baf pathway? These are not easy experiments in *C. elegans*, which we are currently performing.

Wilson: There's another issue, since BAF is capable of regulating gene expression. Whether it does so at this early stage of embryogenesis is unknown. However embryonic death almost certainly indicates that BAF has structural roles involving chromosomes and mitosis. In *Xenopus* egg extracts, where there is no transcription, all the structural components needed to assemble nuclei and chromatin are present. When extra BAF is added to these nuclear assembly reactions we get very potent structural effects, either enhancing chromatin decondensation, or if you add too much then potent compression of chromatin structure. We conclude that BAF has direct structural roles that involve interactions with LEM domain proteins and lamins, but nothing that would depend on gene expression. However in somatic cells, BAF could have roles in both structure and gene expression.

Gruenbaum: We should probably also test the effect of BAF on transcriptional activity.

Wilson: It could also be a DNA replication problem.

Starr: It could also be specific to *C. elegans* in which mitosis is different because it is holocentric: MTs are connecting along the entire length of the chromosome. I have seen very similar anaphase-bridge phenotypes with RNAi against kinetochore proteins (Scaerou et al 2001). The anaphase bridges could be a secondary result of an earlier problem in development.

Stewart: Is it possible to inactivate BAF at a later stage of development or in specific tissues?

Gruenbaum: You have raised an excellent point, which is the role of BAF during the later stages of development. It is much more complicated to perform these experiments for example expressing BAF dsRNA driven by stage- or tissue-specific promoters. Interestingly, when we looked at embryos outside the most potent windows of RNAi there were many abnormalities at later stages but we have not analysed them yet.

Malone: One reason that RNAi is so powerful in *C. elegans* is that you can deplete the protein from the gonad. You effectively have freshly packaged oocytes with defined protein levels. You don't have to watch the cell deal with lower and lower levels of protein.

Wilson: It may be that you get distinct phenotypes when there is less BAF than when there is no BAF, so watching the cells run out might be good.

Gerace: How many copies of BAF are there per nucleosome? Also, even though it is known to be a DNA binding protein, how does it interact with chromatin structures?

Wilson: BAF binds histones specifically. BAF binds to the tail of certain linker histones and also histone H3. We have started to map the binding regions in BAF for histones, and vice versa. Our model is that BAF interacts with nucleosomes. We have no idea how this influences its binding to LEM domain proteins.

Gruenbaum: In *Xenopus*, Dr Wilson has measured the concentration of BAF and found that it is an abundant protein.

Gerace: I am trying to get a sense of the relative abundance of BAF in relationship to nucleosomes.

Wilson: The concentration of BAF is 12 μM in *Xenopus* eggs. What is the content of histones?

Gerace: There's one octamer for every 200 base pairs in a somatic cell.

Gruenbaum: *Drosophila* and *C. elegans* embryos contain a large pool of maternally deposited BAF so the amount of active BAF can vary at different stages. Somatic *Xenopus* cells are probably the best place to measure the ratio between BAF and histones.

Wilson: We need to understand the BAF-histone interaction. We speculate that BAF might regulate the higher-order structure of nucleosomes, and it wouldn't need to be present at a 1:1 level with nucleosomes to do this.

Stewart: What happened to BAF's binding to naked DNA?

Wilson: BAF still binds naked DNA. This was the first thing to be discovered by Bob Craigie. Each BAF dimer can bind two pieces of DNA non-specifically. This was the basis for his model that BAF might 'bridge' DNA in retroviral preintegration complexes. If our model is correct that BAF binds nucleosomes, remember that linker DNA is naked.

Gerace: What do you mean by non-specifically? Is it saturable binding?

Wilson: BAF has no sequence specificity, but does require double-stranded DNA. This was shown by Bob Craigie. We haven't measured its affinity for DNA.

Gruenbaum: An important experiment would be to FRAP cells and see whether BAF is mobile or not.

Wilson: GFP–BAF is extremely mobile, but still had meaningful direct interactions with emerin at the nuclear envelope (Shimi et al 2004). However more work is needed because GFP might affect its mobility. Tokuko Haraguchi carefully tested several GFP-fusion constructs, and the FRAP studies were done with GFP fused to the N-terminus of BAF using a relatively long linker. This protein behaves like wildtype BAF in terms of its binding properties. BAF is so small (10 kDa). If you attach anything to its C-terminus (even a Myc tag), BAF function or dimerization appears to be affected, and we've backed her conclusions up biochemically. BAF is so small and sensitive that it needs to pack closely with various partners. If you attach anything to its C-terminus it will disrupt the dimer. It looks like it is changing BAF into a conformation that allows certain binding interactions but not others. Even a Myc tag will do this at the C-terminus.

Goldman: Then the GFP experiment may not be meaningful at all.

Wilson: We don't know yet. The N-terminal fusion behaves normally by other criteria. The high mobility of GFP–BAF is similar to the mobility of linker histones.

Capeau: What is the proportion of BAF attached to the nuclear membrane?

Wilson: About 50% of BAF is in the nucleus, and most of that is near the nuclear envelope.

Gruenbaum: This is probably also correct for early *C. elegans* embryos.

Ellis: In your emerin knockouts, have you looked to see whether MAN1 or any other nuclear envelope protein is up-regulated?

Gruenbaum: It's a good experiment. We can do it by Western blot analysis. By immunofluorescence we don't see a significant increase in MAN1 intensity as compared to wild-type, but this is not a quantitative method.

Ellis: Have you looked at the post-translational modifications of either emerin or MAN1 in *C. elegans*? The mammalian one is phosphorylated during mitosis.

Gruenbaum: We have not. Mitosis in *C. elegans* is very different from that in vertebrates because membrane proteins stay in a spindle envelope until mid-late anaphase. The nuclear pores disassemble earlier during prometaphase and molecules can get in and out of the nucleus. *C. elegans* lamin doesn't have the cdc2 site. I haven't found any obvious cdc2 sites on either MAN1 or emerin.

Starr: In your RNAi treatments do you see any earlier phenotypes, such as in the gonad or in the position of the meiotic spindle?

Gruenbaum: There is a reduction in fertility, so we should probably also look at the gonads.

Malone: With regards to the chromosome bridges, have you imaged embryos during pronuclear migration and the first mitosis to see whether other defects might cause these bridges as well?

Gruenbaum: We found chromosome bridges first during the embryonic mitosis, but not earlier.

Capeau: You said that there was a chromatin condensation. Do you see increased apoptosis of these cells?

Gruenbaum: When we analyse embryos with down-regulated lamin by thin-section EM, the nuclei look apoptotic. However, we see no difference in lethality when lamin is down-regulated in ced-3 or ced-4 mutant strains and TUNEL assay does not label nuclei in embryos with down-regulated lamin. Interestingly, in *C. elegans*, one of the first apoptotic events is CED-4 translocation from the mitochondria to the nuclear envelope. Also interesting is that in *C. elegans*, lamin is probably not a substrate for the CED-3 caspase and is degraded very late in apoptosis.

Gasser: I'd like to return to the question of the anaphase bridge. If I understand correctly, neither lamin, emerin, Man1 or BAF is actually associated with the late mitotic chromosome. This is all an effect on something that has happened earlier in the cell cycle.

Wilson: There is BAF on the chromatin surface during late anaphase and telophase.

Gasser: What fraction of cells has the anaphase bridge? It seems that one is observing interference with proper replication, segregation and/or cohesion loading. In yeast, anything that interferes with this (proper firing of origins, loading of cohesion and decatenation) gives this phenotype. Many times you can compensate for that by slowing down S phase. That is, if you give the cell more time, it can resolve the anaphase bridge. In yeast you do it by adding low levels of HU or nocadazole; this slows down mitotic progression allowing the cell time to sort out the chromosome entanglement. I know you haven't yet looked at origin firing or cohesion loading, but you can get a feeling for whether this is going on in every cell, or whether it is partially stochastic. If it is stochastic then you might want to slow down mitosis by testing the effects at a lower temperature.

Gruenbaum: We can see embryos in all nuclei undergoing the anaphase bridges. We never tried a very low temperature. The lowest we tried was 16 °C and we see the same thing. It is always possible that a lower temperature would resolve the bridge. In addition, we see very early on condensation of chromatin. We concluded that both mitosis and global chromatin structure are affected by down-regulating the pathway.

Gasser: But there is lots of condensed chromatin in a nucleus, and it doesn't normally interfere with mitosis. I don't see the link between the two phenotypes.

Hutchison: I wonder whether you are thinking about this the wrong way round. What is curious to me in the double knockouts is that you are not actually getting any membranes forming around those chromosomes. There has to be something mediating the interaction between the membranes and the chromatin at anaphase. Kathy Wilson showed years ago that if there are no membranes in a *Xenopus* system you get no chromatin decondensation. I wonder whether what you are actually seeing results from limiting the amount of decondensation at anaphase because you are limiting the amount of membrane interacting with that. Is this a non-standard cut phenotype or is it normal?

Gruenbaum: This is the classical cut phenotype. Maybe the cut phenotype represents a variety of different cell cycle stages. The other thing that is striking here is that the membranes form next to segregated and decondensed chromatin, but not around the chromatin in the bridge. The latter contains the mitotic marker phosphohistone H3. We can not conclude — from the data that I presented — a role for the nuclear envelope in helping nuclei to get out of mitosis, but would suggest a role for the nuclear membranes in the decondensation of chromatin.

Collas: I would argue against that. In HeLa cell-free systems, we can induce some chromatin decondensation independently of membranes. This requires protein phosphatase 1 activity controlled by an, as yet, unknown nuclear PP1 regulatory subunit. This would correlate with your phospho H3 data, suggesting that there is a highly localized phosphatase activity. My guess would be that the lack of decondensation that you are seeing is independent of membrane binding.

Gerace: Is there any way to examine whether the length of S phase is increased in the cells in which you have down-regulated inducing the cut phenotype?

Gruenbaum: We are starting to do metabolic labelling experiments which we hope will give us some idea about what fraction of nuclei is undergoing S phase and even tell us something about the length of S phase. Another way to look at S phase would be to follow PCNA, which stains differently during different stages of S phase.

Wilson: Has anyone ever examined the cut phenotype in *C. elegans* cells to see whether the nuclear envelope reassembled? I thought the cut phenotype involved breaking chromosomes but there is still a whole nucleus involved. It is quite striking that there is some membrane attachment to chromatin, which is fairly condensed in the pictures that Yossi showed. This suggests that other types of nuclear membrane proteins mediate contact with the chromatin surface, but these are insufficient to decondense chromatin. I think these data are starting to reveal the relative contributions of LEM domain proteins.

References

Scaerou F, Starr DA, Piano F, Papoulas O, Karess RE, Goldberg ML 2001 The ZW10 and Rough Deal checkpoint proteins function together in a large, evolutionarily conserved complex targeted to the kinetochore. J Cell Sci 114:3103–3114

Shimi T, Koujin T, Segura-Totten M, Wilson KL, Haraguchi T, Hiraoka Y 2004 Dynamic interaction between BAF and emerin revealed by FRAP, FLIP and FRET analysis in living HeLa cells. J Struct Biol 147:31–41

Mutations in the mouse *Lmna* gene causing progeria, muscular dystrophy and cardiomyopathy

Serguei Kozlov, Leslie Mounkes, Dedra Cutler, Terry Sullivan, Lidia Hernandez, Nicolas Levy* Jeff Rottman† and Colin L. Stewart[1]

*Cancer and Developmental Biology Laboratory, CCR, NCI at Frederick, PO Box B, Frederick. Maryland 21702, *Inserm U491 Genetique Medicale et Developpement Faculte de Medecine de la Timone 13385 Marseille Cedex 05 France, and †Department of Internal Medicine Vanderbilt University School of Medicine, Nashville, TN 37232, USA*

Abstract. At least ten different diseases have been linked to mutations in proteins associated with the nuclear envelope (NE). Eight of these diseases are associated with mutations in the lamin A gene (*LMNA*). These diseases include the premature ageing or progeric diseases Hutchinson-Gilford progeria and atypical Werner's syndrome, diseases affecting striated and cardiac muscle including muscular dystrophies and dilated cardiomyopathies, lipodystrophies affecting white fat deposition and skeletal development and a peripheral neuropathy resulting in motor neuron demyelination. To understand how these diseases arise from different mutations in the same protein, we established mouse lines carrying some of the same mutations found in the human *LMNA* gene, as both mouse and human lamin genes show a very high degree of sequence conservation. We have generated mice with different mutations resulting in progeria, muscular dystrophy and dilated cardiomyopathy. Our mouse lines are providing novel insights into how changes to the nuclear lamina affect the mechanical integrity of the nucleus and in turn intracellular signalling, such as the NF-κB pathway, as well as cell proliferation and survival, cellular functions that, when disrupted, may be the basis for the origin of such diseases.

2005 Nuclear organization in development and disease. Wiley, Chichester (Novartis Foundation Symposium 264) p 246–263

In all metazoans, a thin protein meshwork, the lamina, underlies the nuclear face of the inner nuclear membrane (INM). Because of its role in maintaining nuclear envelope (NE) integrity and providing anchoring sites for chromatin domains, the nuclear lamina is a significant determinant of interphase nuclear architecture

---

[1]This paper was presented at the symposium by Colin L. Stewart to whom correspondence should be addressed.

246

and function. The major components of the lamina are intermediate filament proteins, the nuclear lamins. Most adult mammalian somatic cells contain three major lamins, A, B1 and C, as well as several minor lamins (B2 and AΔ10). These various forms are grouped into two classes, A-type (A, AΔ10 and C) and B-type (B1 and B2). Separate genes encode lamins B1 and B2, while a single gene, *LMNA*, encodes the A-type lamins, which arise through alternative splicing of a common transcript (Burke & Stewart 2002).

The nuclear lamina has important roles in regulating DNA synthesis, RNA transcription and in the organization of chromatin (Goldman et al 2002). Transcriptional co-factors also associate with the lamins, indicating that the lamina and NE are important co-factors in transcriptional regulation. The lamins are also developmentally regulated, with all cells expressing one or other lamin B, whereas A-type lamins are absent from early embryonic stages of development and certain adult stem cell populations (Rober et al 1989, 1990, Stewart & Burke 1987).

The functional significance of the lamins has undergone a major re-evaluation, primarily due to the finding that at least eight inherited diseases, ranging from muscular dystrophies to premature ageing, are caused by mutations in the *LMNA* gene. The eight congenital diseases linked to mutations in A-type lamins can be clustered into three groups: the first and largest group predominantly affects striated muscles and includes the autosomal dominant form of Emery-Dreifuss muscular dystrophy (AD-EDMD), dilated cardiomyopathy (DCM) and limb girdle muscular dystrophy 1B (LMG1B). The second group of diseases has minimal if any effect on muscle but does affect white fat distribution. This group includes familial partial lipodystrophy (FPLD) and mandibuloacral dysplasia (MAD). In addition to the effects on fat distribution, MAD also results in cranio-facial abnormalities and osteolysis of the digits. A peripheral neuropathy, Charcot-Marie-Tooth type 2B (CMT2B), resulting in demyelination of the motor nerves, is also inherited as a rare recessive mutation in the *LMNA* gene (Mounkes et al 2003a, Worman & Courvalin 2002). The most recently identified group of diseases includes premature ageing or the progeric disease, Hutchinson-Gilford progeria syndrome (HGPS) and some cases of atypical Werner's syndrome. These diseases are primarily caused by amino acid substitutions in the A-type lamins. The exception is the most common mutation causing HGPS, G608G, a C→T base change that leads to a splicing defect in exon 11 of *LMNA* (De Sandre-Giovannoli et al 2003, Eriksson et al 2003).

With these seemingly diverse diseases having tissue-specific defects, one of the most immediate questions has been, how can so many different diseases arise from mutations in the same protein that is expressed in the majority of adult cells? Structural studies on the lamin A protein have provided some clues as to how the different diseases might arise. The mutations causing the muscular dystrophies and

DCM are distributed throughout the A-type lamins. Many of these mutations disrupt assembly of the lamins, and their incorporation into the lamina. These mutations also affect nuclear morphology and the localization of other proteins, such as emerin, to the nucleus. Mutations resulting in FPLD and MAD are clustered in the carboxy-terminal globular domain. They are predicted to have minimal effects on lamin structure and do not appear to disrupt lamin assembly or emerin localization. Precisely how the FPLD and MAD mutations result in disease is unclear, although it has been suggested that these mutations may either disrupt or enhance interactions of the A-type lamins with other nuclear proteins (Burke & Stewart 2002).

Our approach to understanding how mutations in *LMNA* result in the different disorders has been to derive mouse models of the different diseases, by introducing some of the same mutations found in the human gene into the mouse ortholog. Having a mouse model for each disease provides a valuable resource to understanding the physiological consequences and molecular basis of each disease. This article summarizes the current insights such mouse mutants have provided in understanding the laminopathies.

Lmna-deficient mice

In mammals, expression of A-type lamins is developmentally regulated while B-type lamins are found in all nucleated somatic cells. In the mouse, A-type lamins are absent from all pre-implantation stage embryonic cells (including ES cells), with their appearance commencing at about embryonic day 9 within the visceral endoderm and trophoblast (Stewart & Burke 1987). Subsequently, A-type lamins appear asynchronously in various tissues with certain cell types not acquiring these proteins until after birth (Rober et al 1989). A few cell types, notably cells of the immune system, pancreatic islets and Purkinje cells, only express B-type lamins. These findings indicate that at the cellular level, A-type lamins are non-essential. However, the significance of these observations to embryogenesis and how they relate to lamin function remains uncertain, although it has often been suggested that A-type lamins are involved in terminal differentiation, and are possible determinants of chromatin organization.

Surprisingly, *Lmna* deficient mice develop quite normally from birth (Sullivan et al 1999). However by the weaning stage the mice showed retarded growth, muscle weakness, locomotory problems and they had all died by 7–8 weeks of age. Death was associated with extensive muscular dystrophy, particularly in the muscles of the limb-girdles, intercostal musculature, and perivertebral muscles. The oesophageal, tongue and jaw muscles were also dystrophic. The heart developed dilated cardiomyopathy with left ventricular dilation and abnormal nuclear morphologies. The exact cause of death was unclear, although the effects on the jaw and oesophageal muscles made eating and drinking difficult. The affected mice had very little white fat, but they were not lipodystrophic (Cutler et al 2002).

Male mice also showed defective spermatogenesis with very few if any mature sperm being produced, apparently due to meiotic arrest and apoptosis at the zygotene stage. These results indicated that the minor germline specific form, lamin C2, which is expressed in the male germline may be important to gametogenesis. In contrast, in the female germ line, mature oocytes express the A-type lamins. However, A-type lamin deficient oocytes are fertilized and develop normally (Alsheimer et al 2004).

In humans, complete loss of A-type lamin expression has been reported in one instance, and was associated with late gestational lethality. Nuclei in cell lines derived from the fetus exhibited many of the structural changes seen in nuclei from the lamin deficient mice. In addition, one case of AD-EDMD was caused by a premature stop codon resulting in the affected allele producing only the first 6 amino acids of the A-type lamins. This individual would be equivalent to a functional heterozygote. The lamin deficient mice, therefore, do not correspond to any known case in humans (Muchir et al 2003). Nevertheless, the fact that they develop muscular dystrophy and show a DCM pathology makes them a useful animal model for these diseases (Nikolova et al 2004).

Emerin-deficient mice

Emerin is a transmembrane protein primarily localized to the inner nuclear membrane (INM) of the nucleus. Emerin interacts with other nuclear factors such as the transcription factor germ-cell-less (*gcl*), the DNA binding protein BAF and splicing associated factor YT521-13 (Holaska et al 2003,Wilkinson et al 2003). Emerin localization to the INM depends on the A-type lamins, as loss of A-type lamin expression results in substantial redistribution of emerin to the endoplasmic reticulum (Sullivan et al 1999). The functions of emerin in the nucleus, apart from binding to the proteins listed above, remain unclear. However, mutations in the X-linked gene *STA*, which encodes emerin, result in X-linked Emery-Dreifuss muscular dystrophy (EDMD) in humans (Bione et al 1994). Patients with EDMD show contractions of the Achilles and elbow tendons, with progressive muscle wasting in the humero-peroneal region. Patients are prone to cardiac conduction defects and dilated cardiomyopathy, which is often lethal (Maidment & Ellis 2002).

We derived mouse lines in which the X-linked *Emd* gene was deleted resulting in the complete loss of emerin. To date the mice have not exhibited any overt signs of muscular dystrophy or atrophy, nor has ECG analysis of their hearts revealed any evidence of cardiac abnormalities. Such differences between mice and human patients were unexpected, and it is unclear why mice are tolerant of the loss of emerin.

Muscle affected by dystrophy attempts to compensate for muscle fibre loss by a process of regeneration, in which satellite cells proliferate and replace dead or damaged muscle cells. Adult mouse muscle tissue is able to regenerate after damage, with regeneration re-capitulating many of the features of myogenesis, including satellite cell activation, proliferation, differentiation, fusion of the myoblasts into myotubes, and subsequent formation of myofibres (Zhao & Hoffman 2004). GeneChip array analysis has been used to stage the changes in patterns and levels of gene expression over the 3–4 week regenerative process. Preliminary studies from the emerin-deficient mice revealed that, compared to their normal siblings, muscle regeneration was slightly delayed, with the emerin-deficient mice showing elevated levels of MyoD and myogenin expression 3 days after the start of regeneration, indicative of the emerin-deficient muscle compensating for a defect in regenerative pathways. Furthermore, expression of myogenic factor regulated genes, such as the acetylcholinesterase receptor and embryonic myosin-heavy chain, was delayed. These results suggest that X-linked EDMD may arise as a consequence of defective regenerative and repair processes in muscles (Melcon et al 2005).

Mice with dilated cardiomyopathy

Dilated cardiomyopathy is one of the major pathologies associated with mutations in *LMNA*, and we have established a mouse model for this cardiac pathology. A missense mutation, N195K, was one of the first identified *LMNA* mutations associated with DCM (Fatkin et al 1999). We introduced the same amino acid substitution into the mouse *Lmna* gene. As heterozygotes, these mice grow at the same rate as their wild-type siblings, are overtly normal and have shown no indications of cardiac abnormalities. However, mice homozygous for the N195K mutation die by 12–14 weeks of age. Apart from a slight retardation in growth rate, the mice do not exhibit any overt pathology in their tissues with the exception of severe cardiac abnormalities, including extensive extracellular fibrosis, chamber dilation, and an increase in heart mass. ECG measurements on the mutant hearts revealed the mice developed arrhythmic bradycardia before succumbing. Overall these results indicate that the N195K mice die from DCM (Mounkes et al 2005) (Fig. 1).

Progeria and the lamins

Hutchinson-Gilford progeria syndrome (HGPS) is a rare dominantly inherited disease, in which patients show symptoms of premature ageing (progeria),

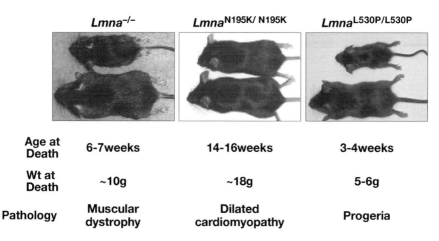

	Lmna$^{-/-}$	*Lmna*$^{N195K/N195K}$	*Lmna*$^{L530P/L530P}$
Age at Death	6–7weeks	14–16weeks	3–4weeks
Wt at Death	~10g	~18g	5–6g
Pathology	Muscular dystrophy	Dilated cardiomyopathy	Progeria

FIG. 1. A summary of the different phenotypes, ages at death and phenotypes of the mice carrying different mutations in the *Lmna* gene.

including severe growth retardation, loss of subcutaneous fat, alopecia, loss of bone density and poor muscle development (Sarkar & Shinton 2001). The average age of death in HGPS is 12 to 15 years, usually due to myocardial infarction or stroke (Sarkar & Shinton 2001). Patients, however, do not show any increase in tumour susceptibility, cataract formation or cognitive degeneration, features often associated with normal ageing. HGPS has, therefore, been referred to as a segmental progeroid syndrome, because it only partially reproduces the ageing processes (Martin & Oshima 2000).

In humans, the majority of HGPS cases are associated with a splicing defect in exon 11 of the *LMNA* gene, resulting in a truncated lamin A protein lacking 50 amino acids of the carboxy-terminal globular domain (Chaouch et al 2003, Eriksson et al 2003). The mutant A-type lamin has been tentatively assigned the name progerin. A mouse progeria model was derived from a splicing defect in exon 9 of *Lmna*, leading to a set of phenotypes remarkably similar to HGPS (Mounkes et al 2003b). Loss of subcutaneous fat, decreased bone density, poor muscle development, and growth retardation are some of the most striking features of the mouse model (Mounkes et al 2003b). In addition, the mice die prematurely by 4 weeks of age, and have craniofacial deformities similar to both MAD and HGPS. The extent to which these phenotypes reflect a normal ageing process is unclear, but the overall similarities in terminal phenotypes of the mice and patients is striking (Table 1).

A second premature ageing syndrome, Werner's syndrome, is inherited as an autosomal recessive due to mutations in a 3'-5'RecQ DNA helicase-exonuclease

TABLE 1. Comparison of the pathologies between children with HGPS and the progeric mouse line carrying a splice mutation

Human	Mouse
Severe growth retardation	Severe growth retardation
Short stature; failure to thrive	Short stature; failure to thrive
Mean death at 12–15 years	Death at 4–5 weeks
Craniofacial disproportion; micrognathy	Micrognathy
Abnormal dentition	Abnormal dentition
Parchment thin skin; loss of subcutaneous fat	Loss of subcutaneous fat
Decreased eccrine, sebaceous glands	Decreased eccrine, sebaceous glands
Scleroderma	Increased collagen deposition in skin
Alopecia, onset at ∼1 year	Decreased hair follicle density
Hyperkeratosis in some patients	Hyperkeratosis
Bone hypoplasia, resorption, osteoporosis	Decreased bone density; thin trabeculae
Hypoplasia/resorption of clavicles	Malformation of scapulae
Resorption of hip girdle joints; shuffling gait	Waddling gait
Congestive heart failure; decrease in vascular smooth muscle	Heart pathology; subtle changes consistent with pulmonary hypertension
Incomplete sexual maturation	Hypogonadism
Hypoplastic facial bones	Not determined
Thin diaphyses	Thinner femur; other diaphyses not studied
Osteolysis of terminal digits	None observed
Protruding ears; prominent eyes	Protruding ears
Myocardial fibrosis	Increased cardiac collagen & fibrocyte number
Atherosclerosis	No obvious defects in aorta, small vessels
Poor muscle development; atrophy	Poor muscle development and/or atrophy

that unwinds DNA and cleaves nucleotides from DNA termini. The disease, which maps to the *WRN* locus, manifests a high incidence of cancers, early onset cataracts, atherosclerosis, diabetes, premature graying of hair and early death, usually in the late 40s from myocardial infarction (Fry 2002, Hickson 2003, Oshima 2000). Unlike HGPS, Werner's syndrome is associated with an increased risk of neoplasms (Mohaghegh & Hickson 2001), although the mean age of death (47 years) in Werner's is much greater than HGPS, possibly allowing the accumulation of mutations that might enhance risk of unchecked cell proliferation.

The majority (83%) of Werner's patients have defects in the *WRN* locus. A small, but significant, number of cases of Werner's syndrome do not carry mutations in *WRN*. These are known as atypical cases, and recently mutations in

LMNA were found in 15% of these atypical Werner's cases (Chen et al 2003). These patients do not die in their teens, but have short stature, alopecia, osteoporosis, lipodystrophy, diabetes and muscle atrophy (Chen et al 2003).

Charcot-Marie-Tooth 2B mice

Another line of mice we have derived is a model for the peripheral neuropathy, CMT2B. We substituted the arginine residue at 298 for cysteine (R298C) and derived mice homozygous for this mutation. To date, the mice have shown no overt phenotype or any indication of muscle weakness that may be caused by the neuropathy. Furthermore, the nuclear morphology of fibroblasts established from R298C homozygous embryos is indistinguishable from wild-type nuclei. The reason why no evidence of neuropathy has been found in these mice is unclear, and the lack of phenotype is particularly puzzling as the *Lmna* null mice do show demyelination of motor neurons (De Sandre-Giovannoli et al 2002).

Disease mechanisms

We have introduced a range of mutations into the mouse *Lmna* gene, as well as a null mutation for the gene encoding the protein emerin. Of the mouse lines we have generated, the $Lmna^{-/-}$, N195K-DCM, progeric mice and, tentatively, emerin-deficient mice all develop distinct phenotypes, consistent with the diseases associated with the respective human mutations.

Many of the *Lmna* mutations result in significant alterations to nuclear structure and morphology. Although these alterations to nuclear structure have been some of the most obvious consequences of *LMNA* mutations, the significance of these morphological alterations and their involvement in disease outcome is less clear, as there is no consistent relationship between nuclear morphology and a particular laminopathy. Two principal hypotheses, the 'gene expression' and 'mechanical stress' hypotheses have been invoked to account for how the different laminopathies arise. In the 'gene expression' hypothesis, disruption to the lamina and/or NE affects chromatin organization and the expression of specific genes (Worman & Courvalin 2002). Evidence supporting the gene expression hypothesis includes the fact that some transcriptional factors, such as Oct-1, Gcl, Smads and pRb, associate and interact with the lamins and/or other proteins in the NE (Imai et al 1997, Kennedy et al 2000, Nili et al 2001, Osada et al 2003). Lamin mutations affect chromatin organization, which may in turn disrupt tissue specific gene expression resulting in the different pathologies. The 'mechanical stress' hypothesis, however, suggests that lamin mutations weaken the structural or mechanical integrity of nuclei, making cells more susceptible to damage caused by physical stress. Such a hypothesis may be valid for the disorders affecting mechanically stressed tissues such as skeletal, cardiac and smooth muscle. The role of mechanical stress in the lipodystrophies, and perhaps some of the tissues affected in the progerias, is less obvious (Burke & Stewart 2002).

Direct measurements on the deformability of nuclei from *Lmna* null mice have revealed that the nuclei are indeed more deformable, or less 'stiff,' when subjected to mechanical stretching. Intriguingly, the cytoplasm was also less rigid, indicating possible interactions between the nucleoskeleton and cytoskeleton. One of the outcomes from these studies is that *Lmna* null cells are more prone to apoptosis and necrosis following mechanical stress than their normal counterparts. Furthermore, loss of the lamins also compromised the ability to activate the NF-κB pathway, and in turn, upregulate the expression of genes such as the potential anti-apoptotic factor IEX-1 (Lammerding et al 2004).

In addition to these effects on the mechanical integrity of nuclei, we found that some of the lamin mutations affect cell proliferation. Disruption of the lamins and other NE associated proteins in *C. elegans* and *Drosophila* results in abnormal mitosis, incomplete chromosome segregation at mitosis, and eventual cell death. Cells derived from both embryonic and adult lamin null mice show a reduction in growth rate and eventual growth cessation that is more acute than that seen in normal cells, although immortalized *Lmna* null cells may eventually appear. In the progeric mice the effect of the mutation on proliferation is even more acute, and in contrast to the *Lmna* null mice, is developmentally regulated. Fibroblasts from progeric E13 embryos show no difference in short or long-term proliferation compared to cells from wild-type embryos. However, fibroblasts from 3–4 week old postnatal progeric mice had severely decreased rates of proliferation resulting in rapid death of the cells when placed in culture (Fig. 2). Together these results reveal that different lamin mutations can dramatically compromise cell proliferation, albeit in different ways (Fig. 1).

The observations that the laminopathies involve defective tissue regeneration and that progeric children (and mice) show significant growth retardation, suggest that cellular proliferation may be a significant factor contributing to these disorders. Impaired proliferation would not only reduce the ability of an affected tissue to regenerate, it would also, if more widely distributed in adult tissues, affect overall growth of the individual. In progeric children, development of the nervous system is apparently relatively unaffected, whereas most affected

---

FIG. 2.   Different lamin mutations affect fibroblast proliferation at different stages of development. (A and B) Fibroblasts from postnatal tissues of the progeric mice show severe growth retardation and death. (C) In contrast, fibroblasts derived from progeric E13 embryos have growth rates indistinguishable from wild-type embryos. (D) Fibroblasts from E13 *Lmna*$^{-/-}$ embryos show retarded proliferation and premature senescence.

A

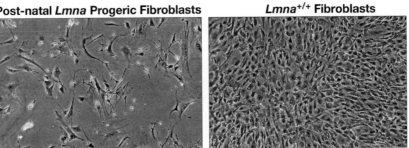

B Proliferation of post-natal progeric fibroblasts

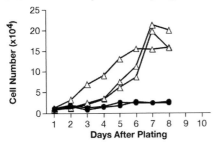
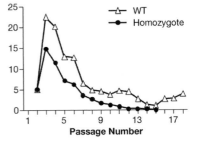

C Proliferation of embryonic E13 progeric fibroblasts

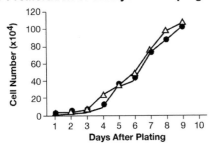
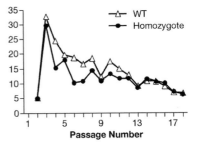

D Embryonic *Lmna*⁻/⁻ fibroblasts

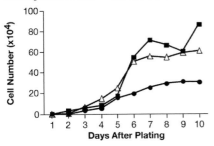
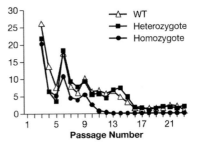

tissues, such as the skeletal, muscular and cardiovascular systems are of mesenchymal origin. At birth, almost 75% of the human neural system has formed and during postnatal development growth of the neural system largely involves myelination. In contrast, growth of most other organ systems including those of mesenchymal origin involve extensive increases in cell number (Sinclair & Dangerfield 1998), and it would be these tissues that are most vulnerable to any defect in proliferation. Here we wish to propose that some of the particular disorders in the laminopathies may arise due to a combination of cellular structural weakness, which leads to cellular stress, subsequent cellular death, and attempts by the affected tissues to replace lost cells. Replacement of the lost cells requires cell proliferation of tissue stem cells. However, our results suggest that proliferation may be compromised as a consequence of improper function(s) of the mutated A-type lamins in DNA replication and/or mitosis. Such defects may only become apparent postnatally, affecting those tissues in which there is extensive proliferation or regenerative capacity. Tissues would become subjected to the day-to-day physical stresses such as movement. In the uterus, the fetus may be protected from such stresses by being cushioned by the amniotic fluid.

Conclusions

The similarities between the phenotypes and tissues affected in patients with FLPD, MAD, EDMD, and progeria, as well as during the normal ageing process, are some of the more striking and provocative findings that have emerged from studies on the laminopathies. The normal process of ageing is accompanied by a loss of muscle mass and decreased muscle function, a process called sarcopenia (Hepple 2003), which is similar to the muscle wasting and atrophy that characterize EDMD. A similar degenerative pathway has been described in ageing adipocytes *in vitro* (Yu & Zhu 2004). 'Metabolic syndrome,' which is associated with ageing, includes a decline in glucose tolerance, increased insulin resistance, and dyslipidemia, all of which often accompany the changes in fat redistribution prevalent in ageing, such as loss of subcutaneous fat from extremities and accumulation of fat in truncal regions (Björntorp 1998). Accordingly, one could view FPLD as an accelerated version of these particular aspects of ageing. Defects in A-type lamins may also result in an altered capacity of cells to differentiate, or maintain the appropriate differentiated state in muscle, fat and bone lineages leading to phenocopies of premature ageing. Both muscle and bone lineages can transdifferentiate into adipocytes, and myoblasts from 'old' mice more readily form adipocytes than myoblasts from younger mice. Whether, alterations in lamin expression and/or function are part of the normal ageing process is a provocative and challenging issue requiring future research.

References

Alsheimer M, Liebe B, Sewell L, Stewart CL, Scherthan H Benavente R 2004 Disruption of spermatogenesis in mice lacking A-type lamins. J Cell Sci 117:1173–1178

Bione S, Maestrini E, Rivella S et al 1994 Identification of a novel X-linked gene responsible for Emery-Dreifuss muscular dystrophy. Nat Genet 8:323–327

Björntorp P 1998 Etiology of the metabolic syndrome. In: Bray GA, Bouchard C, James WPT (eds) Handbook of obesity. Dekker Inc., New York, p 573–600

Burke B Stewart CL 2002 Life at the edge: the nuclear envelope and human disease. Nat Rev Mol Cell Biol 3:575–585

Chaouch M, Allal Y, De Sandre-Giovannoli A et al 2003 The phenotypic manifestations of autosomal recessive axonal Charcot-Marie-Tooth due to a mutation in Lamin A/C gene. Neuromuscul Disord 13:60–67

Chen L, Lee L, Kudlow BA et al 2003 LMNA mutations in atypical Werner's syndrome. Lancet 362:440–445

Cutler DA, Sullivan T, Marcus-Samuels B, Stewart CL Reitman ML 2002 Characterization of adiposity and metabolism in Lmna-deficient mice. Biochem Biophys Res Commun 291:522–527

De Sandre-Giovannoli A, Chaouch M, Kozlov S et al 2002 Homozygous defects in LMNA, encoding lamin A/C nuclear-envelope proteins, cause autosomal recessive axonal neuropathy in human (Charcot- Marie-Tooth disorder type 2) and mouse. Am J Hum Genet 70:726–736

De Sandre-Giovannoli A, Bernard R et al 2003 Lamin a truncation in Hutchinson-Gilford progeria. Science 300:2055

Eriksson M, Brown WT, Gordon LB et al 2003 Recurrent de novo point mutations in lamin A cause Hutchinson-Gilford progeria syndrome. Nature 423:293–298

Fatkin D, MacRae C, Sasaki T et al 1999 Missense mutations in the rod domain of the lamin A/C gene as causes of dilated cardiomyopathy and conduction-system disease. N Engl J Med 341:1715–1724

Fry M 2002 The Werner syndrome helicase-nuclease — one protein, many mysteries. Sci Aging Knowledge Environ (13):re2

Goldman RD, Gruenbaum Y, Moir RD, Shumaker DK Spann TP 2002 Nuclear lamins: building blocks of nuclear architecture. Genes Dev 16:533–547

Hepple RT 2003 Sarcopenia — a critical perspective. Sci Aging Knowledge Environ (46):pe31

Hickson ID 2003 RecQ helicases: caretakers of the genome. Nat Rev Cancer 3:169–178

Holaska JM, Lee KK, Kowalski AK Wilson KL 2003 Transcriptional repressor germ cell-less (GCL) and barrier to autointegration factor (BAF) compete for binding to emerin in vitro. J Biol Chem 278:6969–6975

Imai S, Nishibayashi S, Takao K et al 1997 Dissociation of Oct-1 from the nuclear peripheral structure induces the cellular aging-associated collagenase gene expression. Mol Biol Cell 8:2407–2419

Kennedy BK, Barbie DA, Classon M, Dyson N Harlow E 2000 Nuclear organization of DNA replication in primary mammalian cells. Genes Dev 14:2855–2868

Lammerding J, Schulze PC, Takahashi T et al 2004 Lamin A/C deficiency causes defective nuclear mechanics and mechanotransduction. J Clin Invest 113:370–378

Maidment SL Ellis JA 2002 Muscular dystrophies, dilated cardiomyopathy, lipodystrophy and neuropathy: the nuclear connection. Expert Rev Mol Med 4:1–21

Martin GM Oshima J 2000 Lessons from human progeroid syndromes. Nature 408:263–266

Melcon M, Kozlov S, Cutler DA et al 2005 Loss of emerin at the nuclear envelope disrupts the Rb and E2F developmental pathways during muscle regeneration, in preparation

Mohaghegh P Hickson ID 2001 DNA helicase deficiencies associated with cancer predisposition and premature ageing disorders. Hum Mol Genet 10:741–746

Mounkes L, Kozlov S, Burke B Stewart CL 2003a The laminopathies: nuclear structure meets disease. Curr Opin Genet Dev 13:223–230

Mounkes LC, Kozlov S, Hernandez L, Sullivan T Stewart CL 2003b A progeroid syndrome in mice is caused by defects in A-type lamins. Nature 423:298-301

Mounkes L, Kozov S, Rottman J Stewart CL 2005 Mice expressing mutant A-type lamins develop cardiac conduction system defects leading to sudden death, in preparation

Muchir A, van Engelen BG, Lammens M et al 2003 Nuclear envelope alterations in fibroblasts from LGMD1B patients carrying nonsense Y259X heterozygous or homozygous mutation in lamin A/C gene. Exp Cell Res 291:352–362

Nikolova V, Leimena C, McMahon AC et al 2004 Defects in nuclear structure and function promote dilated cardiomyopathy in lamin A/C-deficient mice. J Clin Invest 113:357–369

Nili E, Cojocaru GS, Kalma Y et al 2001 Nuclear membrane protein LAP2beta mediates transcriptional repression alone and together with its binding partner GCL (germ-cell-less). J Cell Sci 114:3297–3307

Osada S, Ohmori SY Taira M 2003 XMAN1, an inner nuclear membrane protein, antagonizes BMP signaling by interacting with Smad1 in Xenopus embryos. Development 130:1783–1794

Oshima J 2000 The Werner syndrome protein: an update. Bioessays 22:894–901

Rober RA, Weber K Osborn M 1989 Differential timing of nuclear lamin A/C expression in the various organs of the mouse embryo and the young animal: a developmental study. Development 105:365–378

Rober RA, Sauter H, Weber K Osborn M 1990 Cells of the cellular immune and hemopoietic system of the mouse lack lamins A/C: distinction versus other somatic cells. J Cell Sci 95:587–598

Sarkar PK, Shinton RA 2001 Hutchinson-Gilford progeria syndrome. Postgrad Med J 77:312–317

Sinclair D, Dangerfield PH 1998 Human growth after birth. Oxford University Press, Oxford, p 1–251

Stewart C, Burke B 1987 Teratocarcinoma stem cells and early mouse embryos contain only a single major lamin polypeptide closely resembling lamin B. Cell 51:383–392

Sullivan T, Escalante-Alcalde D, Bhatt H et al 1999 Loss of A-type lamin expression compromises nuclear envelope integrity leading to muscular dystrophy. J Cell Biol 147:913–920

Wilkinson FL, Holaska JM, Zhang Z et al 2003 Emerin interacts in vitro with the splicing-associated factor, YT521-B. Eur J Biochem 270:2459–2466

Worman HJ Courvalin JC 2002 The nuclear lamina and inherited disease. Trends Cell Biol 12:591–598

Yu YH Zhu H 2004 Chronological changes in metabolism and functions of cultured adipocytes—a hypothesis for cell aging in mature adipocytes. Am J Physiol Endocrinol Metab 286:E402–E410

Zhao P Hoffman EP 2004 Embryonic myogenesis pathways in muscle regeneration. Dev Dyn 229:380–392

DISCUSSION

Collas: Can you derive ES cells from the $Lmna^{-/-}$ and the progeric mice?

Stewart: We have derived ES cells from the $Lmna^{-/-}$ mice and we are starting to work with these. The derivation of ES cells from the progeric mice has not yet been done, although it is high-up on our list of things to do. It should be quite feasible. We have used this trick of crossing some of the mutations we make onto the p16/p19 mutant background. p16 and p19 are Cdk inhibitors or regulate p53 and Rb, depending on how they are spliced. We found that on this background mouse cells don't go through this typical senescence profile. When we crossed our *Lmna*$^{-/-}$ mice onto that background, while p16/p19 null mice start to develop tumours at 6–7 months, we found mice heterozygous for *Lmna* in addition to being homozygous for mutant p16/p19 start to develop tumours as early as two months. It looks as though even in the heterozygous state the disruption is enhancing the predisposition to form tumours.

Collas: It is possible to grow primary fibroblasts under reduced oxygen conditions and extend lifespan. Have you tried doing this?

Stewart: We've just bought the gas cylinder that has the low oxygen so we can do this experiment. If you grow mouse cells under normal conditions (5–10% CO_2 in air), it is the 20% oxygen level in air that is the prime cause for the senescing of these mouse cells.

Gruenbaum: You talked about the developmental switch between embryonic and adult fibroblasts. When you compare the embryonic to adult fibroblasts, the adult ones have already undergone many more cycles of proliferation. So can you really differentiate between a developmental switch and some kind of threshold switch?

Stewart: That is a possibility. But we have cultured the embryonic cells up to passage 20 and they are behaving just like the wild-type cells.

Oshima: Have you looked at the processing from pre-lamin to lamin in the cryptically spliced HGPS mice? This might be the cause of the disorder.

Stewart: Not yet.

Young: I have a question about the L530P mice. When you do a Western blot for lamin, what do you see, given all the abnormal splicing?

Stewart: That is a good question. We didn't see much on the Western, although with the immunofluorescence it looks as though at least lamin A is there. Then, having heard in earlier discussions about the technical difficulties of doing proper Westerns, we may have to readdress this to get a conclusive answer.

Young: So you don't really see any lamin C.

Stewart: It is very faint. We would like to do pulse labelling with ^{35}S methionine and then do some immunoprecipitations.

Fatkin: I was wondering about your experiments with muscle cell differentiation. The textbook understanding is that lamin A is expressed in differentiated cells. Do you believe that lamin A is only a feature of the postnatal heart?

Stewart: No. We know that myoblasts in our hands are expressing lamins. Even the C2C12 myoblasts do this. It is not as though A-type lamins come up with the differentiation into the myotubes. Interestingly, in Eric Hofman's profiling of the regenerating muscle, there is a short lag before A and C-type lamin expression. I can't remember which day they come up on. The embryonic expression of A-type lamin is that we begin to see it happening in the trophoblast and visceral endoderm. In the developing heart it is there by day 10.

Capeau: Why in general are laminopathies delayed? And why is this delay dependent on the tissue that is affected? One possible explanation could be that there is a stock of stem cells that could regenerate. Do you think that your experiments with embryonic fibroblasts which showed that growth was normal when the fibroblasts were derived from the embryos and was abnormal when they were derived from other tissues could exclude the possibility that the delay is due to the different stocks of stem cells in body tissues?

Stewart: That is a very sophisticated suggestion. With the primary cells I am not sure that we see any profound differences in proliferation at the outset. I am intrigued by this stem cell idea, and the possibility that lamins may have some effect there. We have noted in these primary muscle cultures from either the progeric mice or the lamin deficient mice, that the primary cultures have very high numbers of adipocytes. Are we looking at some stem cell phenotype, where the stem cell can't make up its mind to become fully committed to muscle (which would be one of the more intriguing aspects), or is it simply that in these tissues by the time we put cells in culture there are already a large number of adipocytes. We would have to do some sort of clonal assay to work this one out.

Julien: Have you tried to get the cell-type-specific knockout in order to rescue the phenotype?

Stewart: We now have a conditional allele of the A-type lamin deletion. We are generating another version that is in a pure BL6 background. We have been trying to do a rather obscure experiment. We wanted to delete the A-type lamins from Sertoli cells because there was some question about the role of lamins in male gametogenesis. We are not sure how well this has worked but we do have conditional alleles for the lamin deletion, so these experiments will be possible.

Julien: You will also be able to use these to address the problem about fibroblasts. You can delete the gene in fibroblasts after a certain number of passages.

Levy: You have shown skin sections from the L530P knock-in mouse. You showed that there is hyperkeratosis. Is this associated with extensive dermal fibrosis?

Stewart: We haven't looked.

Goldman: Hyperkeratosis is an overgrowth of the keratinocytes, which are the epithelial cells.

Young: With Karen Reue in Los Angeles we have made lamin B1 knockout mice (Vergnes et al 2004). Fibroblasts from these mice also show premature senescence and fairly striking chromosomal abnormalities, namely an increased incidence of aneuploidy and polyploidy. This occurs very early on in cell culture.

Goldman: But do the nuclei appear normal?

Young: No, the nuclear shape is strikingly abnormal.

Stewart: Some of the first spectacular results with RNAi technology came from deleting lamins in cells. If you knock out lamin B1 by RNAi in culture the cells die.

Young: These cells can be immortalized quite easily.

Goldman: Do they have abnormal nuclear morphology? What happens to the mice?

Young: Yes, incredibly abnormal. The mice die very late during embryonic development.

Gruenbaum: In the lamin B1 silencing with RNAi, in HeLa cells we get this apoptosis that you mention. This can be delayed by adding caspase inhibitors to the culture. If we go to a different cell line we don't always get it. It seems that HeLa cells are almost primed to undergo apoptosis.

Stewart: Philip Collas should come in here. He had amazing results with preventing lamin B1 reassembly into the nuclear envelope.

Gruenbaum: Wasn't it B1 and B2?

Collas: The antibody we used didn't allow us to distinguish between them. We found that if you prevent the targeting of B-type lamins to the periphery in HeLa cells by messing around with phosphatase targeting you induce apoptosis quickly, early on in G1. In the lymphoid cell line that doesn't express lamin A (KE37 cell line), if we prevent the targeting of B-type lamins to the nuclear envelope, normal looking nuclei are formed and they actually assemble A-type lamins. They switch on this gene which is normally switched off. The lamin A assembles and the cells can't cope, so they go into apoptosis, but this is delayed compared with what is seen in HeLa cells.

Stewart: Have you seen this in any other cells?

Collas: No. We haven't looked.

Gruenbaum: When we looked in the lamin B1 knockout we didn't see compensation by lamin A.

Stewart: Steve Young, have you looked at A-type lamin expression in lamin B1 knockout mice?

Young: There doesn't appear to be any change in the A-type lamins, but preliminary studies suggest that B2 may be increased.

Davies: Changing tack, your emerin-null mice have no phenotype. Have you crossed these with any of these other mutant mice to see whether they modify the phenotype?

Stewart: We crossed them against the lamin-deficient mice to see whether there is escape of more emerin into the endoplasmic reticulum that might have been contributing to pathology. Most of the double nulls from the lamin A and emerin died like the lamin A nulls. We haven't done this yet with other mutations. We did put the progeric mutation over the lamin A null: the mice may have a slight prolongation of viability by a few days.

Gasser: Colin Stewart, I didn't understand your interpretation of the cultured cells from the progeric mutant mouse model. You said that you could immortalize these easily, and then they grow indefinitely. Is that correct?

Stewart: Primary mouse cells in culture are very sensitive to the ambient oxygen concentration. If you grow mouse cells in low oxygen tension (e.g. 1%) then they will grow more effectively and don't go into a senescent phase. Under normal tissue culture conditions mouse cells switch on growth arrest genes such as p16 or p19Arf, and this is what causes them to go through this senescence phase. Then you get immortalized derivatives due to random loss or deletion of these genes, so that they start to take off. We have taken mice that are already deficient in some of those growth arrest genes and put the mutation into that background. With the lamin A nulls, we can get much better proliferation. We have just got the first adult postnatal fibroblasts from the progeric mice on a p16 null background in culture, and it is too early to say what effect that is having. We have crossed the progeric mice into a p53 null background because there is debate about p53 possibly being a modifier. If you have up-regulated p53 you have fewer tumours but you die earlier. What is intriguing is that we haven't yet had any double nulls (progeric and p53 null) born. The frequency in the p16/p19 double null background is less than what we would have expected. *In vivo* it doesn't look as though losing these genes (that profoundly affect proliferation) is having a great effect.

Gasser: So do the embryonic fibroblasts (EFs) from the progeric nulls, without the others crossed in, senesce even faster than the lamin A nulls?

Stewart: No, the EFs from the progeric mouse show no difference in their growth rate compared to wild-type EFs. It is only postnatally that we see a difference. We have done array analysis on the progeric EFs and we couldn't find a single gene that is statistically significantly different. It doesn't look as though the progeric mutation is having a profound effect on gene expression patterns.

Wilson: I want to consider the stress hypersensitivity in the mice. This is a common theme and it is a huge physiological response, i.e. death. The injury itself can't be causing the death. Is this telling us something about hormonally induced stress responses triggering pathways that lead to death?

Stewart: That's a good point. Their cardiovascular system is compromised, and if you pick up a mouse it will feel stressed.

Julien: It would be interesting to make mouse chimeras with these knockouts.

Stewart: We are doing this. We are making ES cells from these mice with a view to seeing just how well there is rescue of any progeric cells within the tissue.

Julien: In Cleveland we made chimeras in different ways, by injection into blastocysts or by aggregation. We got different results. By aggregation we got a good mosaic distribution, but by injection we got more patches.

Stewart: There is a long literature in the mouse embryology field of the combination of different strains. If you combine different strains as chimeras one strain often predominates over another. The intriguing thing may be to use tricks that would allow you to get one donor type to colonize a particular tissue.

Courvalin: Have myoblastic cell lines been established from these mouse models?

Stewart: We did some preliminary experiments with the progeric mice to try to get myoblasts. There we found a huge number of adipocytes when we tried to culture.

Courvalin: This is frequently observed with myoblast cultures. I have been told by colleagues who produce mouse myoblast cell lines that you first have a proliferation of adipocytes. That is the time when you and I, we give up. But we are wrong, because it is exactly the time, according to these experts, when myoblasts start to proliferate.

Stewart: We never really saw that in the wild-type controls.

Courvalin: My friends at the Pasteur Institute in Paris, who are used to these problems, have already proposed to help us to establish these cell lines.

Capeau: We began with the mechanical stress response, and now perhaps we have to also consider the general stress response. which could play a role in the immune system and also be responsible for the increased levels of some cytokines and hormones. Has anyone got data from patients or mouse models which could indicate that there is a general stress response with increased cortisol or TNFα in those patients or models?

Stewart: I think there may have been some attempts to look at immune function in some patients. These patients are so rare and in such demand that they have become reluctant to be biopsied and fooled around with. This is quite understandable.

Capeau: I wasn't just thinking about the progeria patients, but all the laminopathies.

Bonne: I don't think this has been done.

Reference

Vergnes L, Péterfy M, Bergo MO, Young SG, Reue K 2004 Lamin B1 is required for mouse development and nuclear integrity. Proc Natl Acad Sci USA 101:10428–10433

The nuclear membrane and mechanotransduction: impaired nuclear mechanics and mechanotransduction in lamin A/C deficient cells

Jan Lammerding*, Richard T. Lee*†[1]

*Biological Engineering Division, Massachusetts Institute of Technology, 77 Massachusetts Avenue, Cambridge, MA 02139, and †Cardiovascular Division, Brigham & Women's Hospital, Partner's Research Facility, 65 Landsdowne St, Cambridge, MA 02139, USA

Abstract. Mutations in the lamin A/C gene cause a variety of human diseases including Emery-Dreifuss muscular dystrophy, dilated cardiomyopathy and Hutchinson-Gilford progeria syndrome. The tissue specific effects of lamin mutations are unclear, in part because the function of lamin A/C is incompletely defined, but the many muscle specific phenotypes suggest that defective lamin A/C could increase cellular mechanical sensitivity. Lamin A/C deficient fibroblasts were subjected to mechanical strain to measure nuclear mechanical properties and strain-induced signalling. We found that lamin A/C deficient fibroblasts are characterized by impaired nuclear mechanics and mechanotransduction, reflected by increased nuclear deformations, increased nuclear fragility, attenuated expression of mechanosensitive genes, and impaired transcriptional activation, leading to impaired viability of mechanically strained cells. Lamins and other nuclear envelope proteins can thus affect several levels of the mechanotransduction cascade, altering nuclear and cytoskeletal mechanics as well as playing an important role in mechanically activated gene regulation. Individual mutations in the lamin A/C gene could potentially selectively interfere with any of these functions, explaining the tissue-specific effects observed in the laminopathies.

2005 Nuclear organization in development and disease. Wiley, Chichester (Novartis Foundation Symposium 264) p 264–278

Lamins are structural components of the nuclear lamina, a protein network underlying the inner nuclear membrane that determines nuclear shape and size (Liu et al 2000, Aebi et al 1986). Lamins bind to chromatin and to several nuclear

---

[1]This paper was presented at the symposium by Richard T. Lee to whom correspondence should be addressed.

envelope proteins such as emerin, recruiting them to the nuclear envelope, and lamins have been shown to contribute to the correct spacing of nuclear pore complexes (Hutchison 2002, Holt et al 2003, Sullivan et al 1999). Mammalian cells contain two types of lamins: A-type lamins (lamin A, C, AΔ10 and C2) are encoded by a single gene (*Lmna*) and are developmentally regulated and expressed in differentiated cells. B-type lamins (B1 and B2/B3) are encoded by two distinct genes (*Lmnb1*, *Lmnb2*) and are constitutively expressed in all cells (Liu et al 2000, Biamonti et al 1992, Lin & Worman 1995, 1993, Furukawa et al 1994, Machiels et al 1996, Harborth et al 2001). Mutations in the gene encoding A-type lamins and their binding partners have been associated with a variety of human diseases, including Emery-Dreifuss muscular dystrophy (EDMD), dilated cardiomyopathy (DCM), Dunnigan-type familial partial lipodystrophy (FPLD) and Hutchinson-Gilford progeria syndrome (Bonne et al 1999, Fatkin et al 1999, Cao & Hegele 2000, Shackleton et al 2000, De Sandre-Giovannoli et al 2003, Eriksson et al 2003). The molecular mechanisms underlying the varied phenotypes are unknown, but the many muscle specific phenotypes suggest that abnormal lamin A/C function could increase cellular mechanical sensitivity. This increased mechanical sensitivity could arise from direct structural deficiencies in the nuclear lamina of lamin A/C deficient cells, leading to an elevated risk of nuclear rupture, but it could also be a consequence of deficient cellular mechanotransduction, i.e. cellular signalling in response to mechanical stimulation. Although the precise molecular mechanisms of mechano-transduction remain elusive, it is clear that mechanotransduction plays a major role in maintaining the normal function of tissues, particularly in mechanically active tissues like myocardium and skeletal muscle.

Our current studies explored two potential mechanisms of the lamin associated diseases: the 'structural hypothesis' suggests that lamin mutations lead to increased nuclear fragility and eventual nuclear disruption in the mechanically strained tissue; the 'gene regulation hypothesis' proposes that the mutations result in tissue-specific gene regulation defects caused by altered interaction of lamin A/C with transcriptional regulators and changes in chromatin organization. These hypotheses are not mutually exclusive and could in fact be interrelated. Here we present a brief description of our findings; detailed results along with a precise description of the applied methods can be found in Lammerding et al (2004).

Results

Embryonic fibroblasts derived from lamin A/C null mice were used for all experiments. Lamin A/C null (*Lmna*$^{-/-}$) mice are indistinguishable from their littermates at birth but develop severe muscle wasting and contractures similar to

EDMD by 3–4 weeks and die by eight weeks (Sullivan et al 1999). Cells derived from lamin A/C null mice have mis-shaped nuclei and obvious ultrastructural damage (Sullivan et al 1999, Raharjo et al 2001), similar to fibroblasts from patients with EDMD or FPLD and cells from *C. elegans* with reduced lamin levels (Liu et al 2000, Vigouroux et al 2001).

Measurements on nuclear mechanics were obtained by using a novel technique to measure nuclear deformation in wild type and lamin A/C null fibroblasts plated on elastic silicon membranes subjected to bi-axial strain. This method yields quantitative measurements of nuclear stiffness compared to cytoskeletal stiffness in living cells without having to isolate the nuclei. In wild-type cells, the nucleus is significantly stiffer than the surrounding cytoskeleton and shows only minor deformations under strain. In contrast, lamin A/C null fibroblasts showed significantly increased nuclear deformations, indicating that lamin A/C deficient nuclei have impaired nuclear mechanics.

Since nuclear deformation in intact cells is a function of nuclear mechanics as well as strain transmission through the cytoskeleton, we investigated cytoskeletal stiffness using two separate techniques previously described by Bausch et al (1998). In the first method, the displacement of small, fibronectin-coated paramagnetic beads attached to the cell surface in response to an applied magnetic force is used as an indicator of cytoskeletal stiffness. In the second method, the induced displacement of fibronectin coated non-magnetic polystyrene beads on the cell surface in proximity to the magnetic bead is included in the analysis to evaluate the cytoskeletal stiffness independent of potential artifacts caused by variations in the degree of magnetic bead attachment. Both experimental techniques independently revealed that lamin A/C null fibroblasts have significantly decreased cytoskeletal stiffness compared to wild-type cells. Therefore, the observed increased nuclear deformation in lamin deficient cells is unlikely to be due to altered force transmission to the nucleus, as the softer cytoskeleton would in fact result in an underestimation of the nuclear stiffness. Based on these findings we conclude that lamin A/C deficient cells have decreased cytoskeletal stiffness and impaired nuclear mechanics.

Strain-induced damage to the more fragile nucleus could provide one explanation for tissue-specific effects of lamin A/C mutations in mechanically active tissues like cardiac and skeletal muscle. To examine nuclear envelope integrity, we monitored the subcellular localization of microinjected fluorescently labelled 70 kDa dextran. This high molecular weight dextran is too large to cross the intact nuclear envelope and is thus excluded from the nucleus when injected into the cytoplasm. Accordingly, when injected directly into the nucleus, it is retained in the intact nucleus and does not diffuse into the cytoplasm. In our experiments, dextran injected into the cytoplasm was excluded from the nucleus in both wild type and lamin A/C null cells, indicating that nuclear integrity is not significantly impaired in lamin A/C null cells under resting conditions. However, when microinjecting dextran directly into the nucleus, nuclear integrity appeared temporarily compromised in a significantly larger number of lamin A/C null cells. In these cells, fluorescently labelled dextran could temporarily escape into the cytoplasm, indicating an increased rupture potential of lamin A/C deficient nuclei. For both cell types, the number of intact nuclei decreased with increasing pressure, and sufficiently high pressure led to nuclear rupture in all cells.

To determine if lamin A/C deficient fibroblasts are thus more susceptible to mechanical strain, cells were subjected to prolonged cyclic, bi-axial strain. Following strain application, lamin A/C null fibroblasts had a significantly increased fraction of dead cells compared to wild-type cells, whereas differences in cell viability between unstrained control and lamin A/C null cells were not statistically significant. Dual labelling with propidium iodide and a FITC-conjugated annexin V antibody revealed that the decrease in viability was due to an increase in both necrotic and apoptotic cell fractions compared to wild type cells.

The concomitant increase in the apoptotic cell fraction of mechanically strained lamin A/C null cells indicates that necrosis through nuclear rupture alone can only in part explain increased sensitivity to mechanical stimulation, and that, in fact, strain activated protective signalling pathways are affected in the lamin A/C null cells. Consequently, we found that expression of the mechanosensitive genes *egr-1* and the anti-apoptotic gene *iex-1* in response to mechanical stimulation was impaired in lamin A/C null cells, whereas expression of the mechanically unresponsive genes *thioredoxin-1* and *glyceraldehyde 3-phosphate dehydrogenase* (*GAPDH*) was unaltered, indicating that transcription is not impaired in a nonspecific manner and instead suggesting impaired mechanotransduction signalling in the lamin A/C null cells.

Because *iex-1* is an NF-κB dependent survival gene (De Keulenaer et al 2002), and because NF-κB can be biomechanically activated (Inoh et al 2002), we examined if biomechanical signalling through NF-κB is disturbed in lamin A/C null cells. Our analysis revealed that nuclear translocation and transcription factor binding of the NF-κB subunits p65/RelA were not negatively affected in the lamin A/C null cells, but NF-κB dependent activity of a luciferase reporter gene in response to cytokine stimulation was significantly impaired in the lamin A/C deficient cells. These results indicate that the deficient response of lamin A/C null cells to mechanical or cytokine stimulation is based on a role of lamin A/C in transcriptional activation following transcription factor binding.

Discussion

These results establish the importance of lamin A/C for nuclear stability and highlight its role in transcriptional regulation in response to mechanical or chemical stimulation. Based on these observations, the dominant effect of laminopathies on muscle tissue can arise from two mechanisms:

- Impaired nuclear mechanics render the lamin A/C deficient or mutant cells more susceptible to nuclear rupture, with damage accumulating over time in mechanically strained tissue. Evidence of fragmented nuclei has been reported in skeletal muscle fibres from emerin-deficient EDMD patients and in fibroblasts from FPLD patients following heat shock treatment (Markiewicz et al 2002, Vigouroux et al 2001)
- Impaired mechanotransduction reduces adaptive and protective responses in mechanically strained tissue

Here, mechanotransduction can be defined as the signalling response of cells to mechanical stimulation. Changes in the cellular mechanical environment, including the extracellular matrix, the cytoskeleton and the nucleus, can be translated into conformational changes of yet unidentified mechanosensor proteins. These conformational changes are in turn translated into a biochemical response that triggers a complex network of signalling pathways, leading to the activation of mechanosensitive genes and transcription factors.

It is important to notice that lamin A/C can affect the mechanotransduction chain at several levels. Structural deficiencies in the nuclear lamina affect the cellular mechanical microenvironment, and compensatory changes of the cytoskeleton can affect the cellular mechanical environment, leading to effective changes in the strain level of the nucleus. The nucleus itself could serve as a mechanosensor, and this function could be altered in laminopathies. Furthermore, mutations in lamins could directly interfere with the transcription factor binding and transcriptional regulation. These aspects are discussed in more detail below.

Nuclear and cytoskeletal mechanics

Nuclear deformations and adaptation in response to changes in the mechanical environment have been observed in a variety of cells, including compressed chondrocytes (Guilak 1995) and stretched endothelial cells (Caille et al 1998). In wild type cells, the nucleus is 5–10 times stiffer than the surrounding cytoskeleton and shows only small deformations under mechanical strain (Caille et al 1998, 2002, Guilak 1995, Maniotis et al 1997), but these deformations could be sufficient to affect chromatin structure and possibly activate nuclear mechanosensory elements. Recent studies report desmin intermediate filament mediated changes in chromatin in response to mechanical strain and hypothesized that stretch-induced changes in chromatin can lead to activation of hypertrophy-associated genes (Bloom et al 1996). Immunogold electron microscopy studies of lamin A/C null cardiac myocytes revealed disorganization and detachment of desmin filaments form the nuclear surface with progressive disruption of the cytoskeletal desmin network (Fatkin et al 2003). Ultrastructural analysis of these cells showed severe alterations in nuclear shape and size with central displacement and fragmentation of heterochromatin (Fatkin et al 2003). Nuclei in left ventricular cardiac myocytes showed more severe morphological changes compared to nuclei from left atrial myocytes or hepatocytes (Fatkin et al 2003), suggesting that increased mechanical strain in the ventricular myocytes can lead to nuclear structural damage. Furthermore, mechanical coupling between integrins, cytoskeletal filaments and the nucleus have been demonstrated by micromanipulation with microbeads and micropipettes in endothelial cells (Maniotis et al 1997). Mutations in nuclear envelope proteins such as lamin or emerin could interrupt some of these connections and impair nuclear mechanotransduction pathways. The impaired nuclear and cytoskeletal mechanics observed in the lamin A/C null cells could lead to significantly altered nuclear strain distribution, which could result in impaired nuclear mechanosensing. The observed alterations in cytoskeletal stiffness could arise as a compensatory mechanism to protect a more fragile nucleus, but could also contribute to the pathophysiology of the disease. Altered cytoskeletal mechanics not only affect the transmitted force to the nucleus under applied strain, but can also play an important role in cell shape, migration, and other critical functions with direct consequences on the affected tissue.

Mechanotransduction and transcriptional regulation

In addition to affecting the mechanical environment and nuclear mechanics, lamins can play a role in the signal transduction itself. Accumulating experimental findings suggest that the three-dimensional structure and distribution of genes and protein binding sites in the genome have to be taken into account to understand the complex mechanisms governing transcriptional control, replacing the old paradigm of gene regulation completely described by the one-dimensional sequence of promoter and transcription factor binding sites. In this context, it becomes obvious that nuclear deformations and changes in chromatin structure and organization can lead to direct consequences in transcriptional regulation.

Lamins are not constrained to the nuclear envelope, but also form stable structures within the nucleoplasm (Moir et al 2000), and interactions between lamin and chromatin as well as nuclear transcription factors have been previously demonstrated *in vivo* and *in vitro* (Dreuillet et al 2002, Lloyd et al 2002). Furthermore, lamin A/C speckles have been proposed to mediate spatial organization of splicing factor compartments and RNA II polymerase transcription (Kumaran et al 2002), which could affect gene regulation after the initial mechanotransduction event. In our experiments, NF-κB regulated luciferase activity was impaired in lamin A/C null cells despite normal NF-κB translocation and an increase in transcription factor binding, indicating an important role of lamin A/C in transcriptional activation with an unknown role of lamin A/C in the assembly of enhancesomes or as a scaffolding protein for transcription factors and co-activators. In addition to these nucleoplasmic effects, lamins and lamin-binding proteins can have an important function at the nuclear envelope. Emerging experimental evidence suggests a potential role for the nuclear envelope as a filter for signalling and transcription factors entering and exiting the nucleus. Thus lamins can play an important role not only in the signal transduction from mechanical deformations to biochemical responses but can also affect the downstream gene regulation through its direct and indirect interaction with transcription factors and splicing factors.

Outlook

While the experiments presented here were performed in fibroblasts that completely lack lamin A and lamin C, most human laminopathies arise from heterozygous lamin mutations, with mutant lamin expressed at similar levels to wild-type lamin and often apparently as stable (Ostlund et al 2001). In many cases, the mutation can lead to a structurally impaired form of lamin A/C that can act as a dominant negative and lead to cellular mechanical deficiencies as observed in the lamin A/C null cells. Other mutations can affect the binding of lamin to other proteins or chromatin with fewer effects on the structural role of lamin itself, resulting in a partially functional protein that might only affect specific signalling pathways. Combinations of mechanical and transcriptional regulation defects could result in complex phenotypes affecting several tissue types in diseases such as progeria. Studying the individual levels of mechanotransduction, i.e. cell/ nuclear mechanics, signal transduction, gene regulation, will not only help to understand the effect of lamin mutations on mechanically active tissue but can also lead to new insights into the normal function of lamin. We have here presented a first approach to develop independent tests for measuring the structural and gene-regulatory functions of lamin A/C, thereby helping to clarify the effects of specific mutations. Additional experiments can be designed that investigate the effects of lamin A/C mutations on cell mechanics and gene regulation in more detail. These techniques will include three-dimensional imaging of cells and nuclear structures *in situ* in tissue affected by laminopathies, such as the heart, to observe the changes in cytoskeletal and nuclear structure during the process of the disease. Furthermore, cells derived from different tissues from laminopathy patients as well as from mice bearing specific lamin mutations can be cultured and subjected to strain and microinjection experiments to study the effect of specific mutations on nuclear stability and fragility, and nuclei isolated from these cells can be studied using atomic force microscopy (AFM) to examine nuclear mechanics more directly. Depending on the force application mode, these methods can reveal new insights into chromatin structure as well as the transmission of forces through the nuclear envelope to the nucleus. Applying these techniques to cells derived from different tissues and different mutations in humans and mice may then provide insights into why lamin A/C mutations cause diverse phenotypes.

Summary

Lamin A/C deficient fibroblasts are characterized by impaired nuclear mechanics and mechanotransduction, reflected by increased nuclear deformations, increased nuclear fragility, attenuated induction of mechanosensitive genes, and impaired transcriptional activation, leading to decreased viability of mechanically strained cells. The tissue-specific effects observed in several laminopathies may thus arise from two mechanisms:

- The impaired nuclear stability renders mechanically strained tissue more susceptible to cellular damage and
- Abnormal transcriptional regulation impairs adaptive and protective pathways. Individual mutations in the lamin A/C gene could potentially selectively interfere with any of these functions, explaining the diversity of observed phenotypes.

References

Aebi U, Cohn J, Buhle L, Gerace L 1986 The nuclear lamina is a meshwork of intermediate-type filaments. Nature 323:560–564
Bausch AR, Ziemann F, Boulbitch AA, Jacobson K, Sackmann E 1998 Local measurements of viscoelastic parameters of adherent cell surfaces by magnetic bead microrheometry. Biophys J 75:2038–2049
Biamonti G, Giacca M, Perini G et al 1992 The gene for a novel human lamin maps at a highly transcribed locus of chromosome 19 which replicates at the onset of S-phase. Mol Cell Biol 12:3499–3506
Bloom S, Lockard VG, Bloom M 1996 Intermediate filament-mediated stretch-induced changes in chromatin: a hypothesis for growth initiation in cardiac myocytes. J Mol Cell Cardiol 28:2123–2127
Bonne G, Di Barletta MR, Varnous S et al 1999 Mutations in the gene encoding lamin A/C cause autosomal dominant Emery-Dreifuss muscular dystrophy. Nat Genet 21:285–288
Caille N, Tardy Y, Meister JJ 1998 Assessment of strain field in endothelial cells subjected to uniaxial deformation of their substrate. Ann Biomed Eng 26:409–416
Caille N, Thoumine O, Tardy Y, Meister JJ 2002 Contribution of the nucleus to the mechanical properties of endothelial cells. J Biomech 35:177–187

Cao H, Hegele RA 2000 Nuclear lamin A/C R482Q mutation in canadian kindreds with Dunnigan-type familial partial lipodystrophy. Hum Mol Genet 9:109–112

De Keulenaer GW, Wang Y, Feng Y et al 2002 Identification of IEX-1 as a biomechanically controlled nuclear factor-kappaB target gene that inhibits cardiomyocyte hypertrophy. Circ Res 90:690–696

De Sandre-Giovannoli A, Bernard R, Cau P et al 2003 Lamin A truncation in Hutchinson-Gilford progeria. Science 300:2055

Dreuillet C, Tillit J, Kress M, Ernoult-Lange M 2002 In vivo and in vitro interaction between human transcription factor MOK2 and nuclear lamin A/C. Nucleic Acids Res 30:4634-42

Eriksson M, Brown WT, Gordon LB et al 2003 Recurrent de novo point mutations in lamin A cause Hutchinson-Gilford progeria syndrome. Nature 423:293–298

Fatkin D, MacRae C, Sasaki T et al 1999 Missense mutations in the rod domain of the lamin A/C gene as causes of dilated cardiomyopathy and conduction-system disease. N Engl J Med 341:1715–1724

Fatkin D, Nikolova V, Leimena C et al 2003 Nuclear structure and function defects promote dilated cardiomyopathy in lamin A/C-deficient mice. Circulation 108:449

Furukawa K, Inagaki H, Hotta Y 1994 Identification and cloning of an mRNA coding for a germ cell-specific A-type lamin in mice. Exp Cell Res 212:426–430

Guilak F 1995 Compression-induced changes in the shape and volume of the chondrocyte nucleus. J Biomech 28:1529–1541

Harborth J, Elbashir SM, Bechert K, Tuschl T, Weber K 2001 Identification of essential genes in cultured mammalian cells using small interfering RNAs. J Cell Sci 114:4557–4565

Holt I, Ostlund C, Stewart CL et al 2003 Effect of pathogenic mis-sense mutations in lamin A on its interaction with emerin in vivo. J Cell Sci 116:3027–3035

Hutchison CJ 2002 Lamins: building blocks or regulators of gene expression? Nat Rev Mol Cell Biol 3:848–858

Inoh H, Ishiguro N, Sawazaki S et al 2002 Uni-axial cyclic stretch induces the activation of transcription factor nuclear factor kappaB in human fibroblast cells. FASEB J 16:405–407

Kumaran RI, Muralikrishna B, Parnaik VK 2002 Lamin A/C speckles mediate spatial organization of splicing factor compartments and RNA polymerase II transcription. J Cell Biol 159:783–793

Lammerding J, Schulze PC, Takahashi T et al 2004 Lamin A/C deficiency causes defective nuclear mechanics and mechanotransduction. J Clin Invest 113:370–378

Lin F, Worman HJ 1993 Structural organization of the human gene encoding nuclear lamin A and nuclear lamin C. J Biol Chem 268:16321–16326

Lin F, Worman HJ 1995 Structural organization of the human gene (LMNB1) encoding nuclear lamin B1. Genomics 27:230–236

Liu J, Ben-Shahar TR, Riemer D et al 2000 Essential roles for Caenorhabditis elegans lamin gene in nuclear organization, cell cycle progression, and spatial organization of nuclear pore complexes. Mol Biol Cell 11:3937–3947

Lloyd DJ, Trembath RC, Shackleton S 2002 A novel interaction between lamin A and SREBP1: implications for partial lipodystrophy and other laminopathies. Hum Mol Genet 11:769–777

Machiels BM, Zorenc AH, Endert JM et al 1996 An alternative splicing product of the lamin A/C gene lacks exon 10. J Biol Chem 271:9249–9253

Maniotis AJ, Chen CS, Ingber DE 1997 Demonstration of mechanical connections between integrins, cytoskeletal filaments, and nucleoplasm that stabilize nuclear structure. Proc Natl Acad Sci USA 94:849–854

Markiewicz E, Venables R, Mauricio Alvarez R et al 2002 Increased solubility of lamins and redistribution of lamin C in X-linked Emery-Dreifuss muscular dystrophy fibroblasts. J Struct Biol 140:241–253

Moir RD, Yoon M, Khuon S, Goldman RD 2000 Nuclear lamins A and B1: different pathways of assembly during nuclear envelope formation in living cells. J Cell Biol 151:1155–1168

Ostlund C, Bonne G, Schwartz K, Worman HJ 2001 Properties of lamin A mutants found in Emery-Dreifuss muscular dystrophy, cardiomyopathy and Dunnigan-type partial lipodystrophy. J Cell Sci 114:4435–4445

Raharjo WH, Enarson P, Sullivan T, Stewart CL, Burke B 2001 Nuclear envelope defects associated with LMNA mutations cause dilated cardiomyopathy and Emery-Dreifuss muscular dystrophy. J Cell Sci 114:4447–4457

Shackleton S, Lloyd DJ, Jackson SN et al 2000 LMNA, encoding lamin A/C, is mutated in partial lipodystrophy. Nat Genet 24:153–156

Sullivan T, Escalante-Alcalde D, Bhatt H et al 1999 Loss of A-type lamin expression compromises nuclear envelope integrity leading to muscular dystrophy. J Cell Biol 147:913–920

Vigouroux C, Auclair M, Dubosclard E et al 2001 Nuclear envelope disorganization in fibroblasts from lipodystrophic patients with heterozygous R482Q/W mutations in the lamin A/C gene. J Cell Sci 114:4459–4468

DISCUSSION

Julien: Have you examined heat shock proteins (HSPs) in your fibroblasts?

Lee: A set of HSPs always goes up with mechanical stimulation. It is part of the same stress response. We haven't studied this in lamin-deficient cells yet

Stewart: Following on from the signal transduction, if you were to 'rack' the cells and treat them with PMA, would you see a reduction in necrosis or apoptosis? What might you expect to happen if you were to do the racking on cells deficient in the p65 component of NF-κB? Would the cells deficient in NF-κB signalling be more susceptible to stress-induced apoptosis?

Lee: Quite possibly. One of the only ways you can get a mechanically stimulated cell to undergo apoptosis is to overexpress I-κB to knock out the pathway and then come in with some other major stimulus.

Gruenbaum: Can you do the same experiments on cells that don't normally express lamin A, such as lymphoblasts?

Lee: That's a good question. We haven't done this yet but we should.

Julien: Is the level of vimentin changed in the lamin knockout fibroblasts? This could explain some of the change in the cell stress.

Lee: We don't know yet. The effect seems reproducible: the cytoskeleton is softer in these cells.

Goldman: We have looked at the vimentin network in these cells: it looks reasonably normal.

Gerace: What is the main determinant of the property of cytoplasmic 'softness' that you measure? Is it the cross-linked actin meshwork? Is there any way to visualize this?

Lee: It probably is the actin network. If you look at actin staining you don't see any differences.

Gasser: What is the arrangement of nesprin in lamin A knockout or progeric mice?

Bonne: We have screened and analysed cells of the human lamin A/C knockout. Nesprin is like emerin: it stays in the ER. Once you give back lamin A or C by transfection, nesprin 1α goes back to the nucleus.

Goldman: Is that the full length nesprin?

Bonne: We used the antibody against nesprin 1α prepared by Elisabeth McNally's group.

Goldman: This would recognize all nesprins. There are many different splice variants. We don't know whether this is a cytoplasmic nesprin or a nuclear nesprin.

Bonne: With this antibody we used, in control cells the staining is only at the nuclear envelope. If you look at knockout human cells lacking lamin A/C, you don't see any staining of the nucleus, just proteins in the ER. Then, if those cells are transfected with lamin A or C the staining goes back to normal in the nuclear envelope. So this antibody might detect an isoform specific to the nuclear envelope and what we observed in those cells may reflect some reality.

Goldman: Have you blotted with the same antibody? Have you done total cell extracts?

Bonne: The problem is that the antibody does not work very well on Western-blots, at least not for us, but Elisabeth McNally has well characterized this antibody (Mislow et al 2002).

Goldman: With all these different forms of nesprin we need to be extremely cautious. When cells are fixed one can easily remove soluble proteins, or those that aren't as stable. Different fixatives give different results.

Wilson: Have you tried introducing wild-type lamin A into those cells? Once they start expressing A-type lamins, how long does it take to recover the nuclear versus cytoplasmic stiffness? Could this be something that takes two days to re-establish because it requires changes in gene expression patterns, or is it re-established the minute lamin A re-enters the system?

Lee: We haven't done the experiment, although it is a good idea. I would imagine that it would take some time. It is most likely going to require some transcription.

Capeau: What is known about this stress response in other cell types such as muscle or adipose tissue? Is this protein induced by the same cytokine? Is something known about the protective gene?

Lee: I showed just one: there are many genes that are mechanically responsive. If you mechanically stimulate almost any cell they will respond with some of these cytoprotective molecules. We haven't looked at adipose cells, but it would surprise me if we didn't see the same effect.

Capeau: Is it generally cytokines that are inducing these genes?

Lee: Many mechanically-induced genes are also cytokine-responsive.

Fatkin: I would like to support Richard Lee's proposal that a combination of structural defects and the altered gene regulation contributes to the cardiac dysfunction in these mice. We have found both changes in cytoskeletal desmin organization and a failure of compensatory hypertrophic gene activation. The nuclear softness Richard sees gives us a potential explanation for the predominance of muscle phenotypes with *LMNA* mutations. 'Soft' nuclei are likely to be more susceptible to the push and pull effects of the muscle contraction. Hence, changes in nuclear shape may be more severe in contractile tissue than in non-contractile tissue.

Bonne: Did you have any way to check that the softness was the same in muscle cells or heart cells from the mice? Is it possible to check this?

Lee: Yes, but we haven't done this.

Gerace: When you increase apoptosis by stretching the lamin A knockout fibroblasts, do you think this is due to breaching the nuclear–cytoplasmic boundary, or is it rather just that the cytoskeleton doesn't respond properly and signals to the nucleus to induce apoptosis?

Lee: I don't know. It could be a combination of the two.

Gerace: If the nucleus is breached does apoptosis follow? I know the nuclear envelope repairs itself when you microinject cells.

Gasser: It doesn't necessarily induce apoptosis.

Lee: Those cells will keep living indefinitely as far as we know.

Bonne: Did you also test heterozygote fibroblasts? Is the nucleus less soft than the homozygotes? It would also be interesting to check fibroblasts in a knock-in mouse for a specific mutation, where it is assumed to have the same level of protein expression, to see whether the change in softness would be similar to what is observed in the homozygous lamin knockout. Is this a global mechanism of softening of the nucleus, whatever mutation there is?

Lee: That is a good question. We were studying homozygous null cells. We need to study heterozygous cells more carefully.

Goldman: One should also consider the organization of lamins. You don't need to get rid of them; disorganizing them will have the same effect. Richard Lee, you showed an image of a nucleus that was highly distorted in the lamin A knockout. But you can look around the field and see other nuclei that aren't quite as distorted or look quite normal. When you look at the more normal looking nuclei, are they softer too?

Lee: Yes.

Goldman: So lamin B can't compensate. B type lamins are the only ones that are expressed during early development. It would be interesting to look at a cell or nucleus that doesn't have any lamin A in it.

Lee: Perhaps in early development cells don't need this kind of protection.

Goldman: Or, the A type defect is also having a profound effect on B. This could be the case.

Hutchison: Using this technology are you able to look at the way the cell deforms when you apply pressure as you are doing it? The reason for the question is that in a polarized cell such as a fibroblast, one might expect that the A type lamins are interconnecting with the cytoskeleton. So, when you have them there you might expect that the nucleus deforms in a polarized way when you interfere with it. If you break those connections in the absence of A-type lamins, then you might expect more radial deformation. Can you do this using your technology?

Lee: Yes, and they do deform three dimensionally. If we take a normal cell and stretch it uniformly on a membrane, then the nucleus will stretch out and flatten down a bit.

Hutchison: What is the comparison between normal and lamin A-deficient cells?

Lee: We haven't done this three dimensionally.

Capeau: When we looked at fibroblasts from FPLD patients we were able to show that the nuclear envelope was more fragile. Lamin and emerin were easily extracted from the nucleus. The question is, have you tested fibroblasts from patients?

Lee: No. Howard Worman has suggested that we study this. I don't want to suggest that the idea of mechanical coupling is a unifying explanation, but perhaps it is part of the diversity of laminopathy symptoms in mechanically active environments such as the heart.

Young: Have you checked the softness of the nuclei at early passage versus much later on? Colin Stewart showed that they started growing off well but then after a certain number of passages they slowed down. Have you correlated that growth disturbance with softness of the nuclei or the cells? Is there any increase in softness with time?

Lee: We haven't tried this. All the ones we have done have been early passage.

Gasser: What is the natural variation in your measurements? If you take 100 nuclei, what is the standard deviation for deformation?

Lee: If we do atomic force microscopy we get a lot of variation because we are looking at one single point of the membrane.

Gasser: Does this depend on the cell cycle stage?

Lee: We would guess that it would. We are usually looking at serum-starved cells, so probably cell division rates are low.

Gruenbaum: If the lamin A knockout cells are more likely to be in G1 while the rest of the cells are more likely to be in S or G2, it may cause cell cycle-specific responses to mechanical stress.

Gasser: We have done a lot of diffusion studies that are appropriate to this, looking at chromatin in the nucleus at different stages of the cell cycle and comparing nuclei of different volume. Small changes in nuclear volume result in dramatic changes in the mobility of proteins or DNA within the nucleus. This would say that as you replicate DNA if the nuclear volume doesn't increase proportionately at different points of the cell cycle, you would have different density and viscosity just within the nucleus. This is not to say that the structural aspect isn't there — I am sure that it is — but it would be interesting to see whether size is not another source of variation.

Wilson: Does this relate to yeast or mammalian cells?

Gasser: Yeast.

Goldman: It is known that nuclei change in size going from early G1 to S phase. It is possible that A-type lamin deficient cells might spend a longer time in S phase than normal cells.

Goldman: What about the amount of DNA?

Lee: This would change things if you put in so much DNA that you forced the material onto a different part of its stiffness curve.

Gasser: This is assuming that you are only measuring the flexibility of the envelope itself, and that this is not influenced by the material within.

Lee: This will depend on whether you are pushing or pulling. If you are pulling on the nucleus from the outside then what is inside will be less relevant. If you are pushing from outside then it will be dominated by what is on the inside. In most cases we are pulling from the outside.

Gerace: The way chromatin is attached to the inside could be non-uniform. We shouldn't forget this.

Lee: I was very interested by the chromatin attachment. Perhaps this is why when we walk across the surface we are hitting the spots where the chromatin is attached or not attached.

Gerace: Do the pores cluster more in the knockout?

Burke: This appears to be the case, yes.

Gerace: The pore-containing regions might have different properties to those next to them.

Goldman: Someone said that there were no pores in the blebs in the mouse lamin A knockout cells.

Gasser: I would expect that the distribution of pores would have a major effect on the flexibility of the envelope.

Goldman: This is a new area of interest. It is a fascinating observation that the mechanical properties in the entire cell are changed by deleting one nuclear protein. Ultimately, this will become recognized as a very important finding.

Giles: I agree. Recently, our interest has moved to ventricular fibroblasts exhibiting excitation–secretion coupling, and through this fundamental mechanism being able to change the whole paracrine milieu of the endocardium if not the entire heart. The myofibroblast has the ability to secrete things such as ANF and endothelin. The clear demonstration of the deformability, regardless of the exact molecular explanation, provides a new way of looking at the fibroblast as an endocrine organ, as part of endocardoid excitation–secretion coupling.

Goldman: It has now been quite well established that the cytoskeleton is exquisitely sensitive to small perturbations in mechanical stress. I have heard that actin changes very rapidly; we certainly known that IFs are very sensitive to such changes. Ultimately these findings may help to explain the movement of signals between the cell surface and the nucleus.

Gerace: The induction of gene expression by mechanical stress on the cytoskeleton by no means excludes the classical mechanism of release of transcription factors from tethering sites in the cytoplasm, entry through pore complexes and their activation by classical mechanisms. It does not need to involve pulling on genes through direct mechanical connections, although this remains possible.

Wilson: I remember Tom Pollard saying that the actin cytoskeleton responds dynamically to pressure, and if you deform very slowly it will give. However, if you push fast and hard, it resists. Are the forces you are using in your measurements slow, allowing a deformation response, or fast?

Lee: We oscillate and do it at different frequencies. We do look at these viscoelastic responses. The forces that we are using are within the normal range of what a muscle cell might experience in terms of frequency, but they are far below the forces a muscle cell generates. We are using 2 nano newtons, and a heart cell will generate much more force with a typical contraction.

Gruenbaum: I believe that Glen Morris had an interesting observation that the emerin minus cells are proliferating faster than wild-type cells. Is that correct?

Morris: If you grow mixed populations from a female carrier which contain emerin-positive and negative cells, gradually the emerin negative cells will take over the population (Morris 2004).

Gruenbaum: Did you try doing a similar experiment in mice? It would be interesting to see whether emerin has an effect on cell proliferation.

Hutchison: We already know the answer to that: emerin knockout human cells have an up-regulation of the beta catenin pathway. That is why they do it.

References

Mislow JM, Kim MS, Davis DB, McNally EM 2002 Myne-1, a spectrin repeat transmembrane protein of the myocyte inner nuclear membrane, interacts with lamin A/C. J Cell Sci 115:61–70

Morris GE 2004 Protein interactions, right or wrong, in Emery-Dreifuss Muscular Dystrophy. In: Evans DE, Hutchinson CJ, Bryant JA (eds) Communication and Gene Regulation at the Nuclear Envelope. BIOS Scientific Publishers, Oxford, in press

Summing up

Robert D. Goldman

Department of Cell and Molecular Biology, Feinberg School of Medicine, Northwestern University, 303 E Chicago Avenue, Chicago, IL 60611-3008, USA

It's difficult to summarize in the few remaining minutes, the remarkable amount of experimental data that has been presented over the past few days. We started out by discussing how descriptions of the general structural features of the nuclear surface have changed over the past 50 years or so. We have come to realize that the molecular composition of each of the major structural components of the nuclear envelope, the lamina, nuclear pores and the inner and outer membranes, is remarkably complex. For example, we have learned at this meeting, that in addition to the known components of the inner nuclear membrane, there may be as many as 67 novel membrane-associated proteins, many of which are potential lamin-binding partners. Furthermore, these inner membrane proteins have been derived exclusively from liver. It is therefore possible that we are only scratching the surface and that there are many more of these nuclear proteins specifically expressed in other tissues. In contrast, we haven't learned much about the outer nuclear membrane. In this regard, I don't think that we should continue to assume that this membrane is merely an extension of the endoplasmic reticulum. It is likely that we will hear, in the not too distant future, that there are unique outer membrane proteins and receptors. In this same vein, we further discussed the likelihood that there are connecting proteins between the lamina and the outer surface of the nucleus, providing links between the nucleoplasm/chromatin and cytoplasmic components such as those comprising the cytoskeleton. One candidate for providing such linkages is nesprin which, either in its full length form and/or one or more of its isoforms, is depicted as a long protein interacting with a SUN domain protein in the lamina and connecting to the cytoskeleton. Primarily we discussed the possibility of nesprin binding to cytoskeletal actin. However, no one has really looked for other binding domains on nesprin, which due to its very large size could possess many domains that could bind to other cytoplasmic components. It would be extremely interesting if one of these turned out to be an intermediate filament binding site, as it is well established that large numbers of IF are closely associated with the outer membrane of the nuclear envelope. It is important to note that the nuclear lamins which comprise the bulk of the lamina, are Type V IF proteins. So, in the case of nesprin, we have a potential structural link between the nuclear interior and the cytoplasm. Another possible nuclear/cytoplasmic cross-talk protein which has numerous splice variants is plectin. One of these variants has been reported to be present in close association with lamins in the lamina of nuclei. Plectin possesses an IF binding domain, in addition to actin and microtubule binding domains. Although much more work is required to convincingly support the role of such linker proteins in nuclear/ cytoplasmic interactions, there is little doubt that this area of research is emerging as one of the most challenging and exciting areas of cell biology.

One issue that has arisen frequently throughout our discussions relates to the proliferation of names. If we simply add the new components of the inner nuclear membrane, the need for a sensible nomenclature system becomes clear. Nuclear architecture is an emerging field, so it is important that we have a consensus on the names used to describe the enormous number of proteins already discovered and the new proteins which undoubtedly will be discovered over the next few years. We already have examples of this in the case of nesprin which has too many names to remember. We must find a way to come to grips with this situation soon, or else we will become mired in utter confusion.

The major theme of the meeting has been disease linked to alterations in nuclear lamin A. There are now numerous examples of mutations throughout the length of lamin A which cause a wide range of devastating diseases. Since my own interest has long been focused on lamin structure and function, these mutations are helping us sort out the different functional domains of lamins and how they interact with each other to form complex networks within nuclei. These mutations are also beginning to reveal how lamins interact with their binding partners such as the Lamin Associated Proteins (LAPs). This is important as very little is known about specific binding sites within the lamin A protein chain.

As seen throughout this meeting, in most laminopathies reported to date, the nuclei of cells derived from patients are mis-shapen. Although such aberrantly shaped nuclei are frequently considered the pathological hallmark of these diseases, there is little known about the relationship between abnormal nuclear shape and nuclear dysfunction. It is therefore critical to understand how alterations in nuclear shape change nuclear function. In this regard we and others have shown that normal nuclear lamin organization is required to maintain nuclear shape and mechanical integrity. In addition, normal organization of lamins is required for nuclear assembly and disassembly, as well as DNA replication and transcriptional activity related to RNA polymerase II. Obviously, the presence of mutant lamins could have profound effects on one or more of these processes.

In closing, I thank you all for your participation in what has been a very exciting meeting. I also wish to thank the Novartis Foundation and all of its staff who have made our stay here in the heart of London a most enjoyable experience.

Index of contributors

Non-participating co-authors are indicated by asterisks. Entries in bold indicate papers; other entries refer to discussion contributions.

Subject index

Page numbers in *italics* indicate figures and tables.

A

A-kinase anchoring protein 149 (AKAP 149) 6
A-type lamins
  embryogenesis 84
  emerin 52–53
  Emery-Dreifuss muscular dystrophy 52–53, 73
  expression 36
  inner nuclear membrane binding 37
  interaction with B-type lamins 11, 48
  tumour progression 46
acanthosis nigricans 168, 172
actin
  emerin 55–56, 60–61
  F-actin 55, 58–59
  lamin binding 62
  multiple forms 58
  nesprin binding 62
  RBD-2 52
α-actinin 105
adipokines 167–168
adiponectin 167–168
adipose tissue 166–168
ageing
  heart 133, 134
  normal 256
aggresomes 115–116
AGPAT2 168
agrin 101
ALS2 184, 187–188
Alsin 184, *185*, 187–190, 193–194
amyotrophic lateral sclerosis 183–192
  ALS2 184, 187–188
  Alsin 184, *185*, 187–190, 193–194
  dynactin gene 184, *185*
  dynamitin 184, *185*
  dynein 184, *185*, 196
  intermediate filaments 184, 186–187
  NF-H gene 184

  peripherin 184, *185*, 186–187, 192
  SOD1 183, 187
  spheroids 186
  tubulin cofactors 184
anaphase bridges 237–238, 243
anc-1 209, 212, 214
Anc-1 6
ANC-1 25, 212–215, 217
'anti-anchor' 147
'anti-silencing' 147
apoptosis, lamins 16
autoimmune disease, lipodystrophies 170

B

B-type lamins
  embryogenesis 84
  emerin 52
  essential nature 18
  genes encoding 36–37
  inner nuclear membrane binding 37
  interaction with A-type lamins 11, 48
  nuclear assembly 11–12
Barraquer-Simons partial lipodystrophy 170
barrier to autointegration factor (BAF) 232–233, 240–242
  DNA binding 54–55, 233, 241–242
  gene expression 54–55, 240
  identification 233
  LEM domain 6, 52, 54, 232
  mitosis 55
  nuclear assembly 54–55, 233
  nuclear membrane 242
  transcription repression 54
Berardinelli-Seip congenital lipodystrophy (BSCL) 168, *169*
BLOC1 103
bone morphogenetic protein 39, 42
BSCL1 168
BSCL2 168
Btf 53